The challenge of ecology

The challenge of ecology

Clair L. Kucera

Division of Biological Sciences,
University of Missouri,
Columbia, Missouri

With Chapter 10 by
Stephen Chaplin and **John Faaborg**

Division of Biological Sciences,
University of Missouri,
Columbia, Missouri

SECOND EDITION

with 156 illustrations

The C. V. Mosby Company

Saint Louis   1978

SECOND EDITION

Copyright © 1978 by The C. V. Mosby Company

Previous edition copyrighted 1973

Printed in the United States of America

Distributed in Great Britain by Henry Kimpton, London

The C. V. Mosby Company
11830 Westline Industrial Drive, St. Louis, Missouri 63141

Library of Congress Cataloging in Publication Data

Kucera, Clair L
    The challenge of ecology.

    Bibliography: p.
    Includes index.
    1. Ecology. I. Chaplin, Stephen, joint author.
II. Faaborg, John, joint author. III. Title.
QH541.K83   1978   574.5   77-14596
ISBN 0-8016-2802-4

GW/CB/CB   9 8 7 6 5 4 3 2 1

To LIZ

Prologue

These words have a prophetic ring as we view an ever-changing and depreciating environment. People are the dominant species on earth. Through our power and technology, we are able to manipulate, change, even destroy our habitat. Today we know that large-scale modifications are in process, and they continue at an ever-faster pace. The oceans, despite their vastness, are virtually nowhere free of pollutants. These include oil spills, toxic chemicals, and radioactive wastes. Some of our deepest lakes show steady deterioration in water quality. Their self-cleansing biological processes can no longer cope with the constant intrusions of industrial wastes. Our rivers, literally, are open sewers. The air is no longer clean. Smog hands in a pall-like shroud over major cities, often dimming the sun for days at a time. Whole landscapes are being destroyed, stripped of vegetation and viable topsoil. Pesticides, among other chemicals, have appeared and continue to appear in our food and water supplies. The list of environmental injuries is long, persistent, and commonplace. The human-made revolution going on in the environment today is a sober reminder of the widening gap between expanding technology and shrinking ecological balance. As these impacts continue, it is inevitable that change and instability will also increase. The quality of our lives, if not life itself, is being challenged.

Our earth is indeed a spaceship. It has a limited life-support system for sustaining a human population. Its carrying capacity is finite. Some say we have already exceeded it, if certain goals concerning living standards are to be maintained. Undoubtedly if such goals for all peoples are to match those currently enjoyed by Western industrial countries, the carrying capacity of the earth in terms of available resources is already oversubscribed. In the United States alone, with a little more than 6% of the world's population, we use an average of 30% to 35% of the earth's resources and energy produced each year. The material comforts we have come to accept as daily necessities are not without their price in terms of growing pollution and diminishing resources. Stated in another way, such demands are reflected in the statistic indicating that in the last 30 years more materials were used in this country than in the rest of the world combined.

Throughout human history, we seemingly have taken our natural resources and wild-life heritage for granted, if our treatment and use of them are any guide. Picture the decimation of the bison herds in North America, the actual extinction of the passenger pigeon, or the creation of the Dust Bowl in the Great Plains during the 1930s—all this in less than a century. The exploits of the ancient world are no less impressive. The denudation of Mediterranean lands, the expanding deserts of Africa, and the disappearance of forests everywhere are convincing proof of civilization's influence.

Unfortunately the philosophy of exploitation so characteristic of us and our traditional view of nature persists even today, as resources continue to give way to development. In the United States 5 acres of open space disappear every 1½ minutes under the impact. Around the world the land and waters are undergoing severe pressure to feed and serve more people. Starvation and malnutrition take their toll at the rate of one death every 8 seconds! Yet a conservative estimate places the net growth of the earth's population at 70 million people every year, a population equivalent to one-third that of the United States! With each additional increment in world growth, the quality of human life can only deteriorate. Projections are that in less than 30 years there will be twice as many people for whom food and living space will be needed. The oceans cover about 70% of the earth. They are an important source of protein in the harvest of commercial fisheries. Ironically, however, less than one tenth of the ocean supports food chains that can produce usable food for human consumption. The sea is not as large as it would seem to those who would rely on it to feed an unrestricted growth of human populations.

An irrefutable fact about our environment, then, is its limited resources. There is just so much usable water. Oxygen, minerals, and space for growth of whatever form, too, are measurable only in finite quantities. The environment is also fragile, vulnerable to the effects of industrial and technological development, and always in a state of delicate balance. Our planet in effect is one vast working system of interlocking functions and processes. We might ask how much longer, short of complete breakdown, the environment can sustain the pressures of use and the impacts of disruption that arise from expanding economies and the needs of growing populations. The day is here when we should recognize these ecological restraints on our management plans.

The universality and oneness of our plight in a finite and constantly degrading environment would seem an effective impetus for concerted action in solving mutual problems. A litany of facts, figures, and statements about a dismal future, however, is not enough. Alone, such a jargon may do a disservice to constructive action, for it may instil a fatalistic attitude toward the total environmental picture. There are other considerations. If we are to save our environment, arrest its further decline, and mend the damage already done, awareness, concern, and new attitudes toward nature are needed. We must bring about a reordering of priorities in our traditional approach to natural wealth. Our philosophy of continuing growth and corporate bigness as synonyms for progress should be replaced by a planetary concern for the quality of human living. What, then, should be done to ensure a native environment that provides not only the basic necessities for survival but also those intangibles such as biotic diversity, natural beauty, open spaces, and quiet surroundings that enrich human life?

Constructive measures applied through forward-looking and far-reaching educational programs are the main basis of hope in meeting the ecological challenge. These programs would combine innovations and interdisciplinary approaches. All areas of scholarship and learned endeavor in science, the humanities, and the social sciences must be involved. So often ecology and the environmental crisis are equated in cognate fashion, but erroneously so. The environmental crisis deals with people. It is caused by people, and a reciprocity in kind is implied when helpful dialogue and pragmatic solutions are sought. Thus implementation of corrective proposals will require the support of all segments of society. Only through instilled appreciation and an understanding of human-related causes and their interaction with ecological processes can meaningful progress be made.

For example, would a broader understanding of species diversity as a functional concept in maintaining ecosystem stability help change attitudes toward predators? Would a more broadly based knowledge of environmental resistance as a mechanism in control of a native species, thereby ensuring its future, assist in changing attitudes about our own population growth? Such examples are numerous. Much basic information is already available concerning environmental relationships across a wide front of ecological observations. It is true that the studying, the measuring and sampling, and the empirical effort should go on in the testing of hypotheses and the advancement of knowledge. But the extrapolation and extension of this knowledge into new domains are also needed. Neither is it a one-way street. The interactions that develop should provide a catalyst for synergistic accomplishment.

What are the real and significant confrontations between ecological processes and the environmental crisis? The purpose of this text is to show such relationships wherever they occur. It is hoped that this presentation will provide for beginning students, our learned colleagues in other disciplines, and nonprofessionals generally a better understanding of the interphase between such ecological values as diversity and stability and man's impositions that have led to the present situations. New dialogues and exchange of viewpoints are keys to awareness, to concern, and, hopefully, to a reorientation in public attitudes toward the use of resources. Without a change in our thinking, there can be no real abatement of the environmental problem. Our efforts become a series of post-factum skirmishes between groups sympathetic to environmental quality and vested interests as each emergency arises.

As our general condition worsens, a situation may arise from which there is no redress. An overview policy at high government levels, given impetus by an informed citizenry, might provide the legislation that could result not only in an early cure of the damage already effected but more importantly in the prevention of future alterations.

In the first edition I envisioned hope for the future regarding the conservation of resources and the preservation of our surroundings. The Environmental Protection Agency (EPA) was formed in 1970 by federal statute. The Council on Environmental Quality was also created, with its broadly based role as advisor and coordinator at the national level. In the international sphere, the United Nations sponsored a conference held in Paris in 1971 entitled "Man in the Biosphere." Here, problems were addressed to population and resources in the global perspective. A year later the United Nations Conference on the Human Environment was held in Stockholm, again dealing with global issues. In 1973 an international meeting on endangered species of animals and plants was convened in Washington, D.C. Since the publication of the first edition, limited gains have been made on various fronts. Our environment in the overview, however, still continues to deteriorate from long-standing threats as well as from ones not even known 5 years ago.

Our problems are exacerbated by the energy crisis, of which we suddenly became aware in 1973 with the oil embargo, and by the persistent worldwide food shortage, particularly with regard to protein needs, in developing countries with explosive population growth. The United Nations World Population Conference held in Bucharest in 1974 was a disappointment. Clearly there is no universal agreement concerning the urgency of controlling population on a worldwide scale. On the contrary, some developing countries with high birthrates view with suspicion the affluent nations for advocating birth control measures in their countries. Yet if the human species is to prevail on this planet within a civilized framework and a social order, we must arrive at some accommodation with global resources. This can only be achieved by a reduction in population growth and by conservation of resources, leading ultimately to an equilibrium condition between man and nature. The choice is ours to make. The time is late, but not too late. If we choose not to implement our words with positive action, we invite armed conflict and an uncertain future for all. If, however, we start now to cope realistically with expanding populations and shrinking resources, we can and will, like Prometheus, yet prevail.

Many persons contributed to the preparation of this book. I wish to express a deep appreciation to all for their assistance. Special thanks are extended to Rosemary Crane and Cindi Williams who typed the manuscript and to Ruth Dalke. Liz Hollis made additional drawings. Jim Alexander gave many hours in reference assistance. In addition, numerous correspondents kindly provided photographs or granted permission for the use of diagrams and tabular data from various publications. Several agencies should be noted, especially the Missouri Conservation Commission, the United States Forest Service, the Soil Conservation Service, the United States Geological Survey, the California Water Resources Board, the Alaska Cooperative Wildlife Research Unit, the State of Alaska Department of Environmental Conservation, the Serengeti Research Institute, and the Environmental Protection Agency. My appreciation is extended to Drs. Stephen Chaplin and John Faaborg for their preparation of Chapter 10 and to Dr. Arthur Witt for his assistance with Chapter 2, dealing with freshwater systems. Last but not least I wish to thank my graduate students and several faculty members for evaluation and critique of certain parts of the manuscript.

Clair L. Kucera

Contents

PART ONE

Introduction

1 A common ecology

MAN IN THE ENVIRONMENT

People have always had a vital interest in their environment. Their mode of living depended on it. The tools they developed, the types of dwellings they built and occupied, the foods they gathered, and the forage they used for domesticated animals have of necessity reflected an awareness of the resources available to them. Local environments were important to their survival, and they adjusted accordingly. As their skills improved, the number of artifacts increased to more effectively cope with their environment. Until approximately 10,000 years ago, they were fruit gatherers and later hunters. Then followed in successive order their roles as nomadic pastoralists and agriculturists, or tillers of the soil. The first food crops were probably grown in the Middle East, and it is here that archeological evidence suggests the construction of the first villages. Such early attempts at urbanization had their roots in these primitive human communities.

With improved methods of agriculture and more effective storage, both the amount of available food and the reliability of its supply increased. Transportation systems became more efficient, and food could be shipped from distant sources. Populations grew gradually and began to concentrate in cities. Crafts of increasing number and complexity emerged as more time became available for specialized tasks in the making of goods and the delivery of services. No longer was the day-to-day involvement with gathering food a person's sole existence. This succession of events in human activities resulted in national industrialization, becoming most advanced in western Europe and the United States.

Accompanying these changes in human life-style was a growing impact on the environment, measurable in terms of greater attrition of natural resources such as soil fertility, increasing pollution of the water and air, and a diminishing number of wildlife species.

Underlying these effects of human endeavor, today there is one basic, all-pervading stimulus—that of increasing densities of human populations. This, coupled with a greater affluence and the desire for even more material comforts, provides an incontestable basis for ever-mounting pressures being exerted on the environment. Unless these pressures on space and food resources can be curbed and brought into line with the earth's carrying capacity, the status of our habitat will continue to depreciate. *A finite system devoid of checks is also one without balance.*

WHAT IS ECOLOGY?

Ecology is a relatively new science. It was accepted as an organized discipline with a body of accumulated knowledge at the beginning of the century. Its foundations, however, were laid much earlier, with the classification of plant and animal life, the development of the science of biogeography, and the acceptance of Darwin's theory of evolution. What conditions might warrant such recognition for this or any other field of scientific interest? Concepts are based on stores of relevant information derived from patient and sustained observation. This process of gathering information and organizing it rationally is never ending. The synthesis of generalizations fashioned from details, facts, and particulars is an integral part of the scientific method. In turn, the inductive process serves as the means by which new ideas are generated and tested and old ideas are reevaluated. As we advance the scope and depth of our knowledge and relate it to information provided by other interests and professions, the whole provides a more comprehensive understanding of natural phenomena. The very necessary human preoccupation with environmental relationships for thousands of years has provided us in recent times with a unified approach to scientific inquiry.

Ecology is the study of interrelationships among organisms and their surrounding habitat. Specifically we might ask how plants and animals interact with one another. What are the effects and limitations of the physical environment in relation to such processes as photosynthesis, nitrification, and organic decay? Are there different degrees of biotic tolerance to such factors as temperature, salinity, and moisture? To what extent is the habitat modified by the interaction of these factors? What are the biotic changes associated with successional maturity?

Through this complex of relationships runs a common thread of interdependency. A factor initiated at one point in the environment may produce a chain reaction whose final effect is felt far from the source. There is abundant evidence of this cooperative relation between biological and physical processes. We have seen it in reports of DDT in Antarctic penguins, mercury in swordfish, or radioactive strontium in our milk supply. Still other examples might be mentioned such as the polychlorinated biphenyls (PCBs) whose subtle presence in food chains has only recently come under close scrutiny as a human health hazard. It is literally true that no species, including man, lives in complete isolation from any other species. Neither can any species live apart from its physical environment, although migration, hibernation, and dormancy are adaptive mechanisms that help organisms to avoid extreme or otherwise intolerable stresses that occur on a periodic basis. The biosphere is an "ecological web," linking all life and its external medium of air, water, and soil in a common whole (Fig. 1-1). These binding interactions do not

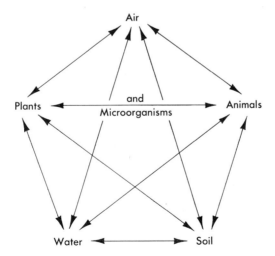

Fig. 1-1. Schematic diagram of interrelationships of the biological and physical components of the ecosystem.

always work to the advantage of the organisms in a given system. The examples just cited show that toxic or harmful substances as well as life's essentials often move through the biosphere with equal facility.

COMMUNITY ORGANIZATION

The worldwide complex of life presents a heterogeneous but orderly and generally continuous pattern of plant and animal assemblages. These assemblages are biotic communities, each characterized by a mutualism among its members and certain distinctive features that permit its identification and differentiation from other communities. The Alaskan tundra is a type of community. It suggests to the initiated an array of plant species and just as readily a list of animal associates. To know one is to have some measure of understanding of the other. So, too, with a Colorado grassland or a rain forest in Costa Rica.

Species organization is thus a key feature of most communities, obviating randomness of relationship, particularly in those systems that are generally mature or in the later stages of development. However, species composition often approaches randomized assortments in communities that are pioneering new areas such as a recently created volcanic deposit or a river sandbar. Such assortments result from the haphazardness of plant dispersal, particularly that which results from the scattering effect of wind-borne spores or seeds. In time successive invasions and species replacements fill bare areas, and competition for resources becomes a more exacting determinant in an emerging species structure.

Biotic communities differ in the number and variety of plants and animals that occur in a given habitat. Some communities are relatively diverse and contain numerous species, whereas others have but a few. Diversity indices have been devised to provide a mathematical basis for making comparisons between communities. Some examples will be discussed in Chapter 4. Communities vary in the way in which they are structured. The spatial arrangement of species and their densities, size and age distributions, and life form are all aspects of community structure. The characteristic size of a grazing herd of ungulates or a school of fishes or the spacing of plants in the desert are expressions of community organization.

LIFE FORM AND ADAPTATION

Life form is a striking property of community structure. We see differences immediately between the low tundra vegetation of the Far North and the tall grass prairies of the Midwest, between the deciduous forests of New England and the semitropical evergreen aspect of southern Florida. The following simple classification for plants has been devised by the Danish botanist Raunkaier:

Phanerophytes	(P)	Trees, shrubs, and vines; buds, more than 0.5 meter above ground, aerially exposed throughout year
Chamaephytes	(Ch)	Perennial herbs or low shrubs with buds less than 0.5 meter above ground
Hemicryptophytes	(H)	Herbs dying back to ground level; renewal buds at ground level developing from a rosette of basal leaves

Cryptophytes (geophytes)	(Cr)	Herbs dying back to ground level; renewal buds or organs below soil surface such as bulbs or corms
Epiphytes	(E)	Aerial plants occurring on other species; roots exposed and not attached to soil; this group sometimes placed with P
Therophytes	(Th)	Plants regenerated each year exclusively through new seed that remains dormant during periods of stress such as cold and/or dryness

This system of classification illustrates some of the adaptive and structural differences representing the total range of climatic conditions under which plants are found. All classes may be represented in a particular community, but almost universally the majority of species are found only in one or possibly in two.

The life-form spectrum of a given community consists of the percentage of various species of particular classes found. Some examples are shown in Table 1-1. Thus deserts are well populated by annual species, and grasslands and tundras by low shrubs, grasses, and herbs. In thick forests, on the other hand, ground cover has minimal representation on a percentage basis. At this early stage of our reading, we should realize that nature does not submit readily to generalities. Other adaptive characteristics that are not taken into account by Raunkaier's classification include such manifestations as succulence in cacti and African spurges, microphylly (small leaves) and the deciduous habit in drought-tolerant trees and shrubs, and the high osmotic values of cell sap (causing greater water retention) in various halophytes, or salt-tolerant plants. Fig. 1-2 depicts various desert plants that demonstrate a trend toward greater drought tolerance, as manifest in their succulent and deciduous habits. However, as cold or dryness or both intensify, there is a trend toward the development of structures that reduce the exposure of regenerative organs to environmental extremes.

Some wide-ranging species of warm-blooded animals, or homoiotherms, demonstrate variations in size or surface-volume ratios in different geographical regions. Those species found in cold regions may show a decreasing ratio, with a larger adult size and relatively less surface exposure per unit of body mass. This variation is known as Bergmann's rule, after the German zoologist who proposed such a hypothesis many years ago.

Table 1-1. Life-form spectra for several plant communities existing under a wide range of climatic conditions*

COMMUNITY	P	Ch	H	Cr	Th
Alaskan tundra	0%	23%	61%	15%	1%
Missouri prairie	6%	2%	62%	16%	14%
Minnesota forest-prairie border	35%	3%	46%	16%	0%
North Carolina Piedmont forest	60%	0%	36%	5%	0%
British Guiana rain forest	88%†	12%	0%	0%	0%
California desert	26%	7%	18%	7%	42%

*Modified from Buell, M. 1948. Ecology **29**:352-359; Dansereau, P. 1957. Biogeography; an ecological perspective. The Ronald Press Co. New York; Richards, P. W. 1964. The tropical rain forest. Cambridge University Press. London; and Kucera, C. L. Unpublished data.
†Includes epiphytes and climbers.

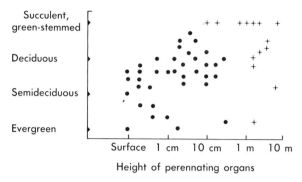

Fig. 1-2. Growth-form patterns in relation to height of perennating organs. The crossed points indicate spinose plants. (Redrawn from Whittaker, R. H. and W. A. Niering. 1965. Ecology **46**:429-452.)

The basis for the relationship is that surface area varies as the square, and volume as the cube, of the radius. Increasing the radius of a sphere twofold, for example, illustrates this relationship:

Radius	X	2X
Surface area	X	4X
Volume	X	8X

Doubling the radius increases the surface area four times and the volume eight times. The surface area of a sphere is equal to $4\pi r^2$; the volume of a sphere is equal to $\frac{4}{3}\pi r^3$. The same relationship between volume and area would be apparent if real values were used for the radius.

Increasing size associated with colder climates is exemplified in an interesting study of the native woodrat. These relationships are shown in Fig. 1-3. Note that weights are used to depict increasing size.

Some zoologists, however, dispute the Bergmann rule. Although larger animals living in cold climates may have a physiological advantage in that their greater body volume subjects them to a lesser amount of stress in terms of loss of body heat, they are at a disadvantage ecologically because larger animals have greater food requirements that they must try to meet in a less productive habitat. The other variables that are also involved include shape of the extremities such as the ear, hair covering, coloration, thickness of skin, and sweat glands. The African elephant, weighing as much as 3 tons or even more, is our largest land animal and inhabits various tropical areas. This animal faces the problem of dissipating rather than conserving body heat; this is accomplished, at least in part, by the very large surface area of its ears, from which body heat is readily lost. The surface-volume relationship regulates metabolic requirements per unit weight. Smaller animals generally are more selective in their diet preferences, requiring higher energy foods. Since plant parts with greater caloric values such as seeds are less abundant than structural features, the biomass levels in these herbivores, for example, are much less than in larger species with a broader selection of plant foods.

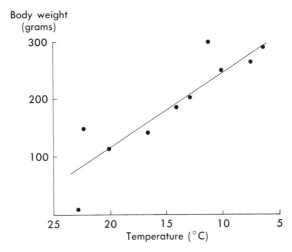

Fig. 1-3. Body weight of adult woodrats *(Neotoma)* increases as climate becomes colder, based on measurements of mean annual temperatures in different parts of their native range. (Based on data from Brown, J. H. and A. K. Lee. 1969. Evolution **23**:329-338.)

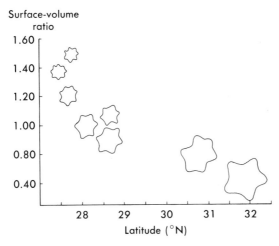

Fig. 1-4. Increasing diameter of stems of the large cactus *Lophocereus scottii* illustrates changes in surface-volume relationships in its natural range. (Redrawn from Felger, R. S. and C. H. Lowe. 1967. Ecology **48**:530-535.)

The principle that an animal of larger size can tolerate colder temperatures better has been extended to plants, but for different reasons and within a narrower frame of reference. In several species of tall, columnar cacti found in the southwest United States and Mexico, there is a distinct shift in size, or diameter, of the trunk. Those subtypes in the northern part of the range demonstrate thicker stems. Because of their greater mass, these are better adapted to withstand more hours of cold temperatures than those with smaller stems found in the warmer parts of the species range. Cross sections of stems of these sub-

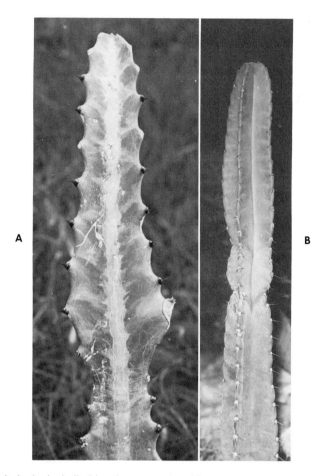

Fig. 1-5. Morphological similarities between the African spurge, **A,** and the American cactus, **B,** an example of convergent adaptation to desert climates. These plants belong to unrelated families but share the traits of succulence, leafless stems, spinescence, and thick cuticle, all of which are characteristic of certain plants that are found in environments with a minimum water economy.

types are shown in Fig. 1-4. This relationship between stem diameter and temperature implies a greater external heat absorption during the day that serves to ameliorate the effects of the dropping ambient temperatures for a longer period at night. It does not, of course, imply a greater generation of heat by the plants themselves, for plants in a sense are poikilotherms, or organisms whose temperatures are not metabolically regulated.

In climates similar to but remote from one another such as the warm deserts of the United States and southern Africa, adaptations similar in form and implied function are observed. The succulent habit of the exclusively American cacti and the African spurges is a suitable example (Fig. 1-5). This phenomenon is called convergent evolution, shown schematically in Fig. 1-6.

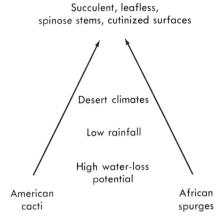

Fig. 1-6. Schematic representation of convergent evolution.

Both plants and animals demonstrate convergence in form under similar types of ecological stress. The kangaroo rat of the southwestern United States and the jerboa, a small rodent of the African desert, for example, are both nocturnal and have similarly developed hind legs for jumping and rapid movement and long tails. Both have the capacity of deriving needed water from the dry seeds that are a principal source of food. Among birds there are also numerous ecological counterparts. An interesting study by David Lack illustrates convergence in beak size and shape between the great tit of the British Isles and the tufted titmouse of eastern North America. Both inhabit broad-leaved woodlands, primarily, and feed near the ground on insects as well as on mast.

THE PLACE OF SPECIES

Each species is an integral part of a given community. It occupies a certain place, or niche, in the total structure. "Niche" refers not merely to physical location, but more importantly to functional role. A community is thus more than a collection in space of ecologically related species. The combined function of all species living together has a synergistic effect on the whole. This quasiorganismic character of the community illustrates the value of maintaining biological diversity, for the success of a community as a viable unit depends on maximum species cooperation. Generally those communities with numerous species and complex structures are characterized by greater functional reliability. Because of their diversity, such communities will be less affected by external influences such as disease organisms or insects than will simpler communities of fewer species.

The dominant species of a community are defined in terms of frequency of occurrence, population densities, size, and so on. In most natural communities, the actual proportion of species that could be classified as dominants is quite small in comparison to the total number of species. This ratio of dominants to the total is an important index of community conditions. (This topic is discussed in detail in Chapter 4.) Generally an increasing

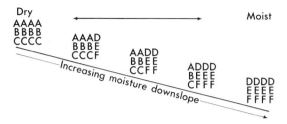

Fig. 1-7. Simplified diagram of a single-factor gradient for moisture. Along the gradient there is a corresponding change in species composition from one community type to another. In nature there are actually many other factor gradients, so that the place of the species in the environment is multidimensional. Its portion of the habitat is also referred to as a hyperspace, or a hypervolume.

dominance on the part of certain species suggests environmental stress and a growing imbalance among species. This point is perhaps aptly shown by the growing numbers and increasing needs of people and the decreasing level of security of their own habitat, as expressed in crowded cities, pollution, and destruction of the native environment.

Adjacent communities differ from one another in varying degrees. Between communities there is frequently a species continuum or transition zone. For the sake of simplicity, we might visualize a single-factor gradient based on moisture. There are idealized communities at the ends of this gradient. Each has three dominant species, depicted as A, B, and C in one, and D, E, and F in the other (Fig. 1-7). At the midpoint of the gradient in this hypothetical scheme, species from both communities are equally represented. At the extremes, however, species of the opposite community are absent, indicating that conditions there exceed their range of physiological tolerance. In nature such a perfect continuum is not likely; nonetheless overlapping of species does occur, particularly in gradually changing environments. In landscapes with irregular terrain and sharply defined habitats, patterns of biotic discontinuity are more common. Whether species changes are gradual or more abrupt, community transitions represent states of flux or relative instability. In this transitional zone, ecological conditions are marginal for both communities, so it is here that species are most sensitive to environmental changes. These zones of stress are important locales for observing biotic tolerances and for analyzing intercommunity relationships. The intertidal zone of sea coasts is an interesting example, for the organisms that inhabit this zone must be adapted to the alternation of wetting and drying, changes in salinity, and a range of temperature conditions. Yet it is no less a community than the land and sea it separates.

THE ECOSYSTEM CONCEPT

The functional aspect of the community implies a working, coordinated system of interacting parts—an ecosystem. Within the system, there is an energy interface between the biotic constituents and the external factors of the environment. A closely knit relationship exists between plant producers and animal consumers. Solar energy, captured and synthesized into food by plants, is transferred through a series of consumer levels that are linked together as a food chain, or more accurately, a food web. This *autogenic*, or *self-driving*, feature is an important property of biotic communities, as it is critical to their ability to achieve and maintain community stability. However, when external influences decimate or eliminate a species, the flow of energy is interrupted, food chains become shorter and more simplified, and the general ability of the system to function is impaired or threatened.

As we have seen, communities differ both structurally and in biotic composition, and zones of species overlap may occur between them. So, too, communities are not isolated units of function. Rather they are integral parts of a collective network in which links of energy join ecosystems together across the landscape. Leaves, twigs, and other debris from a forest community drift into an adjacent stream, where their energy content is utilized by the stream's residents, often at some distance from the forest. Some animals move back and forth through the "edge effect" between communities, relying for their needs on the resources of both communities at different times in their life cycles and seasonal activities. Wind and water currents literally transport usable energy across community boundaries in the form of biological material that is produced in one community and consumed in another. The transport of reusable materials such as mineral nutrients is also associated with energy flow. To be used repeatedly, they must be recycled through the system. Imbalance arises when the transport of such material becomes unidirectional, or nonreturnable. Phosphorus that has collected in ocean sediments is an example. Runoff from cultivated land and municipal wastes entering streams aggravate the difficulties in what is inherently an imperfect cycle.

STUDY OF COMMUNITIES

Autecology is the study of relationships at the individual species level. The study of communities is called synecology, and it implies a comprehensive or holistic approach to the total biotic complex, whereas autecology deals only with separate parts (species or factors) of the system. Measurement of energy relationships in a forest involves the total production of all tree species and is an example of community analysis and interpretation—a synecological study. Observations on the life cycle of the white oak in that forest, its germination, growth, and reproduction, would be an autecological study. Both approaches are complementary to a degree. Data from autecology contribute to the clarification of community interactions. Yet synecological methods are required for a full understanding of community-based relationships. The conclusions based on information derived from community studies should not contradict the findings from population analyses. This dictum obviously applies to any other set of sequential levels in biological organization, from molecules and cells to individual organisms and species populations.

Classification is a valuable and necessary adjunct of biology; nowhere is it more important than in ecosystem interpretations. It is axiomatic that the entity under study must first be identified accurately, according to established criteria, so that it may be properly distinguished from other groups of similar rank. Otherwise data become meaningless. Communities accordingly may be separated and characterized with varying degrees of refinement. The more rigorously the distinguishing criteria are applied, the finer the separations and the more homogeneous the composition and structure of the final community designates.

The largest geographical units of terrestrial communities described primarily on the basis of growth form or physiognomic classification of the vegetation are called plant formations. When the descriptions are more comprehensive and include the constituent animal populations, these units are called *biomes*. Growth form refers to the general habit (characteristic mode of growth, that is, whether woody or herbaceous, deciduous or evergreen, and so on) of the dominant constituents that yield the readily discernible characteristics that differentiate an area such as temperate deciduous forest from tropical evergreen forest, grasslands, or desert scrub. These are extensive community complexes, regional or even transcontinental in their geographical range (Chapter 8).

Temperature and rainfall as general climatic features are the functional determinants of the geographical distribution of biomes. Temperature and rainfall are important not as single independent variables but as interactive influences. The seasonal distribution and the total amount of moisture in relation to the range of temperatures are the significant determinants in the final pattern. Thus under the influence of extended dry seasons, dense tropical rain forests give way to more open woodlands and savannahs. Similarly in cooler climates we find gradations from deciduous forests, in areas that have substantial year-round rainfull, to grasslands, in areas that have less rainfall that occurs primarily during the summer season. Tundra, taiga, and closed conifer forest belts occur in the high, cold latitudes that generally are characterized by little precipitation. Inserted at specific points in the worldwide mosaic of biomes are still other types such as the well-known ''Mediterranean'' woodland or scrub found in warm regions in which the rainy season occurs during the winter months. Superimposed on this general plan of climatic controls is a network of soil types, topographical variation, and anthropogenic factors such as fire, domestic grazing, and cultivation. These factors combine to create a complex mosaic, a complete discussion of which is beyond the scope of this text. The worldwide distribution of the principal biomes or biome types is depicted in simplified form in Fig. 8-2, p. 187.

INTERNATIONAL COOPERATION

Because of the large-scale environmental changes taking place, ecological programs that are regional in scope become increasingly more significant. Biomes are comprehensive geographical units representing particular biotic resources, ecosystem relationships, and environmental conditions. Because of their regional and even continental character, biomes more often than not extend across national boundaries. In order to gain an understanding of how ecosystems function at these levels of integration, it is apparent that ecological cooperation between countries is essential. To this end the International Biological Program (IBP) was instituted in 1964. Fifty-six nations have agreed to participate, and a broad spectrum of studies has been initiated. The primary aim is the development of a scientific basis on which to implement programs of resource management and environmental improvement.

In 1971 The Institute of Ecology (TIE), a consortium of universities and research organizations, was formed under the auspices of the Ecological Society of America. Its broad objectives are to improve communication in the ecological community and to help implement the decision-making process in matters of legislation pertaining to environmental improvement. Both U.S. and foreign institutions in the Western Hemisphere are participants.

Quantitative information concerning ecosystem function and response to change is lacking for many situations. For example, how much is known about thermal pollution, and what will be its impact in the years to come? Is it somehow possible to make predictions about that impact before it is actually felt? The demands we make on the environment increase and proliferate, setting a limit on the amount of time we have to develop an understanding of the probable results. Therefore it is important that we gain some measure of expertise as rapidly as possible in predicting how a given ecosystem would respond to a given impact when a stress factor is applied. According to Barret et al., we presently lack adequate information concerning stressed systems.

In these times of increasing manipulation and devastation of the environment, stress ecology is a new ''growing point'' of the science. With the aid of computers, ecological conditions can be simulated and models of ecosystems can be developed using data that are either available or currently being acquired. Such models would allow us to make assertions about the real world and its environment with varying degrees of confidence (measurable probability). Answers are urgently needed *now* so that we may forestall irreversible changes. The knowledge of what would happen and when it would occur under given stress conditions provides the input necessary for fashioning environmental policy to better cope with future change.

Fig. 1-8. Beer cans littering a remote beach of one of the Galapagos Islands. (Courtesy Joel W. Hedgpeth, Marine Science Center, Oregon State University, Newport, Ore.)

Since the predictive power of the computer is only as good as the data being supplied, it is particularly important to provide as much ecologically comprehensive factor data as possible. All ecosystems should be analyzed. The wide gamut of natural conditions and the changes induced by our technology are a fertile ground for instruction about the human species as an inherent part of nature and a common ecology. Such a learning approach would allow us to correct social indifference and its farflung effects such as those depicted in Fig. 1-8. We could also learn to safeguard the environment against irrevocable change that is detrimental to the biosphere as a whole, to our own well-being, and to the well-being of future generations.

SUMMARY

All life shares the basic needs for energy, water, and oxygen. Biotic communities with their many and varied forms of life are composite expressions of biological adaptation. A main feature of a community is the interdependency demonstrated among its constituent species. In turn, communities are integral parts of ecosystems, which are the functional units of environmental organization. The production of energy, its allocation and use within the system, and the recycling of nutrients are universal processes. The effects of physical factors, species interactions, and molding influences of both biotic and abiotic components are important considerations in the development of ecosystem stability.

Ecology is the study and interpretation of these relationships. People are a significant and powerful force in the total environmental complex. As one species proliferates, another (or others) must decline, for ours is a finite system in which amoebas and humans alike compete for life's essentials. Because all environments in the biosphere are interrelated, a comprehensive approach to ecological problems is required. Such problems often transcend parochial domains and national boundaries. To realize common goals, cooperation and mutual understanding are needed to bring about environmental improvement, control of pollution, and the judicious use of the world's resources.

DISCUSSION QUESTIONS

1. The biosphere is a finite system of interdependent communities. Discuss its "common ecology" as the rational basis for resource management and conservation of natural wealth.
2. What is convergent evolution? What is its most important determinant? Give both plant and animal examples.
3. Discuss "stress ecology" as an important aspect of environmental study. What factors underlie its growing significance? Explain.

REFERENCES

Adams, R. M. 1960. The origin of cities. Sci. Am. **203:**153-168.

Barrett, G. W. et al. 1976. Stress ecology. Bioscience **26:**192-194.

Beals, E. W. 1969. Vegetational change along altitudinal gradients. Science **165:**981-985.

Braidwood, R. J. 1960. The agricultural revolution. Sci. Am. **203:**130-148.

Brown, J. H. and A. K. Lee. 1969. Bergmann's rule and climatic adaptation in woodrats (Neotoma). Evolution **23:**329-338.

Buell, M. 1948. Life form spectra of the hardwood forests of the Itasca Park region, Minnesota. Ecology **29:**352-359.

Cain, S. A. 1950. Life-forms and phytoclimate. Bot. Rev. **16:**1-32.

Cloudsley-Thompson, J. L. 1970. The size of animals. Sci. J. **6:**24-31.

Daubenmire, R. 1966. Vegetation: identification of typal communities. Science **151:**291-298.

Felger, R. S. and C. H. Lowe. 1967. Clinal variations in the surface-volume relationships of the columnar cactus Lophocereus schottii in Northwestern Mexico. Ecology **48:**530-535.

Gause, G. F. 1936. The principles of biocoenology. Q. Rev. Biol. **11:**320-336.

Hammond, A. L. 1972. Ecosystem analysis: biome approach to environmental research. Science **175:** 46-48.

Irving, L. 1966. Adaptations to cold. Sci. Am. **214:**94-101.

Kendeigh, S. C. 1969. Tolerance of cold and Bergmann's rule. Auk **86:**13-25.

Lack, D. 1969. Tit niches in two worlds: or homage to Evelyn Hutchinson. Am. Nat. **103:**43-49.

Langer, E. 1964. Cooperative research: biologists plan international study program. Science **143:**455.

Loucks, O. L. 1970. Evolution of diversity, efficiency and community stability. Am. Zool. **10:**17-25.

McIntosh, R. P. 1970. Community, competition, and adaptation. Q. Rev. Biol. **45:**259-280.

McNab, B. K. 1971. On the ecological significance of Bergmann's rule. Ecology **52:**845-854.

Raunkaier, C. 1934. The life forms of plants and statistical plant geography. The Clarendon Press, Oxford, England.

Richards, P. W. 1964. The tropical rain forest. Cambridge University Press. London.

Robbins, L. H. 1972. Archeology in the Turkana District, Kenya. Science **176:**359-366.

Whittaker, R. H. 1967. Gradient analysis of vegetation. Biol. Rev. **42:**207-264.

Whittaker, R. H. and W. A. Niering. 1965. Vegetation of the Santa Catalina Mts. Arizona: a gradient analysis of the south slope. Ecology **46:**429-452.

ADDITIONAL READINGS

Dansereau, P. 1957. Biogeography; an ecological perspective. The Ronald Press Co. New York.

Daubenmire, R. 1968. Plant communities. Harper & Row, Publishers. New York.

Odum, E. P. 1971. Fundamentals of ecology. W. B. Saunders Co. Philadelphia.

PART TWO

The ecosystem

2 Air and water

A DIFFERENT KIND OF PLANET

Most of us take for granted the air we breathe and our daily use of clean water. Our earth is unique among planets for its plentiful supply of each for supporting life as we know it. To explain how these essentials were produced is to recount a vast segment of our earth's history, dating back in time hundreds of millions of years. Our atmosphere is approximately one-fifth free oxygen (O_2), primarily the result of photosynthesis by green plants. Other gaseous components are nitrogen (N_2), carbon dioxide (CO_2), water (H_2O) vapor, and several others of less significance such as argon (Ar). Nitrogen occurs in the largest amount, about 79%. It is colorless, odorless, inert, and, unlike oxygen, nonsupportive of life in the N_2 state. In Chapter 6 we shall discuss the role of nitrogen-fixing bacteria in the conversion of gaseous nitrogen to forms usable by green plants.

Thus the atmosphere consists mainly of two elements, one reactive and readily combining chemically with other substances and the other, as mentioned, inert and nonreactive within the range of atmospheric temperatures. Together these gases make up 99% of the atmosphere. Carbon dioxide is present in small amounts by comparison, a little over 0.03%. Over the past century, the amount of CO_2 is shown to be *steadily rising* because of the accelerating consumption of fossil fuels, especially in the developed or industrialized nations (see Fig. 5-1, p. 89). Of all the gases of the atmosphere, water vapor is the only one that *fluctuates* in both time and place around the earth. There are seasons when the atmosphere is more moist than at other times, and there are regions where the air is characteristically more or less humid year round.

The following data are estimates of the gaseous components so necessary to the earth's life-support system and their relationship to amounts in the oceans:

ATMOSPHERIC GAS	AMOUNT IN THE ATMOSPHERE (tons)	RATIO TO EQUIVALENT AMOUNT IN THE OCEAN
CO_2	2×10^{12}	1/60
O_2	1×10^{15}	100/1
N_2	4×10^{15}	100/1
H_2O	1×10^{13}	1/100,000

Water vapor and its precipitate in the form of rain are an integral part of vast circulatory systems. Differences in rainfall together with variations in global temperatures are the fundamental bases for climatic characteristics of the biosphere. Solar heating's effects and functional relation to latitude as causal factors in temperature differentials are discussed in Chapter 5.

THE ATMOSPHERIC PROFILE

A slice through the atmosphere shows that gradients of its component parts occur with increasing altitude. About one half of the water vapor is present below the 6000-foot level. At 17,000 feet there is only one half as much O_2 as at sea level. Most of the earth's weather occurs below the 35,000-foot altitude in the middle latitudes. This weather-making zone, reaching to 10 kilometers or so, is called the troposphere. Fig. 2-1 shows an average profile to 50 kilometers above the earth for the middle latitudes. The tropo-

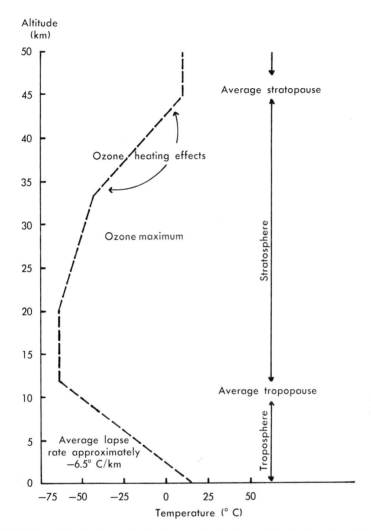

Fig. 2-1. Vertical profile of atmospheric structure to the top of the stratosphere. The broken line represents a typical temperature gradient showing cooling to the tropopause caused by adiabatic expansion of air and increasing distance from stored heat of the earth. The tropopause is the transition layer between the troposphere, or weather-making zone, and the permanently dry stratosphere. The warming trend in the stratosphere is caused by absorption of infrared radiations by the ozone layer. The stratopause marks the upper limits of the stratosphere where effects of ozone heating are maximum.

sphere is thicker in the equatorial regions and thinner at the poles. In the temperate latitudes, the weather-making zone also varies with the seasons, increasing in height in summer and decreasing in winter.

Above this zone is another principal region of the atmosphere, the stratosphere, which extends an additional 40 kilometers or so. It is characteristically devoid of moisture and contains an ozone-rich stratum. Ozone (O_3) at this altitude is produced by the interaction of atmospheric oxygen and incoming ultraviolet (UV) radiation. The high-density ozone layer built up by this photochemical reaction serves as a shield against UV radiations penetrating into the lower atmosphere. Approximately 99% of the UV wavelengths shorter than 300 millimicrons* are screened out by the ozone layer. If the ozone shield remains intact, life in the biosphere is protected against these damaging, shortwave radiations. It has been observed that emissions of oxides of nitrogen from aircraft (SSTs) flying in the stratosphere might reduce the ozone shield through a series of complex photochemical reactions. Even though estimates of oxides of nitrogen emitted by SSTs vary widely, any additions to the normal oxide load of the atmosphere suggest a threat to the ozone layer. The naturally occuring oxides would represent a homeostasis, or equilibrium, with present ozone concentrations.

Recently it has been proposed that this protective shield may be in jeopardy from other synthetic pollutants, including chlorofluorocarbons. Fluorocarbons are one of the ingredients in aerosol sprays. A report by Maugh discussing the findings of several investigators suggests that these compounds migrate into the stratosphere where, in the presence of UV radiation, chlorine atoms are liberated to interfere with ozone formation. At 1973 rates of aerosol use, it is projected that over a period of years a 7% reduction in ozone could occur. Our atmosphere of life-sustaining components and protective strata from dangerous radiations is a relatively fragile part of the biosphere. It is well to ponder its structural features and hence its vital functions, lest we unmindfully impair or destroy it with the wastes and by-products of our own technology.

THE GLOBAL WINDS

As air is warmed over ground and water surfaces in the hot tropics, it rises, causing expansion and cooling with increasing altitude within the troposphere. In this process, referred to as adiabatic cooling, saturated air cools less rapidly than drier air because some of the energy required in expansion is partially offset by the heat of condensation, as water vapor is converted to the liquid state, or precipitation. Thus there are two rates of cooling with ascending air masses, one dry, the other moist. For dry air, the average rate is about 1° C per 100 meters. (It warms at the same rate in descending and contracting.) The cooling rate for air at saturation is approximately one half of the dry adiabatic rate, or 0.55° C per 100 meters. The average lapse rate for the troposphere falls within these extremes (Fig. 2-1).

In regions of persistent cloudiness, we would expect lapse rates to be less than where sunny skies and dry air are common. It should be mentioned that negative lapse rates can

---

*1 millimicron = $^1/_{1000}$ of a micron.

also occur, usually in still air. These take place where cold air is trapped near the ground, below a layer of warmer air. A polluted atmosphere is more obvious under these conditions and persists until fresh air masses enter the area, as in the passage of a frontal system. These inversions are caused when the ground is cooler than the air, and they may occur in all seasons. In winter with deep snow cover, temperature inversions become extraordinarily persistent.

In the global perspective, the rising air aloft that originated over the warm tropics spreads poleward on either side of the equator, some of it descending between 30 and 35 degrees north and south latitude. These descending air masses, warmer and drier in these zones of subsidence (subtropical high-pressure belts), spread outward at the surface, both poleward and back toward the tropics. The latter become the Trade Winds blowing from an easterly orientation. In the Northern Hemisphere, these are called the Northeast Trade Winds; south of the equator, they are the Southeast Trade Winds.

Why do these winds entering the relatively low-pressure systems of the tropics show a deflection to the right in the Northern Hemisphere and to the left in the south? The causal factor is the spin of the earth on its axis. This drifting tendency for any moving substance such as air and water is referred to as the Coriolis effect and is more pronounced with increasing distance from the equator. The same veering effect is shown in those air masses moving poleward at the surface from the subtropical high-pressure belts. These become the basis of the vast circulatory systems known as the prevailing westerlies. In the Southern Hemisphere, with less land mass as impediments to airflow at these latitudes, the westerlies are stronger and more persistent than in the Northern

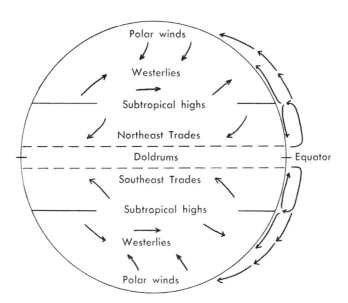

Fig. 2-2. Generalized diagram of atmospheric circulation of the earth.

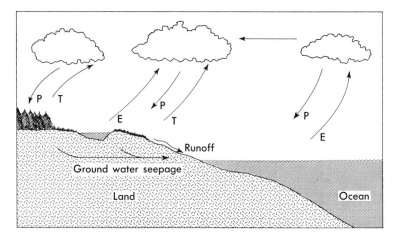

Fig. 2-3. Water cycle of the biosphere. The total amount of precipitation, *P,* is balanced by evaporation, *E,* from physical surfaces such as the soil and water, plus transpiration, *T,* by living plants. Some of the precipitation falling on land is lost as runoff or enters the ground-water system before it is evaporated, indicating in a balanced biosphere that the difference is accounted for by evaporation from the ocean. As surface runoff is increased, the cycle becomes simplified because less water is available for transpiration and evaporation over land and for deep seepage. Therefore if more water is diverted to runoff, its kinetic energy causes more sediment in the form of soil and nutrients to be carried to the sea.

Hemisphere with its greater land masses. Nonetheless the Coriolis influence is responsible for these directional tendencies in atmospheric circulation. These systems are illustrated in simplified form in Fig. 2-2.

On the poleward side of the westerlies, at the ground is a final set of basic wind patterns referred to as the Polar easterlies. In the upper troposphere, however, regardless of latitude, almost all winds blow from a westerly direction.

FRESH WATER

The supply of fresh water for the earth's terrestrial biota is based on the hydrological cycle (Fig. 2-3). Water vapor from the oceans and lakes is carried aloft to condense and precipitate over both land and sea. Periodic inputs of fresh water are needed to sustain life on land. It is stored in the soil and in lakes and reservoirs, with some of it entering underground water supplies and the river systems of the continents. Reservoirs, ponds, and lakes, or those places characterized by still water, are called *lentic* communities; streams and rivers are classified as *lotic* communities. Destructive practices such as clear-cutting of forests, strip-mining, overgrazing, channelization, and road construction reduce the stabilizing features of natural plant cover and hasten the exit of water from the land in its return to the sea. Let us examine a reservoir or lake system regarding seasonal changes in temperature in relationship to its physical and chemical properties.

Temperature relationships

Effect on physical properties. Temperature exerts a controlling influence on the physical and chemical properties of water as well as on the biota within the waters. By way of illustration, a typical lake in the temperate zone is depicted in Fig. 2-4, showing temperature changes during the four seasons. A knowledge of the physical properties of water in relation to temperature is necessary to understand thermal stratification, or layering, in a body of water at different times of the year.

As water is cooled toward 0° C, its density increases. It becomes heavier and begins to sink as it approaches a density of 1.000000 gram per cubic centimeter. Oddly enough, fresh water attains its maximum density at 4° C and, as it continues to cool, becomes progressively lighter. Following are densities of water (in the liquid state) for selected temperatures lowered to the freezing point:

TEMPERATURE	DENSITY (g/cm³)
18° C	0.009622
14° C	0.999271
8° C	0.999876
4° C	1.000000
0° C	0.999868

When water changes from the liquid state at 0° C to the solid state, or ice, its density decreases markedly to 0.917000 gram per cubic centimeter, causing it to float. Similarly, warm, less dense water will tend to float over cold, more dense water. If a lake is warm, cooling of the surface water at night will result in an increase in density, causing it to sink into warmer, underlying water. Layers of water at different temperatures and with different densities will resist the tendency to mix, just as oil and water resist mixing. As water is cooled, its internal friction or viscosity increases to maximum at 0° C. Water at different viscosities also resists mixing.

Water has the greatest specific heat of most substances, that is, 1.0. It takes 1 calorie of heat energy to raise the temperature of 1 gram of water from 15.0° to 16.0° C. This means that water can absorb more heat or energy than air or soil before a similar temperature change will occur. This characteristic of water explains why during the late fall or early winter the ground may already be frozen, but a lake surface is still free of ice. In the late spring, large expanses of ice may persist and are slow to melt relative to above-freezing temperatures of the surrounding air.

Before water can change from the liquid state at 0° C to ice at the same temperature, it must give up 80 calories of energy for each gram of water. Similarly if 1 gram of water is to change from the solid state to the liquid state, it has to absorb 80 calories of heat. Lakes are heated at the surface by radiation from the sun. The various wavelengths are absorbed differentially, with the shorter wavelengths penetrating the farthest in translucent water. Most warming occurs in the first few meters of water. Within the first meter, over 50% of the radiation is absorbed in clear water, and at 10 meters, 99% is absorbed. Water is a poor conductor of heat. The distribution of heat gained at the surface of a lake is accomplished by currents of water being set in motion, ultimately mixing warm surface water with the cooler water below.

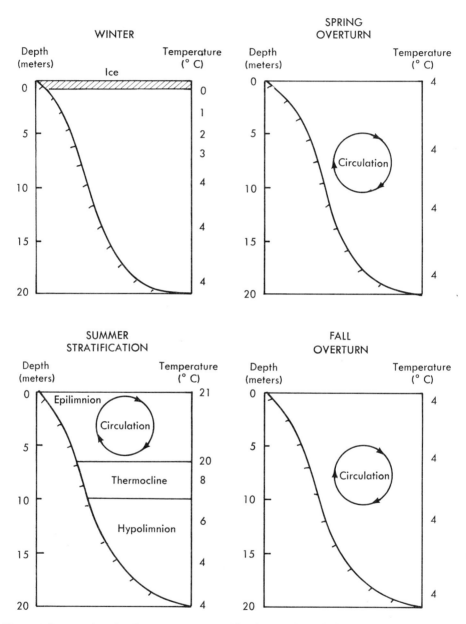

Fig. 2-4. Seasonal cycle of temperature, stratification, and circulation in a typical temperate lake. Depth in meters is at left of each graph, and temperature profile is to the right.

There are two types of currents that cause mixing within a body of water. Wind-driven horizontal currents circulate the water by mixing the surface with lower layers. The other type is the convection current, a vertically oriented flow of water that results from the sinking of cool, dense water at the surface into warmer, lighter water beneath the surface.

In an ice-covered lake during the winter, a vertical temperature profile will show that water immediately below the ice is at $0°$ C, and with increasing depth, the water will warm to $4°$ C and maintain this temperature all the way to the bottom. At this time, the lake is stratified concerning temperature and density. There is a layer of colder but less dense water above a layer of warmer, more dense water. As winter passes and increasing temperatures occur, the ice absorbs heat and melts into water at $0°$ C. Continued absorption of incoming heat by surface water elevates its temperature to $4°$ C, increasing its density and causing it to sink. This displaces deeper and relatively lighter water to the surface. Convection currents are established, and, furthered by wind action, the mixing process results in the transfer of heat to lower levels of the lake. The lake now has a uniform temperature of $4°$ C. All water in the lake is similar in temperature, density, and viscosity. It is easily mixed.

With the increased wind action characteristic of early spring, the water of the lake begins to circulate from top to bottom. This period of circulation is called the *spring overturn*. As the days grow longer with continued heating at the surface, the temperature increases and the density decreases, becoming less than that of the deeper water. The viscosity also becomes less at the surface, and it takes increasingly more wind energy to effect mixing. Eventually the lighter, warmer surface water no longer mixes with the deeper, colder layers, and vertical thermal stratification is established. Typically the lake becomes stratified into three distinct layers: an upper layer called the *epilimnion,* a middle layer, or *thermocline,* and a lower layer, the *hypolimnion.* Summer stratification of a lake is said to be well established when the upper layer is at least $10°$ C warmer than deeper water. In the epilimnion, the temperature decreases slightly, but there is an abrupt drop in temperature in the thermocline layer. The thermocline is defined as that stratum where a decrease of $1°$ C per meter of depth occurs. This layer ends where the temperature decrease is more gradual, and a drop of less than $1°$ C per meter is recorded. This marks the top of the hypolimnion (Fig. 2-4). The water of the epilimnion is warm, with relatively low viscosity and density. It is continuously mixed by wind action. Although the epilimnion circulates during the summer, the thermocline is an effective barrier preventing mixing of surface waters with the hypolimnion. Without access to the lake surface, this lowermost layer of water may become chemically stagnant.

As fall approaches, air temperatures decrease, and the process of surface cooling of the lake begins. The cooled water becomes heavier and sinks, displacing deeper, warmer layers to the surface where they, too, are cooled and sink. Gradually the lake loses heat to the atmosphere. Eventually the temperature of the epilimnion approaches that of the lowermost levels, and the lake begins to mix from top to bottom by wind action at the surface. This is called the *fall overturn,* and temperatures are uniformly $4°$ C throughout the profile.

With winter, temperatures continue to drop, and surface waters become less dense and begin to stratify on top of the warmer, more dense water. At 0° C, ice begins to form. The lake becomes sealed from the atmosphere, and it enters the period of winter stagnation.

Effect on chemical properties. Accompanying the seasonal changes in temperature within the lake are also changes in its chemical properties. The relationships existing among dissolved gases, acidity (pH), and nutrients will be discussed using dissolved oxygen (O_2), carbon dioxide (CO_2), pH, and phosphorus (P) as examples.

During winter stagnation, the lake is covered by ice and snow. Most of the O_2 in the water results from photosynthesis by algae and other aquatic plants. Light is required for photosynthesis, so that a cover of snow over ice reduces the production of O_2. Respiration by both plants and animals and decomposition of organic matter by bacteria use up O_2 and simultaneously release CO_2. In lakes rich in organic matter and with a heavy snow cover, O_2 supplies may be depleted severely, whereas the CO_2 level increases. Carbon dioxide combines with water to form carbonic acid. This weak acid dissociates to form hydrogen ions, and the water becomes more acidic under these conditions. With prolonged winter stagnation, in shallow lakes especially, the depletion of dissolved O_2 and the concomitant increase in CO_2 often cause a loss of fish life.

Phosphorus is an essential nutrient for plants and is recycled by decomposition of organic matter (Chapter 6). In lakes it may become chemically bound to soil particles and settle to the bottom. In winter there is no circulation of water under the ice, and P cannot be distributed throughout the lake for use by algae. At spring turnover, however, the lake "takes a breath." The water is mixed from top to bottom, improving O_2 relations and bringing P and other nutrients to the surface. As temperature and light conditions improve for photosynthesis, CO_2 is absorbed in increasingly greater amounts. In addition, some escapes to the atmosphere because of agitation of the surface waters by wind. The result is a reduction in CO_2 and a shift to less acid conditions. During the spring turnover, plant productivity is accelerated, and biological activity in general is increased. Dissolved gases such as O_2 as well as acidity and nutrients become uniform at all depths throughout the lake.

In summer the circulating epilimnion remains relatively high in dissolved O_2 content because of photosynthesis and wind action. The hypolimnion may stagnate, however, just as the entire lake did in winter. This lower part of the lake is unproductive because of a lack of sufficient light for photosynthesis.

Thermal and chemical stratification is again destroyed in the fall overturn. Dissolved gases, acidity, and nutrients such as P are uniformly distributed. With decreasing water temperatures, however, photosynthesis is less than during the spring overturn.

Thermal pollution

In the foregoing section, we examined the temperature relationships in a lake on a seasonal basis under more or less normal conditions. What, then, is the effect of heated discharges from power plants into aquatic systems? It is only in recent years that so-called thermal pollution, or "waste heat," has gained widespread attention. Such discharges may be made into either reservoirs or streams. Both have one factor in common: all aquatic life becomes more sensitive to stress under increased temperatures. According to Clark, the crayfish, a common invertebrate found in freshwater systems, experiences a rapid increase in heart rate from 30 beats per minute at about 4° C to 125 beats at 22° C. As the temperature is increased further, the heart rate decreases, indicating the approach to a lethal limit in temperature.

As temperatures rise, the water holds less O_2 in solution. At the same time, the demand for O_2 by fish and other aquatic vertebrates tends to increase with increasing temperature of the water. As an example, carp, an unusually hardy species, can survive with 0.5 milligram of O_2 in water just above freezing, but requires three times this amount in waters nearing 37° C. Heating of estuarine waters is all the more critical, since O_2 is less soluble in salt water than in fresh water at the same temperature.

With the inevitable increase in the number of power plants in the latter part of this century, vast quantities of thermal effluent will affect the quality of aquatic environments everywhere. The thermal efficiency of the best operated fossil fuel plants is only 40%. Approximately one half of the energy is dissipated as waste heat in condenser cooling waters; the balance, about 10%, is ejected into the atmosphere. For nuclear power plants, the amount of heat carried away by the water required to cool condensers approaches two thirds of the total energy generated. According to some experts, one fifth of all streamflow in the United States will be cycled through power plants by 2000.

Thermal effluent may raise ambient water temperatures at the point of discharge as much as 15° C. In summer, with temperatures already near a critical maximum for many organisms, the additional influx of heated water can prove lethal for heat-sensitive species. Most fish die when temperatures reach 35° C. According to Cole, temperatures over 30° C on a sustained basis constitute a "biological desert" for most aquatic species. The shift in temperatures invariably causes changes in kinds and numbers of both plant and animal species. As temperatures reach critical levels, diversity decreases, resulting in floras and faunas composed of only the most tolerant organisms. Very often these are the less preferred species. Factors affecting diversity are discussed further in Chapter 4.

SUMMARY

Our planet constitutes a life-support system with an atmosphere and a plentiful supply of water. The atmosphere includes free oxygen with a concentration of 20% by volume. Nitrogen for growth and development of all living things also comes from the atmosphere; however, in its gaseous state, it is not available for biological processes, nor is it chemically reactive within the normal range of atmospheric temperatures. Nitrogen must be converted to available forms for incorporation by green plants and for subsequent use by consumer organisms, including humans. This conversion is achieved in the natural state largely by nitrogen-fixing bacteria. It can also be achieved synthetically, as in the manufacture of nitrogen fertilizer.

The oceans are the basis of the hydrological cycle, supplying fresh water to the earth's terrestrial inhabitants, plants, animals, and microorganisms. Obviously the earth's total water balance is affected by precipitation, with evaporation being equally important. For each hemisphere, divided by the equator, the values of water for input and loss are approximately equal (about 100 centimeters for each hemisphere).

Human technology, its pollution effects, and its overpowering manipulation of the environment are capable of upsetting the atmospheric balance and of disrupting the cycle of evaporation of water and its condensation as fresh water over the continents. Thermal pollution is becoming an environmental issue as more power plants, both fossil fueled and nuclear, are built. These require vast quantities of water for cooling purposes. The effect of waste heat imparted to reservoirs and free-flowing rivers can be destructive to both aquatic plants and animals. To safeguard our environment and its singularly important characteristics of supporting all life, we should treat ecological function to our advantage and prevent what may well be irreversible changes in planetary processes.

DISCUSSION QUESTIONS

1. Discuss the importance of the troposphere to the biosphere. What significance does the stratosphere have for life on earth? Explain.
2. What are the causal factors in the development of the major wind systems of the earth?
3. Discuss spring and fall overturn in ponds and lakes. Under what conditions does a thermocline develop? What is its effect on aquatic life? Explain.

REFERENCES

Cairns, J., Jr. 1970. Thermal pollution—a cause for concern. J. Water Pollution Control Fed. **43:**55-66.

Cairns, J., Jr., A. G. Heath, and B. C. Parker. 1975. Temperature influence on chemical toxicity to aquatic organisms. J. Water Pollution Control Fed. **47:**267-280.

Clark, J. R. 1969. Thermal pollution and aquatic life. Sci. Am. **220:**18-27.

Cole, L. C. 1969. Thermal pollution. Bioscience **19:**989-992.

Cummins, K. W. 1974. Structure and function of stream ecosystems. Bioscience **24:**631-642.

Fan, L. T. and S. N. Fong. 1972. Distributed discharge of cooling water along direction of stream flow. Water Resource Bull. **8:**1031-1043.

Horvath, R. S. and M. M. Brent. 1972. Thermal pollution and the aquatic microbial community: possible consequences. Environ. Pollution **3:**143-146.

Hynes, H. B. N. 1970. The ecology of running waters. University of Toronto Press. Toronto.

Johnston, H. 1971. Reduction of stratospheric ozone by nitrogen oxide catalysts from supersonic transport exhaust. Science **173:**517-522.

Lee, R. E., Jr. 1972. The size of suspended particulate matter in air. Science **178:**567-575.

Maugh, T. H. 1976. The ozone layer: the threat from aerosol cans is real. Science **194:**170-171.

Newell, R. E. 1971. The global circulation of atmospheric pollutants. Sci. Am. **224:**32-42.

Prival, M. J. and F. Fisher. 1973. Fluorides in the air. Environment **15:**25-32.

Schaefer, V. J. 1969. Some effects of air pollution in our environment. Bioscience **19:**896-897.

Scherer, C. R. 1975. On the efficient allocation of environmental assimilative capacity: the case of thermal emissions to a large body of water. Water Resource Res. **11:**180-181.

Snyder, G. R. and T. H. Blahm. 1971. Effects of increased temperature on cold water organisms. J. Water Pollution Control Fed. **43:**890-899.

Van Der Horst, J. M. A. 1972. Waste heat use in greenhouses. J. Water Pollution Control Fed. **44:**494-496.

Wang, W. C. et al. 1976. Greenhouse effects due to man-made perturbations of trace gases. Science **194:**685-690.

Welch, Paul S. 1952. Limnology. McGraw-Hill Book Co. New York.

Werner, R. R. 1972. Population control: *not* a necessity in water resource management. Water Spectrum **4:**37-42.

ADDITIONAL READING

Neiburger, M., J. G. Edinger, and W. D. Bonner. 1973. Understanding our atmospheric environment. W. H. Freeman & Co., Publishers, San Francisco.

3 The soil environment

LIVING SOIL

Soil is one of the more important of our natural resources. It represents a dynamic community of a myriad of life forms. Bacteria, fungi, various soil animals such as earthworms, centipedes, mites, and millipedes, and the root systems of vegetation occupy a variety of niches in a complex trophic relationship. All too frequently we fail to appreciate the "live" dimension of this underground community. Without this life there would be no soil in a functional sense.

When we observe vegetation being cleared, the destruction of habitat and food resources for birds and mammals is readily apparent. Less obvious but equally important is the elimination of thriving communities by the destruction of a soil profile. The number of microorganisms in 1 gram* of soil runs into the thousands or even millions. Representative grassland soils, for example, reveal a tremendous abundance and great variety of organisms. A composite set of data is shown in Table 3-1. Note that the bacteria comprise not only the largest populations but also the greatest biomass. Although other organisms are not as abundant, their specific roles in the total community function are no less significant. Earthworms are especially important.

Many essential ecosystem activities occur in the soil and on its immediate surface. Reduction of plant and animal residues, humification, and mineralization are performed by its inhabitants:

$$\text{Detritus products} \xrightarrow{\text{(Humification)}} \text{Soil humus} \xrightarrow{\text{(Mineralization)}} CO_2, H_2O, \text{ and nutrients}$$

*1 gram = $\frac{1}{454}$ pound or about $\frac{1}{29}$ ounce.

Table 3-1. Characteristic relationships between densities and biomass of soil organisms in 1 m² of topsoil for representative grasslands*†

ORGANISMS‡	DENSITY (numbers)	BIOMASS (g)
Bacteria	1×10^{15}	1000.0
Protozoa	5×10^8	38.0
Roundworms	1×10^7	12.0
Earthworms	1000	120.0
Snails	50	10.0
Spiders	600	6.0
Harvestmen	40	0.5
Mites	2×10^5	2.0
Pill bugs	500	5.0
Centipedes and millipedes	500	12.5
Beetles	100	1.0
Flies	200	1.0
Springtails	5×10^4	5.0

*Modified from Brock, T. D. 1966. Principles of microbial ecology. Prentice-Hall, Inc. Englewood Cliffs, N.J.
†Not all of the organisms listed would be expected to be found in the same soil.
‡The scientific names that were used in the original have been converted to colloquial equivalents based on Barnes, R. D. 1967. Invertebrate zoology. W. B. Saunders Co. Philadelphia.

Humification produces finely reduced, relatively stable organic or humic compounds that are amorphous and dark in color. It is primarily carried out by soil animals such as earthworms that ingest large quantities of leaf litter and other organic detritus in addition to mineral matter. These materials are then egested and become a source of energy for still other soil animals. Earthworms, through their feeding and burrowing activities, improve infiltration and water storage. We have all noticed channels in damp soil covered by a protective mulch of leaves. Burning these leaves or carrying them off in a plastic bag to the city landfill removes an important source of energy for these valuable soil conditioners. Some estimates of the amount of this material egested by earthworms are given in Table 3-2. The relatively small amount of earthworm activity in heath soils is indicative of low densities of earthworms, which characteristically are not very abundant in acid soils.

However resistant to decay the humified products of the soil may be, these, too, eventually disappear in the mineralization process from which nutrients are released. Constant inputs of organic matter are therefore required to sustain thriving populations of soil organisms. The processes of nitrogen fixation, nitrification, and denitrification that will

Table 3-2. Amounts of earthworm excrement added annually to several European soils*

SOIL	AMOUNT (lb/acre)†
Heath	4,930
Mixed forest	19,130
Meadow	38,800
Forest meadow	68,960

*Modified from Schaller, F. 1968. Soil animals. The University of Michigan Press. Ann Arbor, Mich.
†The original values were expressed in grams per square meter.

Table 3-3. Estimated CO emissions in the United States, classified by source*

SOURCE	EMISSION (millions of tons)	% OF TOTAL
Technological sources		
Motor vehicles	59.2	62.7
Other mobile sources, aircraft, etc.	4.6	4.8
Fuel combustion, stationary plants	1.9	2.0
Industrial processes	11.2	11.9
Solid-waste combustion	7.8	8.3
Miscellaneous, man-made fires, etc.	9.7	10.3
Technological subtotal	94.4	100.0
Natural sources, forest fires, etc.	7.2	
Total	101.6	

*Modified from Jaffe, L. S. 1968. In S. F. Singer (ed.). Global effects of environmental pollution. Springer-Verlag New York, Inc. New York.

be discussed in Chapter 6 are just as significant. Recent findings indicate that soil bacteria are important agents in the biochemical conversion of carbon monoxide (CO) to other products, including carbon dioxide (CO_2) and methane (CH_4). This is an important discovery when we consider the tons of CO being released daily into the atmosphere on a global basis. It is estimated that each year more than 250 million tons are produced. The United States is the principal contributor, supplying about 40% of the world total. The main source of atmospheric CO is pollution, and the automobile is most responsible. CO emissions in the United States are categorized according to principal sources in Table 3-3. Note that more than 90% of the total comes from technological sources and the balance from such natural processes as forest fires.

Yet despite the constant intrusion of CO into the air, the level of concentration remains constant and is not increasing, as is the level of CO_2 that results from the combustion of fossil fuels (see Fig. 5-1, p. 89). The explanation that has been advanced is that

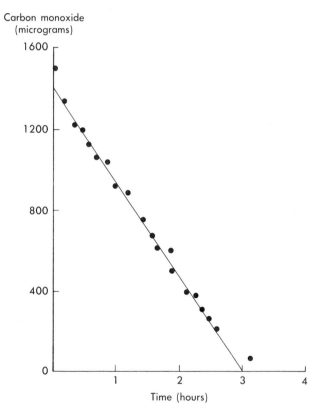

Fig. 3-1. Effect of certain soil bacteria on reducing the amount of CO in the air. In this test the CO was reduced from 1443 micrograms at the beginning of the experiment to 47 micrograms (1 microgram = $^1/_{1,000,000}$ gram) at the end of a 3-hour period. The CO content of air over sterilized soils was not reduced. (Redrawn from Inman, R. E., R. B. Ingersoll, and E. A. Levy. 1971. Science **172**:1229-1231.)

the soil may function as an important sink through the metabolic activities of several bacteria. Experiments utilizing test chambers have shown that CO disappears from the air over biologically active soils, whereas no uptake (disappearance) was measured over soils that had been sterilized. This microbiological effect is shown in Fig. 3-1. Soils having high humus contents were most effective in nullifying CO. These findings imply that microorganisms play a significant role in biospheric function. Certainly the maintenance of these soil-borne activities and the conservation of the organic matter of the soil are critical. How we manage the soil resource with regard to the application of pesticides, herbicides, nutrient fertilizers, and even irrigation water drastically affects its success and survival as a viable system.

The soil is an important element in receiving, storing, and dispensing water for the earth's ecosystems. Of course moisture falling on the land originates over the oceans as evaporation. Carried as vapor, it condenses into rain in generally predictable patterns.

Fig. 3-2. Sheet erosion on cultivated land. It is estimated that 4 billion tons of soil are displaced each year and that half of this amount actually reaches stream systems. (Courtesy United States Soil Conservation Service.)

Ultimately this moisture finds its way back to the sea, and the cycle is repeated (see Fig. 2-3, p. 26). The effectiveness with which the soil receives and retains water, even on a temporary basis, is in part a function of its physical condition. If soil is denuded and sterile, the residence time of moisture in the ecosystem may be relatively short. Erosion is accelerated when runoff is quick to enter the drainage pattern and groundwater systems. Siltation can become a serious problem, as shown in Fig. 3-2.

Under these conditions the ecosystem continues to deteriorate. Productivity decreases, and nutrient dispersal is increased as affected parts of the hydrological cycle are accelerated. The water regimen for the biosphere as a whole depends on a viable soil, one in which the activities of microorganisms, the soil fauna, and root systems alike contribute to maximizing water infiltration, storage capacity, and retention time. Innovative cropping methods to curtail runoff and preserve organic matter in the soil are being employed to an increasing degree. One of these is the no-tillage system of growing row crops, thus preserving a cover of mulch, or crop residues, of the previous season. In place of cultivation, however, chemicals (herbicides) must be employed to control weed growth. Thus there is a trade-off, balancing the benefits of one method against the other, to select the one least harmful to the environment.

In view of the importance of the soil biota, questions regarding the ever-growing volume of herbicides being applied to agricultural soils may be in order. We know that many of these products are short lived and apparently are no threat to soil life; yet the use of others that do constitute a threat (see Table 6-7, p. 140) continues. Even though no large-scale deleterious effects have been demonstrated to date, we cannot say with certainty that the long-range effects of pesticides such as DDT are negligible. Experimental evidence indicates a decline in the number and variety of a wide range of soil organisms when their habitat is subjected to various chemical treatments.

TO MAKE A SOIL

What constitutes the average soil? How is it distinguished from raw parent materials such as volcanic ash, lava flows, loess, river sands, or glacial till? Soil development begins when organic matter is added to these materials. This invariably means that plants, including microscopic organisms such as the blue-green algae, invade new areas. The algae are able to fix atmospheric nitrogen, thus increasing the potential for invasions by still other plants. As these organisms grow, mature, and die, their residues are broken down, humified, and mixed with weathered materials from the earth's crust. The humus residues are a source of energy for a wide spectrum of soil organisms whose activities further the soil-making process, enriching it with their own decay products.

Mineral framework

The mineral or inorganic fraction of the soil consists primarily of silicon, oxygen, iron, and aluminum, which are identified in chemical combination as the iron and aluminum silicates. Other elements that are collectively called bases are also a part of the soil structure. These include calcium, magnesium, potassium, and sodium, among others. The silicates comprise about 90% of the weight of an average mineral soil, the bases

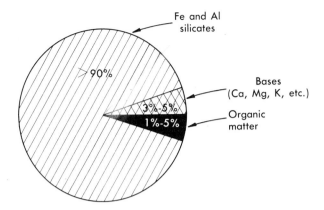

Fig. 3-3. Percentage composition of a typical mineral soil by weight. Deviations in any particular soil can be expected. Some mineral soils contain as little as 1% organic matter; others, between 5% and 10%. Peats and mucks would have much more and are usually classified as organic soils.

another 3% to 5%. The balance is organic matter. Thus the great bulk of the soil is an inorganic or mineral framework into which a relatively small amount of biological material is incorporated (Fig. 3-3). Yet the dynamic importance of organic matter is disproportionately large in comparison to its small fraction of the total soil.

The mineral fraction of soil is most readily recognized in its physical form as varying combinations of sand, silt, and clay. Some soils appear sandy; after a rain, they dry very rapidly. Others have more silt or clay and become sticky when they are wet. We have observed how clay soils swell when wet and shrink as they dry, causing cracks to form. The sand, silt, and clay are the products of physical disintegration and chemical weathering. Freezing and thawing, hydration, hydrolysis, and carbonation tend to break down and even modify the original materials of the earth's crust. In these processes we also see the effects of leaching and the transfer of substances as well as the formation of subsequent products such as oxides, hydroxides, and secondary clay minerals. These include bauxite, kaolinite, and montmorillinite, among others. In some localities, bauxite occurs in commercial quantities as a main source of aluminum. Kaolinite and montmorillinite are hydrous clays that have different capacities for water and nutrient retention. Kaolinite is a more weathered product than montmorillinite, and it occurs in relatively high percentages in tropical regions. Montmorillinites are more prevalent in cooler and usually drier regions where weathering does not occur so rapidly.

The texture of soil is expressed in terms of the relative amounts of sand, silt, and clay and is based on a classification of particle sizes. One commonly used system employs the following divisions:

SIZE CLASS	DIAMETER OF PARTICLES
Sand	0.05 mm or larger
Silt	0.02 to 0.05 mm
Clay	Less than 0.002 mm

These size fractions of the soil are determined by mechanical analysis, based on the principle that particles of different sizes settle in water, for example, at different rates. Obviously the more clay a soil contains, the more material will be in suspension after a given length of time. From this technique we can determine the sand, silt, and clay content of any soil on a percentage basis, which provides a means of comparing the physical properties of soils. For example, the differences between certain soils derived from two different kinds of geological materials in the Missouri Ozarks are as follows:

PARENT MATERIAL	SAND (%)	SILT (%)	CLAY (%)
Sandstone	50	34	16
Limestone	26	47	27

Texture is a significant factor in water-holding capacity and nutrient relationships. Sandy soils have less capacity for water retention and are low in nutrient bases such as calcium and magnesium. In the example just given, the sandy soils are droughty and infertile, able only to support sparse or stunted vegetation. Soil composed of extremely large amounts of clay, however, is also unfavorable to plant growth. Clay soils are generally poorly aerated, and oxygen deficiencies may develop when the soil pores are completely saturated with water. Soils with a high clay content also hold moisture tenaciously because of strong surface-tension forces. As a result, relatively large amounts of water are stored but cannot be extracted by the root systems of plants. This fraction of the total soil moisture thus represents unavailable water and is called the permanent wilting percentage (PWP) of the soil. The soil moisture actually available for plant use is the difference between the field moisture condition at any particular time and the PWP, which is a soil constant. This relationship can be expressed as follows:

$$\text{Available water} = \text{Field moisture} - \text{PWP}$$

As a simple example, if field moisture is equal to 25% of the dried soil weight, and the PWP has a value of 15, 10% of the soil weight could be absorbed as available water. In 1000 grams of soil, the amount of available water is 100 grams.

Heavy clay soils provide a good illustration of the manner in which vegetational patterns in the landscape may be modified. In regions of high annual rainfall where forests might otherwise be expected, this kind of soil frequently supports only grassland or savannah, a phenomenon shared by temperate and tropical regions alike. Grassland and savannah communities are better adapted to minimal soil aeration when saturated conditions occur during the rainy period and to minimal moisture during the dry season— two factors that are deterrents to forest development.

Organic matter

Organic matter is a very effective soil conditioner. It increases the porosity of heavy soils that have a high clay content, improving infiltration rates and thereby reducing runoff and erosion. More water is retained in the soil, increasing the amount of moisture potentially available to plants for longer periods of time. With better aeration, oxygen exchange is also improved.

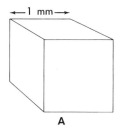

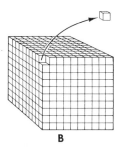

Fig. 3-4. Surface relationships for a given volume of soil based on particle size. **A,** Total surface of a single cube is 6 mm². **B,** Particles one-tenth as large yield a tenfold increase in total surface of the cube; further subdivision would result in corresponding increases in surface area.

In sandy-textured soils in which infiltration is not a problem, other benefits are noted as a result of the addition of organic matter and surface humus. Like clay, this organic material consists of fine particles that are of colloidal dimensions following reduction by soil organisms (humification). Because of the extremely small dimensions of these particles, their presence in a given volume of soil greatly increases the combined surface area, an obvious advantage over coarse, sandy soils. As an analogy, imagine a single cube, 1 mm on each side. This size provides a total surface area of 6 mm². If this same cube is divided into smaller cubes one-tenth as large, the surface area becomes 60 mm². Although each cube has a surface area of only 0.06 mm², there are 1000 of them contributing to the total surface, as depicted in Fig. 3-4.

Increased surface area is a functional property of colloids; it is important in holding soil nutrients against leaching and in the development of a greater water-retention capacity. These soil improvements, in conjunction with more available energy (organic matter), increase the potential for greater soil populations and diversity of species.

When organic matter is lost through cultivation as a result of increased oxidation rates and erosion, population densities and species diversity decrease. Data presented in Table 3-4 indicate that the populations of different soil arthropods (insects, mites, and related groups) are reduced by one-third in cultivated fields as compared to permanent grassland. Reductions are also apparent at lower depths, where the food energy of organic

Table 3-4. Comparison of arthropod diversity and mean population density per square yard of soil*

ECOSYSTEM	DEPTH (inches)	NUMBER OF GROUPS (diversity)	POPULATION DENSITY FOR ALL GROUPS
Permanent grassland	0-6	15	180,000
	6-12	13	75,000
Permanent cultivation	0-8	10	16,000

*Based on data from Jackson, R.M. and F. Raw. 1966. Life in the soil. St. Martin's Press, Inc. New York.

matter is less and oxygen levels are decreased. The buildup of organic matter in the soil reduces fluctuations in soil moisture and surface temperatures, two factors that vary widely in soils with barren surfaces that are low in organic matter. This fluctuation makes the soil habitat less suitable for resident organisms. The more biomass and organic material that can be retained in the system, the more stable it will be, and the development of environmental extremes becomes less likely. Without the influence of vegetation and the decay of plant residues, the geological material is a nonliving entity.

Chemosynthetic bacteria do not depend on the presence of green plants for a source of energy. Functioning as autotrophs, they are able to oxidize inorganic compounds and thus reduce CO_2 in the synthesis of usable carbohydrates. Some bacteria are specific for the oxidation of hydrogen sulfide, others for the oxidation of iron sulfides. The photosynthetic bacteria constitute another important group of autotrophs; they live in environments such as lake muds that are devoid of free oxygen. These are the green and purple sulfur bacteria. As fresh parent materials and new areas become available, autotrophic bacteria may be present in significant densities in the initial stages of soil formation. The extent of their importance in the production of organic matter and in soil development is not known.

SOIL PROFILES

The physical, chemical, and biological characteristics of soils evolve toward a state of equilibrium with the external environment. Like biotic communities passing through transient stages toward a more permanent status, soils also undergo patterns of change from youth to maturity. This aging sequence is evident in an emerging profile development that is marked by varying degrees of horizon or layer differentiation. A mature soil shows strong profile features consisting of distinct morphological horizons. These layers, which compose the profile, differ from each other in color, texture, chemical composition, organic content, and thickness.

Climate is a major influence. Temperature and rainfall are the principal factors in determining the rate of soil development and the kind of soil profile that will ultimately be formed. Vegetation (as a product of climate, the geological substrate, or parent material) and topography are also variables in soil formation. Changing any one of these factors results in a different degree or type of profile development. Other

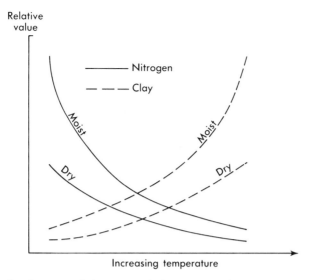

Relative
value

Nitrogen

Clay

Moist

Dry

Moist

Dry

Increasing temperature

Fig. 3-5. Generalized curves of nitrogen and clay content of soil in relation to temperature under given moisture conditions.

factors being equal, high temperatures favor the development of soil profiles with higher clay content and lower nitrogen values. The climatic implications in these two important soil variables are shown in Fig. 3-5. Thus soils of cooler zones such as the midcontinental United States are predictably more fertile than those of tropical areas having comparable terrain and rainfall.

Profile characteristics are the main bases of soil classification and the means for distinguishing between soil types and comparing one type with another. These characteristics include texture, color, organic matter, degree of leaching, internal drainage, and other physical and chemical properties.* A conventional method used by soil scientists in the delineation of profile horizons is a letter designation, with appropriate subscripts when subunit identifications are required. A generalized profile is shown in Fig. 3-6. The top soil is called the A_1 horizon. This is the zone of maximum accumulation of organic matter and the level at which root development is greatest. It generally has the darkest color. The A_1 layer is also the zone of greatest biological activity for all trophic levels in the soil environment. The layer labeled A_2 is a zone of the most intensive leaching of certain chemical compounds and is most apparent under temperate conditions of high rainfall and acidic plant residues such as pine needles. The B horizon is a zone of accumulation, receiving some organic matter, clay particles, and mineral nutrients from above. Below the B horizon is a zone of partially

*Current systems are elaborate and beyond the scope and needs of this text. For more detailed information concerning taxonomy of soils and their classification on a hierarchical basis, the reader should refer to Soil Conservation Service. 1975. Soil taxonomy. Agriculture Handbook No. 436. U.S. Department of Agriculture. Washington, D.C.

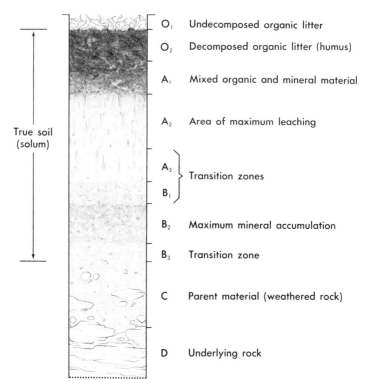

Fig. 3-6. Diagram of a soil profile. O_1 and O_2 , Fresh surface detritus and humified materials, respectively; between them there is a gradation of partially broken-down residues. A_1 , Top mineral layer with most organic matter and greatest root development. A_2 , Zone of leaching, usually lacking an aggregated structure (crumblike) because of a loss of clay and organic matter. B_1 , Zone of leachate accumulation, a blocky structure associated with clay and chemical cementing agents such as iron oxides. C and D, Partially weathered to unmodified material.

weathered to largely unmodified material whose layers are designated as the C and D horizons, respectively. This geological substrate may or may not be the parent material for the developing soil profile overlying it.

At least two layers of organic matter may be distinguished on top of the mineral soil. The first layer, labeled O_1, consists of organic materials recently deposited by local vegetation. It is more or less still intact. The underlying layer, O_2, shows advanced breakdown or an amorphous condition and is the product of humification. This terminology can be employed to discuss and evaluate soil profiles based on the morphological and chemical traits mentioned previously. A typical prairie soil, for example, has a different appearance than a forest soil. Its A_1 horizon is generally deeper and contains more total organic matter. It is a more fertile soil. A forest soil, by contrast, is generally more leached, has a greater acidity, and under certain conditions has a pronounced A_2 zone. Grassland soils typically are important sites of agricultural production in the temperate

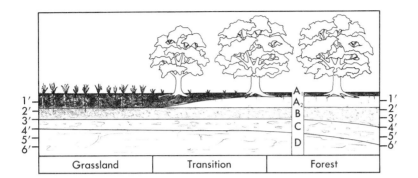

Fig. 3-7. Regional pattern of soil development from forest to grassland. Note that the topsoil, A_1, is deeper under grass than under forest.

Fig. 3-8. Soil profiles developed under tall grass prairie, **A,** and deciduous forest, **B.** Note the darker color of the prairie compared to the forest. (Courtesy Clarence L. Scrivener, University of Missouri, Columbia, Mo.)

regions of the world. A diagram of idealized forest and grassland profiles is shown in Fig. 3-7. Note the gradual transition from one community type to the other. Typical profiles are shown in Fig. 3-8. The light color of the A_2 horizon in the forest-developed profile is conspicuous

Some soils never reach a mature state of profile development. On steep slopes, the products of geological weathering and biological decomposition erode or are washed away as rapidly as they are formed, and distinct profiles have no opportunity to develop. In the desert, dry conditions appreciably curtail the rate of weathering and the contribution plants make to soil formation. Reduced productivity and sparse plant populations result in less organic matter being added to the soil.

In cold climates the biological processes of decay occur at a slower rate, and, as a result, longer periods of time are required to reach organic equilibrium. In the tundra, for example, regenerative processes are much delayed in areas that have been damaged by erosion. Scars remain on the landscape for many years; restoration is slowed because of the limitations imposed on biological processes by climatic factors.

In the tropics, especially in the wetter regions, soils are also particularly vulnerable to deterioration. The reasons involve the faster rate of organic decay, shorter turnover time, and less effective retention of nitrogen and other nutrients. Cycles are completed at a more rapid pace. In addition to higher average temperatures, mineralization also proceeds on a sustained year-round basis because of the frost-free conditions.

The management practices and decisions involving land use that are well established and effective in one climatic zone may not be successful in another. Tropical soils are especially sensitive to modifications that may occur when the original forest is removed. The major pedogenic, or soil-forming, processes will be discussed in more detail to illustrate their differing biological, physical, and chemical characteristics.

SOIL-FORMING PROCESSES

There are three basic types of soil formation. These are calcification,* podsolization, and laterization. A primary aspect of the development of any kind of profile is the leaching of minerals by percolating water. The intensity of the leaching is significantly affected by climatic conditions. The degree of slope is, of course, a modifying influence, since on steeper gradients, the surface runoff will increase at the expense of infiltration. The kinds of leachates involved depend on geological and biological factors. Igneous rocks such as granites, for example, produce acidic substances and thereby increase the solvent action of percolating waters. The weathering of limestone results in alkaline or basic products.

Under coniferous forests, the leachates from decaying needles are more acid than derivatives from deciduous forest litter, which in turn usually are more acid than the decay products of prairie grasses. In a cool, moist climate with adequate drainage sup-

---

*According to a recent glossary of terms, this designation is obsolete; however, it is retained here for convenience. For the interested reader who wishes to pursue the latest changes in classification, refer to Soil Science Society of America. 1971. Glossary of soil science terms. The Society. Madison, Wis.

plied by the sandy-textured soils on which coniferous forests are found, soil profiles are frequently quite acid, as indicated by their low pH values. Very little mixing with the mineral soil takes place because of limited populations of soil animals and less complete breakdown of litter. The soil profile shows strong leaching effects. Such soils are classified as the *mor* type. In soils with less acidity and more active soil populations, greater mixing and incorporation of organic matter with the mineral components takes place. These are called *mull* soils and are more typical of deciduous woodlands and grasslands.

For convenience, the types of soil development can be classified in a broad sense on the basis of the relationship between rainfall and evaporation. Moist climates receive more precipitation than they lose through evaporation and plant transpiration. Excess water is thus available, and it can pass through the profile into the supply of groundwater. In the course of this process, more nutrients and other materials are transported out and away from the ecosystem. Such losses are particularly serious when they occur in conjunction with denudation of ground cover, which allows greater runoff and erosion. In addition, those nutrients that would otherwise be absorbed and bound by the plant community are now subject to being washed away.

From rainfall and evaporation data, we can calculate ratios designated as precipitation/evaporation (P/E) values. Forest climates have ratios approaching or exceeding the value of 1. In grassland and desert climates, values are lower, in most cases considerably so. Under these conditions, the amount of moisture that evaporates potentially exceeds the amount received through precipitation. We say "potentially" because obviously the amount of water that leaves a balanced system cannot exceed the amount that actually enters it. Therefore evaporation is determined experimentally through the use of open pans of water that are refilled as needed. If we find that a sum total of 50 inches of water evaporated during a year in which 40 inches of precipitation was recorded in a rain gauge, the calculated P/E value is $^{40}/_{50}$, or 0.80. This means that on an annual basis the net effect of the real P/E relationship is the creation of a moisture stress in the soil. Plants faced with deficiencies in the water supply for extended periods may exhibit dormancy or reduced metabolic activity. In arid climates, grasses dry and cure readily, thus reducing water needs, but they resume growth with the onset of rains. The climate–vegetation–soil process relationship may be generalized as follows:

CLIMATE	VEGETATION	SOIL PROCESS
Dry (P/E about 0.8 or less)	Grasslands, steppes, and deserts	Calcification
Moist (P/E 0.8 or greater, sometimes exceeding 1)	Forest-grassland transitions and forests	Podsolization (temperate zones) and laterization (tropical zones)

When sufficient data are available, P/E maps can be drawn, and regional comparisons of the major soil groups and vegetation can be made. Such maps have been produced for the United States.

Calcification

In the moister areas in which the pedogenic process occurs, the deep soils that have developed under luxuriant grass are among the most fertile in the world. These are the chernozem soils, from the Russian that literally means "black earth." Moisture that is adequate but insufficient to cause extensive leaching allows for the large accumulations of organic matter from decay of root systems. This advantage combines with the cycling and retention of bases, including calcium, in the profile. In this soil-forming process, calcium carbonate is deposited as a layer of varying thickness and depth in the profile. Carbonic acid produced by the CO_2 derived from respiration plus water provides the solvent action that frees the chemicals that are moved downward in solution. Below the root and moisture zone, these chemicals are precipitated as carbonates, so that the extent of the carbonate development is inversely proportional, within limits, to the amount of precipitation.

Less rainfall supports less vegetation, and under these conditions the carbonate layer forms closer to the surface. Soil profiles produced in regions of persistent dryness are rich in bases because percolation is generally insufficient to carry them into the groundwater system. Inadequate precipitation is a more effective deterrent to plant productivity than lack of nutrients for plant growth. Particularly in the more arid regions in which irrigation is practiced, the excessive evaporation rate causes salts that have been carried in the water to become concentrated in the soil and even form crusts on its surface.

Many soils in the arid western United States are undergoing widespread salinization or impregnation with salts carried to the land through irrigation systems. As the salt concentrations in these areas reach critical levels, it will no longer be possible to grow the original crops. The salt load of the Colorado River, for example, increases manyfold in areas downstream from its headwaters (Table 3-5). When its waters reach the Imperial Valley of southern California, 1200 miles from its beginning, the burden of dissolved materials obtained along the river's route and carried in suspension is equivalent to more than 1 ton per acre-foot of water. If in the course of a given period, 12 inches of water are used

Table 3-5. Downstream increase in salinity of the Colorado River*

MILES DOWNSTREAM	SALINITY (tons/acre-ft of water)†
0 (Hot Sulfur Springs)	0.1
200	0.5
400	0.8
600 (Grand Canyon)	0.8
800	0.9
1000	1.0
1200 (Imperial Valley)	1.2

*Modified from Colorado River Board of California. 1962. Salinity problems in the lower Colorado River area. State of California. Sacramento, Calif.
†An acre-foot is the volume of water covering 1 acre to a depth of 1 ft. Values are rounded to the nearest tenth.

to irrigate 1 acre of desert land, 1 ton of salt is also applied. When the vast amounts of water diverted to the land each year in this arid region are considered, it is possible to visualize the tremendous additions of salt to the soil. Problems of extreme salinity are being encountered in the San Joaquin Valley, one of the world's richest agricultural areas; the problem is particularly acute in the drier southern part, which in addition has no natural outlet. The extreme amounts of salt that may accumulate in an irrigated field are shown in Fig. 3-9. Under these conditions, salinization is always a threat to sustained plant production. Only by drainage systems and flushing operations conducted at specified times during the year can the salt concentration be held at acceptable levels. Problems of a similar nature are also reported from Mexico, China, Russia, and Australia. In Pakistan it is estimated that 30,000 acres yearly are removed from cultivation because of excess salt accumulation.

Very few crop plants can tolerate a high salt content in the soil. However, as human populations continue to expand, it will be necessary to convert more land of a submarginal

Fig. 3-9. Accumulation of soluble salts in an irrigated field in the San Joaquin Valley, California. (Courtesy State of California, Department of Water Resources, Sacramento, Calif.)

nature to food production, and salinization is expected to become more widespread. The soil-forming processes associated with arid conditions, in which the evaporation potential is a key factor, complicates the simple addition of water even when it is abundantly available. Large areas also are salinized under natural conditions. These are usually basins with no outlet that collect water that runs off from the surrounding higher ground. Native species with varying degrees of salt tolerance are found in these habitats. Many areas are perhaps best left to native cover. The constituent species, many of which are true halophytes (salt plants), are more adaptable than artificially introduced crop plants. Furthermore, some provide usable range forage and habitats that as a whole are suitable for wildlife.

Calcification as both calcium (Ca) enrichment in the topsoil of prairie chernozems and carbonate deposition in the lower profile can thus be extrapolated toward increasing aridity. Carbonates build closer to the surface until downward leaching as a net effect is balanced by positive evaporation stresses. Under these conditions, an influx of water invariably results in deposits of soluble salts on the surface of the soil itself. In this ecological gamut of leaching potentials, the soil is saturated with bases; however, compositional differences are apparent. In the chernozem soil, sodium (Na) is either a minor base or is absent because it is more soluble than Ca and would be leached out under these conditions. In drier regions, Na is not leached out as readily and therefore becomes relatively more prevalent under conditions that promote salinization.

Podsolization

Podsolization is a soil-forming process most characteristic of cool, moist climates. Although universally indicative of forested regions of the temperate zones, wet prairies and certain tundra situations may also manifest similar leaching effects to some degree. In this type of profile development, oxides of iron and aluminum, called sesquioxides (Fe_2O_3 and Al_2O_3), are leached out of the surface and deposited at lower levels. Under these conditions, however, silica (SiO_2) becomes immobilized in the upper soil profile. An identifying feature of podsols, then, is a relatively high ratio of silica to iron and aluminum compounds in the upper profile. Calculated on a molecular basis, typical ratios for podsols have values of at least 2, whereas the values for laterites (see following discussion) are characteristically less. The silica-sesquioxide ratio can be expressed as follows:

$$\text{Silica-sesquioxide ratio} = \frac{SiO_2 \text{ (moles)}}{Fe_2O_3 + Al_2O_3 \text{ (moles)}}$$

The following example is used to clarify this relationship. Let us assume in a quantity of soil that we have 100 grams each of SiO_2, Fe_2O_3, and Al_2O_3. The amount of oxide material is based on the clay fraction only. The molecular weights of these oxides are, respectively, 60.1, 159.6, and 101.9. The ratio of the moles of SiO_2 to the combined number of moles of Fe_2O_3 and Al_2O_3 is calculated as follows:

$$\frac{\dfrac{100.0}{60.3}}{\dfrac{100.0}{101.9} + \dfrac{100.0}{159.7}} = \frac{1.65}{0.98 + 0.63} = 1.02$$

This is only a hypothetical example; however, the low ratio would suggest a tropical soil, probably very old and highly weathered, bricklike, and leached of plant nutrients.

Under the generally acid conditions characterizing a podsol, iron (Fe) and aluminum (Al) are soluble and may be leached to lower levels, where they are precipitated or immobilized as oxides. Silica remains relatively insoluble in an acid regime and therefore remains in situ. Thus under humid conditions, there is a mechanism for leaching together with a solubility factor that preferentially releases certain materials for transfer, although others are immobilized.

Acid conditions are found first of all in moist climates, where basic ions tend to be leached from the soil colloid and replaced by acid-forming hydrogen ions. Both carry positive charges, but the latter are held with a greater electrostatic attraction to the negatively charged clay particles. Second, the products of decaying forest litter are themselves acidic, a factor that contributes to the solvent action of percolating water. In regions where podsolization is most intense, the profile is characterized by a strongly leached subsurface horizon. This is the A_2 zone (Fig. 3-6) that was discussed earlier in the chapter.

Fig. 3-10. Soil disturbance in the tundra-permafrost region. The wet tundra is most susceptible to damage from heavy equipment during the thaw period. (Courtesy D. R. Klein and the International Union for Conservation of Nature and Natural Resources, Morges, Switzerland.)

Forest podsols are less fertile than soils produced under grassland. They are less productive as agricultural soils because they accumulate less organic matter in the topsoil and have a lower nutrient saturation. When forests are cleared and cultivation is begun, the residual fertility of this soil is depleted more rapidly. Many abandoned farms in the eastern United States have been "worn out" because of unsuitable early agricultural methods, and today extensive areas have succeeded to second-growth forests. In Great Smokey Mountains National Park, for example, one may see the remains of stone fences and other artifacts of farming beneath dense, secondary forests of native trees such as tulip poplar, various oaks, and sugar maple.

Weakly developed podsols have been noted in certain tundra areas. Cold, generally dry climates, low productivity, and limited biological activity in the soil combine to retard the podsolization process. A characteristic feature of tundra soils is permanent frost at varying levels below the surface. Where moisture is adequate, an ice layer is formed and may extend dozens of feet into the earth. During the brief summer, the ice thaws to a depth of 1 to 3 feet or more. It is during this period that tundra soils are especially vulnerable to erosion and slippage. Denudation of the plant cover, as in the development of roads, accelerates the thawing process. Meltwater then creates an erosion pattern over the tundra surface because the permafrost layer is an effective barrier to its percolation through the ground. Since successional processes are extremely slow (Chapter 7), erosion patterns may continue to deepen and become more serious over an extended period before the detrimental effects can be reversed by an adequate plant cover. Much attention has been given to the tundra in recent years, as it becomes a focal point in the search for oil resources. As these plans are carried forward, the unique condition of these fragile soils should be taken into consideration lest large-scale changes seriously alter the tundra environment (Fig. 3-10).

Laterization

A first impression of a tropical forest, with its luxuriant growth, numerous species, and complex structure, is that here must be a fertile soil, rich in humus and nutrients. However, upland soils of the tropics are generally nutrient poor, with little accumulation of organic matter. Laterization is described by some authors as essentially a geological process that tends to accelerate when the protective plant cover is removed from the soil. The process is unique in that it results in progressive deterioration of the soil environment. Under the impetus of high temperatures and the intercalation of wet and dry seasons, unprotected soils may undergo drastic changes that are essentially irreversible under certain conditions. Without the ameliorating effects of vegetation, soils become hard and intractable, literally baked and rendered agriculturally useless after a few years.

Parent material is a critical factor in laterization. There are productive soils in the tropical zone; most of these, however, are either of alluvial origin, as in the case of river valleys, or are derived from recent volcanic activity. These soils are rich in nutrients that have been supplied by river overflow or by ejecta from the earth itself. However, many tropical regions are old, highly weathered landforms and are particularly prone to laterite development. As a case in point, Hendricks notes that a large section of Java is productive and supports a large population. The soils are of volcanic origin. In neighboring Borneo, however, the soils, derived from weathered shales and sandstones, are laterized in large measure.

In addition to the low organic content and an impoverished nutrient condition, other factors also characterize the laterization process. A key feature in the development of a true laterite is the removal of SiO_2 from the profile. During the rainy season, the SiO_2 load of river waters is greatly increased. Iron and aluminum, however, are immobilized in the profile. The red, bricklike color is caused by the presence of these oxides. Moisture adequate for microbial decomposition and high temperatures promotes rapid decay or oxidation of organic materials and lessened acidification. Silica thus becomes more mobile, and iron and aluminum become less so. The silica-sesquioxide ratio, unlike that for the more acid process of podsolization, tends to be lower (see preceding calculation, p. 50).

Since iron oxides are good cementing agents, laterization over an extended period of time results in a product that is extremely resistant to weathering. When these oxides exceed 25% of the total composition, the material is incredibly hard, water resistant, and essentially immune to further weathering. At this stage of development it is suitable for building purposes, and many man-made structures in the tropics, including the ancient temples of Angkor Wat, are built of laterite.

The expansion of human activities is conducive to laterization. Clearing forests by cutting and fire is widespread throughout the tropics. A particularly ambitious scheme is underway to build roads and new cities and expand agricultural production in the Amazon forests of Brazil. The shortcoming of such activities, particularly in areas where more food is needed to feed growing populations, is that the resulting laterization will lessen the production potential of the soil. Abandoned and eroded lands attest to this fact. As long as new areas are available, the practice of shifting agriculture has been the temporary solution to crop production and expanding populations.

Another practice that has serious implications concerning the soil resource is the defoliation of tropical forests. This activity is carried out by the aerial spraying of various chemical herbicides. Such an operation has been conducted by the United States military in South Vietnam for nearly 10 years. It is estimated that about one half of that country's soils are "potential" laterites. Data supplied by Westing show that more than 6 million acres have been subjected to aerial spraying since 1961 and that some of this acreage has been sprayed more than once. Supporting data are provided in Table 3-6. This acreage comprises about 12% of the total land area of South Vietnam. It should be realized that whether the intention is defoliation or actual clearing, the end result could conceivably be the same. A cursory study by the United States government would suggest that no serious laterization threat exists in Vietnam.

The tropical forests of the world have evolved by selective processes whereby environmental optimization has been achieved. Destruction of these complex communities, their biological influences, and cooperative interactions and symbioses such as direct nutrient cycling produces impoverished soils and increases the possibility of laterite formation. Fig. 3-11 shows the extensive destruction of South Vietnamese mangrove forests caused by herbicide spraying. The probability of laterization in these habitats is undoubt-

Table 3-6. Estimated areas of forest and cropland sprayed in South Vietnam since 1962*

YEAR	THOUSANDS OF ACRES		
	FOREST	CROPLAND	TOTAL
1962	5	1	6
1963	25	<1	25
1964	83	10	93
1965	156	66	222
1966	741	104	845
1967	1486	221	1707
1968	1526	170	1696
1969	1404	115	1519
1970	220	33	253
Total	5646	721	6366

*From Westing, A. H. 1970. In B. Weisberg (ed.). Ecocide in Indochina. Canfield Press. San Francisco. Data com-
piled by A. H. Westing. Windham College, Putney, Vt., from several releases by the U.S. Department of Defense.
These data do not take into account repeated sprayings of the same area, which is estimated at 20%.

Fig. 3-11. Mangrove forests in South Vietnam that have been treated with herbicides
(Courtesy Herbicide Assessment Commission of the American Association for the Advancement of Science.)

edly low, yet the long-range effects of these disruptive influences on the soil, its developmental processes, and its biotic communities are largely unknown.

SUMMARY

The soil is a dynamic community. It consists of numerous and diverse organisms that play essential roles in ecosystem function. Humification conditions the soil by adding organic matter, improving infiltration, and increasing storage capacity. Mineralization is the final process in organic turnover and results in the release of nutrient elements that can then be recycled through the ecosystem. Parts of the nitrogen cycle take place in the soil. Certain bacteria are known to scavenge CO, converting this gas to other products. All these vital activities depend on the biological state of the soil. Foreign substances such as chemicals and long-lived pesticides should be viewed as potential threats to ecosystem function until it is shown that they are absolutely safe for all living communities in the soil.

The profile characteristics of soils vary. Differences are caused by climate, covering vegetation, and parent material, among several other factors. There are three important soil-forming processes or trends. The two processes characteristic of forest areas are podsolization and laterization; calcification occurs under more arid conditions. Forest soils are less productive than those of grassland areas; they are more vulnerable to a decline in fertility and waste of nutrients because the soil contains less organic matter and more acidic residues and thus has a greater leaching potential. The abandoned farmlands in temperate zones that return to forest cover over time attest to early misuse. The current practice of felling and burning tropical forests for a shifting agriculture suggests that soil productivity in these areas declines rapidly.

In drier regions where water loss from soil surfaces is greater than in forest climates, profile developments are quite different. Reduced leaching results in mineral accumulations in the soil profile. Chernozems are grassland soils developed under the calcification process and are among the most fertile soils in the world. Adequate but not excessive rainfall coupled with moderate annual temperatures create a favorable balance between plant growth and humus accumulation on the one hand and minimal loss of nutrients on the other.

The higher rainfall implicit in areas of podsolization and laterization produces more growth, generally in terms of wood and litter, but it also develops soils in which nutrient budgets are marginal and always subject to leaching. Desert soils are especially vulnerable to salinization, a process accelerated by irrigation. Such problems are increasing on a worldwide basis as more submarginal lands in dry climates are being brought into food production. The development of a soil profile is an evolutionary process in which an equilibrium among several basic influences is effected. The physical and biological characteristics of these different soils and the factors that control them are significant considerations in how we use this valuable resource and ensure its biotic productivity and functional permanence.

DISCUSSION QUESTIONS

1. A most valuable but often neglected resource is the soil, with its nutrient capital and accumulated organic matter. Explain the importance of maintaining a viable soil environment.
2. What is laterization? Explain its effect on soil fertility and plant productivity. What factors hasten the process?
3. What are the implications of long-lived chemicals such as pesticides on biological processes in the soil? Explain.

REFERENCES

Belville, L. S. 1971. They are destroying Brazil's paradise. Int. Wildlife 1:36-40.

Brock, T. D. 1966. Principles of microbial ecology. Prentice-Hall, Inc. Englewood Cliffs, N.J.

Brown, J. 1967. Tundra soils formed over ice wedges, northern Alaska. Soil Sci. Soc. Am. Proc. 31:686-691.

California Department of Water Resources. 1965. San Joaquin Master Drain. Bulletin No. 127. State of California. Sacramento, Calif.

Colorado River Board of California. 1962. Salinity problems in the lower Colorado River area. State of California. Sacramento, Calif.

Cook, R. E., W. Haseltine, and A. W. Galston. 1970. What have we done to Vietnam? In B. Weisberg (ed.). Ecocide in Indochina. Canfield Press. San Francisco.

Hendricks, S. B. 1969. Food from the land. In Committee on resources and man. Resources and man. W. H. Freeman & Co., Publishers. San Francisco.

Inman, R. E., R. B. Ingersoll, and E. A. Levy. 1971. Soil: a natural sink for carbon monoxide. Science 172:1129-1231.

Jackson, R. M. and F. Raw. 1966. Life in the soil. St. Martin's Press, Inc. New York.

Jaffe, L. S. 1968. The global balance of carbon monoxide. In S. F. Singer (ed.). Global effects of environmental pollution. Springer-Verlag New York, Inc. New York.

Jenny, H. 1958. Role of the plant factor in the pedogenic functions. Ecology 39:5-16.

Labova, E. V. 1967. Soils of the desert zone of the USSR. (English translation.) U.S. Department of Commerce Clearing House for Federal Scientific and Technical Information. Springfield, Va.

Orians, G. H. and E. W. Pfeiffer. 1970. Ecological effects of the war in Vietnam. Science 168:544-554.

Persons, B. S. 1970. Laterite, genesis, location, use. Plenum Press. New York.

Peterson, J. R., R. S. Adams, Jr., and L. K. Cutkomp. 1971. Soil properties influencing DDT bioactivity. Soil Sci. Soc. Am. Proc. 35:72-77.

Schaller, F. 1968. Soil animals. The University of Michigan Press. Ann Arbor, Mich.

Soil Conservation Service. 1975. Soil taxonomy. Agriculture Handbook No. 436. U.S. Department of Agriculture. Washington, D.C.

Tedrow, J. C. F. and H. Harris. 1960. Tundra soil in relation to vegetation, permafrost and glaciation. Oikos 11:237-249.

Triplett, G. B., Jr. and D. M. Van Doren. 1977. Agriculture without tillage. Sci. Am. 236:28-33.

Tschirley, F. H. 1969. Defoliation in Vietnam. Science 163:779-786.

Weisberg, B. (ed.). 1970. Ecocide in Indochina. Canfield Press. San Francisco.

ADDITIONAL READINGS

Boyko, H. 1966. Salinity and aridity. Monographiae Biologicae. Dr. W. Junk N. V., Publishers. The Hague, Netherlands.

Bunting, B. T. 1965. The geography of soil. Hutchinson University Library. London.

Chapman, V. J. 1960. Salt marshes and salt deserts of the world. Interscience Publishers, Inc. New York.

Eyre, S. R. 1963. Vegetation and soils. Aldine-Atherton Publishing Co. Chicago.

Garrett, S. D. 1963. Soil fungi and soil fertility. Pergamon Press, Inc. New York.

Judson, S. 1968. Erosion. Am. Sci. 56:356.

Kellogg, C. E. 1950. Soil Sci. Am. 183:30-39.

McNeil, M. 1964. Lateritic soils. Sci. Am. 211:96-102.

Soil Science Society of America. 1971. Glossary of soil science terms. The Society. Madison, Wis.

Westing, A. H. 1971. Leveling the jungle. Environment 13:8-12.

Whiteside, T. 1970. Defoliation. Ballantine Books, Inc. New York.

4 Diversity and structure

LIFE IN THE PAST

Diversity is an expression of the different kinds of life that inhabit the earth. Today there are over 1 million living species of animals and at least 250,000 species of plants that are recognized in biological classification. Their evolutionary lineage covers millions of years. From bits and fragments of the fossil record—bones, teeth, pollen grains, wood, leaves, and other preserved materials—ancient life emerges as a series of ever-changing structures. Subject to the vagaries of geological processes and climatic effects, viable habitats were modified and sometimes even destroyed on a periodic basis. Life forms that were affected by these changes perished, migrated away from zones of stress, or were perpetuated as new adaptive forms developed through natural selection. Those traits with survival value in a changing environment were thus transmitted to succeeding generations.

Redwood trees of the early Tertiary period (Table 4-1), related to those found today in California, once grew as far north as Canada and Greenland. Many other temperate forest genera were also found in these high latitudes (the higher the latitude, the greater the distance from the equator). This intermingling of species, which formed a worldwide community, is called the Arctotertiary Forest. The distribution of this ancient temperate forest was accompanied by corresponding northward displacements of tropical trees such as the fig into the continental United States. From this particular story as it is depicted in the fossil record, we might infer that in these ancient times climates were much milder than those presently found in these regions. This hypothesis is more plausible than the assumption that the ecological tolerances of these fossil species differed markedly from those of their contemporary relatives. It is acceptable particularly where morphological similarity and closeness of relationships are most conspicuous.

The evolution of the horse is a study in species adaptation under the selective pressures of changing climate and corresponding shifts in vegetation. The species arose 50 million years ago in North America (and Europe at a time when there were land bridges to the New World) as a small forest animal whose diet consisted principally of browse

Table 4-1. Timetable of periods and epochs of Cenozoic era (older Mesozoic and Paleozoic eras are not included here)*

PERIOD	EPOCH	BEGINNING OF EPOCH (yr ago)
Quaternary	Recent	10,000
	Pleistocene	1 million
Tertiary	Pliocene	11 million
	Miocene	25 million
	Oligocene	40 million
	Eocene	60 million
	Paleocene	70 million

*Modified from Holmes, A. 1965. Principles of physical geology. The Ronald Press Co. New York.

plants such as shrubs and trees. The gradual uplift of the Rocky Mountains in the Miocene epoch cut off much of the moisture flow that had moved inland from the Pacific Coast, creating an increasingly drier climate. Forests were replaced by grasslands, which require less rainfall. As a result, one line of the fossil record shows a progressive change in the horse's bone structure from that of a forest inhabitant to that of a fleet-footed plains animal, much larger in size than its early progenitors. The horse adapted to eating gritty grasses, which typically have a high silica content, for in association with this change in food habits was the development of high-crowned grinding teeth, ridged with enamel, that grew continuously as the surfaces were worn down.

Although the horse disappeared from North America for a period of a few million years for reasons not completely understood, the line was perpetuated in Europe and Asia. Then about 500,000 years ago the species reappeared on the North American continent. The fossil record documents its presence here until approximately 8000 or 9000 years ago, when other animals also disappeared. One theory is that man, a hunter becoming more proficient with weapons and fire, may have been responsible for overkill. Whatever the reason, the horse was able to survive in the Old World and was returned to the Americas by Spanish explorers in the sixteenth century.

Fossils dating from the late Pleistocene epoch give evidence that cold-climate trees such as the spruce grew far south of their present distribution in North America. Pollen of spruce, fir, larch, and pine, an association of conifers of the contemporary Canadian forest, has been found in several bogs in the western Missouri Ozarks. In this region skeletal fragments found in deposits 13,500 to 32,000 years old indicate that the mastodon, horse, musk ox, and giant beaver were also present at some time during this span.

It is in such habitats that pollen grains and other materials escape decomposition and in their preserved form become a documentary record of the life of a given period. Today the paleontologist can extract the fossils from these areas that once were filled in by silt and organic deposits; identify, count, and collate them; and so piece together a biological spectrum. Implicit also in the fossil record is the factor of climatic change, for the study of past life not only provides insights into the relative numbers and kinds of plants and animals that make up communities but also tells us something about the environmental relationship as well.

Even more recent fossil evidence from about 4000 years ago indicates that many savannah-type animals such as the zebra and lion occupied portions of the southern Sahara Desert when the climate there was more moist. Today it is thought that their descendants occupy the plains and savannahs of Tanzania and Kenya in East Africa, having followed the retreating grasslands with the onset of aridity. Many other examples might be cited, and in sum they provide an exciting panorama of past changes in biota, their redistribution and renewal around the world.

It should be clear that vast spans of time were involved in these biological events. Despite decimation and even large-scale extinctions, and there were many, there were opportunities through escape, adaptation, or both for the perpetuation of biotic diversity. Arguments about the possible impending extinction of such contemporary life forms as the brown pelican and the Arabian oryx (see Appendix B) based on these past events are invalid. The geological record, which was thousands and millions of years in the making, is hardly comparable to a fleeting present measured in decades.

VARIATIONS IN BIOTIC DIVERSITY

Biotic diversity across the face of the earth has long been a subject of interest. If there are more plant and animal species in one region than in another, we should inquire into the reasons for these patterns. In general we can say that patterns of diversity for terrestrial plants and animals reflect a trend showing that the number of species increases toward the tropics. Cold climates support relatively few species of plants and animals. However, some regions such as the wet coastal tundras of the far north serve as the nesting ground for a host of migrating birds during the brief summer. Two of the most remarkable migrants are the golden plover and the arctic tern. Both travel thousands of miles to the Southern Hemisphere to spend the winter during their breeding season. Under such circumstances faunal diversity fluctuates with the season. Tramer has shown that the changes in diversity for migrating bird populations are more gradual across wide latitudinal gradients in winter than during the breeding season. Diversity during the latter period increases only within close approaches to the tropics (approximately 20 degrees north and south latitude), and even then the increase is rather abrupt.

In order to make equitable comparisons between regions, observations should be conducted in areas of equal size. As the area under consideration is expanded, more species obviously would be tabulated. Small islands, for example, have fewer species than larger ones, which usually offer more ecological opportunities for the accommodation of evolving forms and their success as species. The tops of high mountains widely separated from each other also serve as islands or units of ecological isolation. Mount Kilimanjaro and other East African block mountains that rise thousands of feet above the semiarid savannah are noteworthy examples. Despite the difficulty of generating equal-area values, such diversity data as are available on a general continental basis are nonetheless very instructive.

One of the regions richest in bird life is Colombia, South America. Farther north, in Guatemala, there are only one third as many species. For every bird species in Greenland, there are 25 in Colombia. Tanzania, with 289 different species of mammals, probably has a greater concentration of such animals than any region of comparable size. Great Britain has but 47 species. The great proliferation of cichlid fishes in several East African lakes is well known. This one family of freshwater fishes derived from a riverine ancestry was remarkably preadapted for lake life. According to data compiled by Lowe-McConnell, this single group of fishes is represented by well over 100 species in some lakes, with almost 200 in Lake Malawi alone. In all of Europe, for all families, the number of fish species totals about 200. Additional examples of this increase in diversity toward tropical regions are presented in Table 4-2.

In the tropical rain forests of the Amazon, there are an estimated 2500 species of trees. When botanical surveys are completed, it may be found that the number is even greater. The deciduous forest of eastern North America contains slightly more than 200 species of

Table 4-2. Animal diversity in different latitudinal zones showing increasing numbers of species toward tropical lands*

TAXA	REGION	NUMBER OF SPECIES
Birds	Greenland	56
	New York State	195
	Guatemala	469
	Panama	1100
	Columbia	1395
Mammals	Northern Alaska	29
	Arizona	135
	Costa Rica	196
Fishes	Great Lakes, North America	172
	Central America	450
	Equatorial Brazil	1383
	Argentina	393
Snakes	Canada	22
	Continental United States	126
	Mexico	293
Ants	Tierra del Fuego	2
	Patagonia	59
	Buenos Aires region	103
	Sao Paulo region	222

*Data from Fischer, A. G. 1960. Evolution **14**:64-81; Cockrum, E. L. 1962. Introduction to mammalogy. The Ronald Press Co. New York; and Lowe-McConnell, R. H. 1969. Biol. J. Linn. Soc. **1**:51-76.

trees, including small understory species such as dogwood, sassafras, and persimmon. Across several thousand miles of latitude, the reduction in species in this comparison exceeds 80%. If we advance further northward into the extensive coniferous forests (communities of cone-bearing trees such as spruce and pine) of Canada, another decrease of similar magnitude is demonstrated. Scarcely more than two dozen trees, including the widespread black and white spruces, are endemic to these high-latitude forests that stretch across the continent.

The grasses are the most cosmopolitan of plant families. A very large subfamily, the panicoids, illustrates species attenuation away from the tropics. Within this natural group, the genus of panicums includes hundreds of species, the greatest number of which are found in tropical America. The second greatest concentration occurs in Africa. In the same subfamily, the bluestems are another large genus. These are massed in the African and Asian tropics, although sizable numbers are native to the Americas as well. Certain other grasses, including the true festucoids, are characteristically cool-region types, and although some genera have numerous species, they do not approach the proliferation encountered in the tropical panicums and bluestems. Ecological sequences of panicoids and festucoids also occur within the tropics in regions of sufficient topographical relief. In Costa Rica, for example, such familiar temperate grasses as bluegrass, brome, redtop, and timothy are restricted to the higher, cooler zones. At these elevations the species-rich

Table 4-3. Comparison of number of native seed plants (excluding such species as ferns and mosses) cataloged for several regions across a wide latitudinal range*

REGION	AREA (sq miles)	NUMBER OF SPECIES
Greenland	1,000,000	400
Arctic Alaska	200,000	700
Sonoran Desert, Baja California	24,100	1500
British Isles	120,000	1600
Indiana	36,000	1900
Central Coast Range, California	24,500	3000
North and South Carolina	80,000	3150
Costa Rica	18,400	8000

*Data from Seward, A. C. 1929. Nature **123**:455-462; Porsild, A. E. 1951. Can. Geog. J. **42**:120-145; and Raven, P. H. 1967. Introduction to the Costa Rican flora. Organization for Tropical Studies. San José, Costa Rica.

panicoids characteristic of the lowlands are less prevalent or do not occur at all. Comparative data on floristic diversity in selected locations are given in Table 4-3. The density ratio of species on an equal-area basis relating tropical Costa Rica and the temperate British Isles is about 35:1. When arctic Alaska is compared with Costa Rica, the ratio is even greater.

The relative richness of tropical regions is also demonstrated by the coevolution of phytophagous (plant-eating) insects and their specific hosts. Insects as a group are notably diverse in the tropics. Their interaction with plants is an interesting aspect of biology and is thought to be a significant factor in the promotion of diversity in both plants and animals. Some of these mutualistic relationships such as those between ants or butterflies and their respective plant hosts have been developed to a high degree of specificity. Janzen has demonstrated that in the case of the ant-acacia tree interaction, the number of obligate associations (combinations involving particular species that cannot effectively be substituted) diminished with increasing distance from the tropics. Greater environmental fluctuation and the less favorable conditions typical of higher latitudes apparently are among the important factors in the decreasing incidence of such animal-plant liaisons.

A corollary of these conclusions is that man-made stresses in all climatic settings also magnify environmental perturbations. Such disturbances can with equal and perhaps greater facility destroy delicately balanced host relationships that have evolved as an intrinsic aspect of community structure and stability. Such interactions between species demonstrate that if we are to understand ecosystem function, we must approach it on a community basis. This approach takes into account the manner in which one part of the system affects another and the fact that all parts (species) are essential to the integrated whole. For example, the grasses characteristically are wind pollinated; yet in dense tropical forests in which air movements near the ground are minimal, various insects have assumed the role of pollinating agents. Indeed both types of pollination have been observed in grasses in the same generic group, whose species are found from the temperate zone to the tropics. The universality of the phenomenon of dual pollination in related groups of grasses awaits further study.

CAUSAL RELATIONSHIPS

Based on the few examples just cited, we may ask why there are more species native to tropical areas. Various explanations and theories have been offered to account for this difference in terrestrial diversity. One possible explanation is that the tropics are the regions on earth that have had the longest continuous biotic occupancy. It might be said that organic evolution has had more time for selecting, testing, and producing new forms. This presupposes, of course, that far northern and southern lands have not as yet reached their full biotic potential, an issue open to question. Nonetheless, increasing cold and advancing ice did not disrupt the tropical biotas. Those glacial oscillations of the Pleistocene common to the higher latitudes, particularly in North America and western Europe, left tropical regions largely unaffected, at least in a destructive sense. Their biotas thus were not subject to the same effects of depletion and possible extinction impressed on other communities such as the deciduous forests of Europe. Tropical Africa probably has been free of biological upheaval longer than any other region on the earth. Furthermore, in regions of complete disruption, biotas had to be reestablished following each glacial retreat by new organic invasions and the time-requiring processes of developing stable communities. In Chapter 7 we shall discuss the various aspects of biotic succession and its functional relationship to the diversity of communities.

Climate is proposed as another causal factor in differing degrees of diversity. Tropical climates are notably distinct from those of the temperate zones. A principal feature is a lack of freezing temperatures, for frost occurs commonly in the tropics only on the higher mountains. Yet elevation allows greater environmental heterogeneity and increases the possibilities for biotic development. The number of potential life zones above sea level obviously is at a maximum in the tropics and diminishes as distance from the equator increases. This relationship is depicted in Fig. 4-1 and assumes, of course, that comparable elevations are possible at all latitudes.

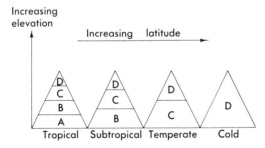

Fig. 4-1. Diagrammatic illustration of relationship of latitude to elevation showing decrease in variety of life forms possible toward the colder latitudes. Tropical climates are indicated by *A,* subtropical by *B,* temperate by *C,* and cold by *D.* In tropical latitudes there is maximum potential for variety. Elevational comparisons with latitude are not strictly accurate, however. One difference is day length, so that although temperatures in the *C* zone, for example, are more temperate-like, it is more accurate to refer to these high-altitude zones as tropical montane, tropical alpine, and so on.

In the tropics there is also a pronounced uniformity in average temperatures on a month-to-month basis. Lack of seasonality, if we exclude the effects of precipitation for the moment, is thus most significant in the tropics. Unlike the extremes sometimes encountered in the temperate zone, especially in continental interiors, the temperature variation throughout the whole year may be a matter of just a few degrees. In tropical Singapore, for example, the annual range between temperatures in the warmest month and in the coolest month is less than 3° C. This compares with 26° C for St. Louis, Missouri and 35° C for Willeston, North Dakota. Other data for comparison are presented in Table 4-4. The very small annual variations experienced in the tropics provide a monotonous but stable environment. This is particularly true of those regions in which high rainfall is

Table 4-4. Mean monthly and annual temperatures (°C) for tropical, temperate, boreal forest, and arctic tundra stations, illustrating greater variation with increasing latitude * †

STATION	J	F	M	A	M	J	J	A	S	O	N	D	AN-NUAL	SEASONAL RANGE
Tropical Belem, Brazil	26	25	26	26	26	26	26	26	26	26	27	26	26	2
Temperate Nashville, Tennessee	4	5	10	15	20	24	26	26	22	16	.9	5	15	22
Boreal Port Arthur, Ontario	−13	−12	−6	2	8	14	17	17	12	6	−3	−10	3	30
Arctic Fort Yukon, Alaska	−29	−27	−18	−6	7	14	16	13	6	−6	−21	−30	−7	46

*Modified from Kendrew, W. G. 1961. The climates of the continents. The Clarendon Press, Oxford, England.
†Data have been converted from Fahrenheit readings. Italicized values represent seasonal extremes.

Table 4-5. Relationship between daily and seasonal temperatures for two tropical stations in Costa Rica—Golfito at sea level (I) and Villa Mills in the alpine region (II) *

		J	F	M	A	M	J	J	A	S	O	N	D
Average daily maximum	I	34	35	34	34	34	33	33	33	33	33	33	34
	II	16	18	19	19	19	18	18	18	17	17	16	17
Average daily minimum	I	23	23	24	24	23	23	25	23	24	24	24	24
	II	4	4	4	5	6	6	5	5	5	5	5	4
Average daily range	I	11	12	10	10	11	10	8	10	9	9	9	10
	II	12	14	15	14	13	12	13	13	12	12	10	13
Average monthly temperature	I	28	29	29	29	29	28	29	28	29	29	29	29
	II	10	11	12	12	13	12	12	12	11	11	10	11

*Modified from Servicio Meteorologico Nacional. 1962. Boletin-meteorologico. San José, Costa Rica. pp. 1-42.

evenly distributed throughout the year. Where pronounced dry seasons occur, the annual temperatures show more fluctuation, as the period of moisture deficit is associated with somewhat higher temperatures (see Chapter 8 for further details).

Another important feature of tropical climates is that these small annual temperature ranges are less than the daily variations. The latter, then, may be of special ecological significance, particularly in the higher elevations that experience considerable cooling at night. The reason for this difference is, of course, that the average temperature for any given day is more like any other day than are the readings between daylight and darkness for any 24-hour period. Data for two tropical stations, presented in Table 4-5, explain these relationships. One station is in the lowland and the other in the high alpine region. Despite differences in absolute values as a result of elevation, the same effect is noted in each.

The conditions necessary for optimal biotic development include favorable temperatures characteristic of the tropical ecosystem together with ample rainfall. In a community sense growth is continuous, even though particular species undergo rest periods or temporary cessation of production. Life cycles overlap, creating a potential for more production and growth and for the building of biological structure. Table 4-6 presents an example of functional overlap as demonstrated in the fruiting periods of 18 species of *Miconia*, a tropical genus of shrubs and small trees. The advantages in such temporal

Table 4-6. Spacing of fruiting period throughout the year displayed by 18 species of the melastome genus, *Miconia,* in Trinidad; species are ecologically and morphologically similar and help provide a reliable food supply for many bird species * †

SPECIES	J	F	M	A	M	J	J	A	S	O	N	D
1		•	X	X	X	•						
2				X	X	•						
3			X	X	X	X						
4				•	X	•						
5				•	X	•						
6						X	X	•				
7						X	X	X	X	•		
8							•	X	X			
9							•	X				
10							•	X	X	•		
11							X	X	X	•		
12								•	X			
13								•	X	X		
14		•						•	X	X	X	X
15									X	X	X	X
16		•							•	X	X	X
17	X	•									•	X
18	X	X	X	•							•	X

*Modified from Snow, D. W. 1964. Oikos **15**:274-281.
†Dots indicate partial months.

Table 4-7. Biomass expressed as weight of total aboveground vegetation in pounds per acre *

COMMUNITY TYPE	VEGETATION BIOMASS (total standing crop)
Tropical rain forest	$400\text{-}500 \times 10^3$
Subtropical forest	$300\text{-}325 \times 10^3$
Temperate deciduous forest	$200\text{-}300 \times 10^3$
Boreal conifer forest	$80\text{-}200 \times 10^3$
Dwarf shrub tundra	$20\text{-}30 \times 10^3$

*Modified from Rodin, L. E. and N. I. Basilevic. 1968. World distribution of biomass production. In Proceedings of the UNESCO symposium on functioning of terrestrial ecosystems at the primary production level. Copenhagen.

divisions of activity are self-evident. Both plant producer and animal consumer are benefited, the first by avoiding competition and the second by having a continuous supply of food (see discussion of niche, p. 75). The cost of community metabolism is also less under mild and constant conditions than it is in cold and fluctuating environments, in which peak food production and storage are confined to abbreviated periods. As we would surmise, this seasonality becomes more emphatic toward colder regions and in areas in which pronounced dry seasons are the rule. Yet, as in all communities, the expenditure of energy is a continuous process, regardless of the food-making potential.

As the biomass of tropical vegetation increases, there is a concomitant increase in structural complexity. More horizontal strata are present, providing for a greater elaboration of subcommunities, or synusiae, within the whole. Epiphytes and aerial plants such as orchids, ferns, and aroids (the latter including such species as philodendrons) grow profusely, forming literally one layer of plants on another. Bird life is also rich and varied; different species are often confined to particular levels aboveground. Thus the total biotic potential is amplified as the habitat possibilities for both animals and plants are increased through more plant growth, more available energy, and greater structural complexity. The total dry-weight biomass of a rain forest may range up to half a million pounds per acre, approximately twice that of a temperate deciduous forest. Comparative data for these and other forest communities are provided in Table 4-7. The exploitation of these biotic resources through specialization minimizes direct competition between allied forms and implements the coexistence of numerous species in a common habitat. Compare the complexity of the rain forest with the relative structural simplicity of the boreal forest discussed in Chapter 8.*

LIMITING ENVIRONMENTS

The presence of a native plant or animal in a stable community attests to the successful interplay between environmental controls and a given range of physiological tolerances. Seen in vertical scale as a segment of the earth's physical environment, these viable habitats make up a comparatively shallow layer of life, the biosphere. The fact that it is

*For a more detailed analysis of causal relationships in tropical diversity, the interested reader should consult Pianka, E. 1974. Evolutionary ecology. Harper & Row, Publishers. New York.

shallow emphasizes its finite nature; the biosphere cannot very effectively be expanded in any direction. In the ocean there are restrictions on the growth of green plants imposed by light-intensity values that decrease with water depth and rapidly become inadequate for photosynthesis. Here also nutrients for plant growth such as nitrates and phosphates generally occur at low concentrations because of the tremendous dilution factor. On the higher mountains, low temperatures, perennially frozen soils, and the resultant lack of available moisture are effective barriers to the development of vegetation. The lack of photosynthesis in these situations obviously also inhibits to a certain extent the development of animal life.

In the Himalayas at elevations above 20,000 feet, several arthropod forms, including springtails and spiders, exist, although no plant life occurs in their immediate vicinity. In these interesting yet sparsely inhabited environments, animals are able to subsist on organic debris that is carried by upward drafts from lower elevations where conditions are more amenable to plant growth. These airborne materials provide the necessary food for the springtails, which in turn are preyed on by spiders. In the lower depths of the ocean, various marine forms subsist on a "rain" of organic particulates that drift down from the upper levels in which light energy is adequate for photosynthesis. These outermost fringes of the biosphere emphasize the tenuous nature of life. More importantly, if the slightest potential for life exists, strategies are developed through time to effect necessary accommodations. The process of exploiting finite environments through selection and adaptation is part of a never-ending contest for survival. Life is not easy to come by.

Increasing human pressure on the environment is a selective process that is often harsh and uncompromising. Our strategy is not always meant to be one of accommodation. All too frequently it is one of drastic change and destruction that inflicts a stress exceeding the tolerance capacity of the affected population. If the population cannot adapt, the change becomes a threat to its continued existence as a species. Present-day modifications such as those induced by the pollution of lakes and rivers or the indiscriminate clearing of rain forests are comparatively rapid. In some cases such as the application of agricultural pesticides, the impact may be worldwide. We can conclude with some assurance that there are no retreats as there were in the geological past, when large-scale movements of biotas to more suitable environments were possible. Today's technology is closing in on every corner of the earth. What effect, for example, will the construction of the 800-mile Alaskan pipeline have on the annual migration of caribou (Fig. 4-2)? What of the atomic testing at Amchitka Island or the proposed military airstrip on Aldabra? This is not to question social progress, but we need to seek alternatives that would achieve the same results for ourselves and at the same time preserve the environment.

Since the biosphere has limits, any drastic modification means an actual reduction in the living space available for some organism or group. Visualize the geographical range of a given animal species. The territory is extensive and contains a continuity of suitable habitats to answer the species' needs for food and cover. As inroads are made, group populations of the species become less continuous and more isolated. Eventually these groups may become so scattered and sparsely distributed that the species' potential as a viable, reproducing unit is placed in jeopardy. A timely example of human encroachment

Fig. 4-2. Caribou in the Alaskan tundra, with the Brooks Range in the background. (Courtesy Atlantic Richfield Co., Anchorage, Alaska.)

is seen in the Serengeti National Park in East Africa. This is a very large area indeed, consisting of approximately 5600 square miles or more than 3½ million acres. Yet this designated sanctuary is not entirely satisfactory as an ecological unit for large numbers of animals. Why should this be so?

Despite the vast acreage involved, various species of herbivores such as the wildebeest move outside the park limits, seeking grass and water during their annual migration. The ancestral (and superior) grasslands have not been set aside in some areas because of the grazing requirements of domesticated animals, so that these wild plains animals must compete with ever-growing numbers of cattle and other livestock. A generalized pattern of migration throughout the year is shown in Fig. 4-3.

Associated with these massive herbivore movements is a contingent of predators, including the lion, cheetah, hunting dog, and even the hyena (once thought to be mainly a scavenger). A pride, or family group, of lions, the only social cat, may have territorial boundaries extending over many square miles. Its movements are more wide ranging during the dry season with some overlapping of territories and accordingly become more integrated with the increased mobility of grazing animals.

As human activities increase, the condition of these plains animals, which must migrate to survive during the 5-month dry season, is expected to become more critical. Similar questions of the survival of given species as viable populations can, of course, be

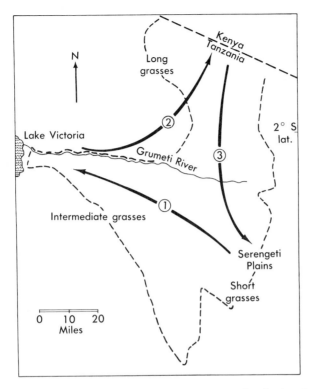

Fig. 4-3. Map of Serengeti National Park showing the generalized migration pattern of the plains herbivores. With the onset of the extended rainless season in June, animals leave the eastern dry plains section for the "corridor" (indicated by arrow *1*), where grass and water are more plentiful. Subsequent movements are also indicated. With the resumption of rains in October and November, the migrants, including wildebeest, zebra, and gazelle, return to the nutritious short grasses on the treeless plains.

posed for many other wild species around the world. Concerns voiced by biologists everywhere apply to industrial societies as well as to those with less advanced development. The responsibility for preserving biotic diversity through adequate allocation of ecological resources should transcend national boundaries and political differences. It is a problem that should receive recognition within a worldwide frame of reference.

DIMINISHING DIVERSITY

Why is there concern for species preservation in natural systems? Cannot social progress be made equally well without certain plants and animals? The dodo is gone; so are the passenger pigeon, the giant moa bird, and the guagga (a zebra), and perhaps it is possible to say that human progress has not been seriously impaired. The depletion of even a few species, however, is symptomatic of a condition that continues to worsen. There are today literally hundreds of species on the brink of extinction. It is estimated that perhaps 120 species of mammals and 150 species of birds have been exterminated during

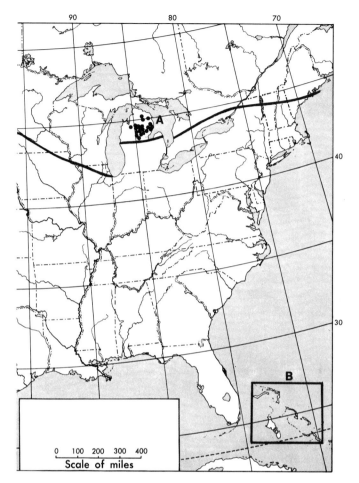

Fig. 4-4. Summer and winter range, *A* and *B,* respectively, of the Kirtland's warbler. The solid line marks the southern limits in the eastern United States for the transcontinental jack pine. (Redrawn from Mayfield, H. 1960. The Kirtland's warbler. Cranbrook Institute of Science Bulletin No. 40. pp. 9-33. Bloomfield Hills, Mich.; base map Copyright © by Denoyer-Geppert Co. Chicago; used by permission.)

the last 350 years. At present 300 species of birds and 375 species of mammals are on the endangered list. If current trends are allowed to continue, these species, too, will disappear from the face of the earth within the next few decades. Appendix B is a list of those birds and mammals that are in the most critical danger of becoming extinct; some of them (the ivory-billed woodpecker, Texas to Florida, for example) may already be lost.

There are perhaps 8500 species of birds and over 4000 species of mammals in the world today. They make up a relatively small fraction of the animal kingdom, even after the arthropods (insects, spiders, mites, scorpions, and so on), which account for about three fourths of all species, are subtracted. However, a disproportionately large share of

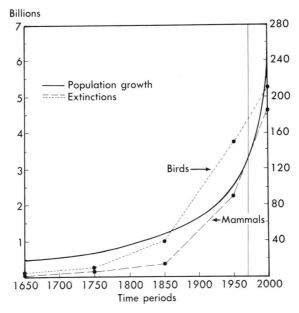

Fig. 4-5. Trends in human population growth and bird and mammal extinctions. Projection of the population curve beyond the vertical line is based on current estimates of population growth rate that indicate a doubling in approximately 35 years. Estimates for additional animal extinctions during the remainder of this century are based on those species whose remnant populations are approaching threshhold values. Included are 67 species of birds and 74 species of mammals. All critically endangered species are listed in Appendix B. (Redrawn from Ziswiler, V. 1967. Extinct and vanishing animals; a biology of extinction and survival. Springer-Verlag New York, Inc. New York.)

the threatened species are birds and mammals. In view of their key role in predator-prey relationships and particularly in view of their effect on the habitat, the impact of such attrition on the ecosystem might be far greater than mere numbers seem to indicate.

A most significant reason for these rapidly growing threats of extinction is the destruction of a given natural habitat, which has as its food and shelter base a plant community of some type. Although factors such as killing of predators, trophy hunting, and harvesting for economic reasons are important, deforestation, wetland drainage, urbanization, agricultural clearing, and pollution are the principal threats. According to Terborgh, the Kirtland's warbler that winters in the Bahama Islands is in serious trouble and may become extinct. Despite efforts to maintain suitable conditions in its summer range in Michigan, American commercial interests have depleted the pine-scrub habitat of the species in certain islands. The summer and winter ranges for the species are shown in Fig. 4-4. This example illustrates the need for cooperation around the world if native species are to be preserved. In California alone more than 250,000 acres of estuarine and tidewater habitat, classified as valuable wildlife lands, have been lost to filling and dredging. This story is being repeated in other states as well. Yet these interesting and unique habitats are valuable not only for maintaining species but also as a source of valuable protein foods and the retention of plant nutrients. This aspect of the problem will be discussed in Chapter 9.

A species is in danger of extinction when mortality, by whatever means, exceeds recruitment for a sustained period of time. Sooner or later under these conditions the population approaches a threshhold number. Beyond this point, recovery as a viable ecological unit may be virtually impossible. Hanging in the balance are such species as the blue whale, the whooping crane, and the Arabian oryx. It may accurately be said that there is an acceleration in the number of species being threatened with extinction. This growing danger, coupled with those extinctions that are already an established fact, is demonstrably related to the expansion of human populations and suggests that such growth is one of the underlying causes of diminishing diversity (Fig. 4-5).

Some may argue that these events do not represent related data and that the recorded extinctions could in fact demonstrate similar trends related to any number of other factors. Obviously such an argument is safe because it would be impossible to produce experimentally validated evidence that could prove a cause-and-effect relationship.

As more and more land is diverted to meet the needs of an expanding technological society, the capacity of wild species to resist the pressures of the human population will decline with increasing rapidity. There is today a mounting list of potential evolutionary cul-de-sacs, dead ends that offer no chance for new species trials in the environmental testing ground. We know that evolution of superior types through extinction of those that are less able to adapt sharpens the competitive edge and ensures biological survival in some form. Extinction without evolutuion, however, is an irreplaceable loss.

DOMINANCE RELATIONSHIPS

Dominance is an expression of relative species importance in the community. It tells us the extent to which a given species or group of species contributes to or influences the total community in some important aspect such as number of individuals, amount of the standing crop, or production of energy. If we could find a community in which all species were equal regarding the chosen parameter, no dominance would be expressed. This is quite an unlikely occurrence in nature. In a community in which all individuals belong to the same species, the dominance trait is maximum. Good examples are monocultures such as grain fields. Because all characteristics of structure and physiological function in this type of community are based on a single species, the degree of sameness is at the highest possible level. We would speculate, too, that dominance is inversely related to diversity, since the latter implies heterogeneous composition or lack of sameness. Thus diversity should be maximum where every individual in the community belongs to a different species, a situation in which there is obviously no dominance. This condition is called *maximum species equitability*. On the other hand, if all individuals are of the same species, it follows that biotic diversity is zero.

In a series of communities in nature in which diversity decreases, the reduction is usually attended by an increase in dominance on the part of certain species. There is "more dominance" expressed in a forest from central Canada than in a tropical forest, and, as we have noted, this dominance implies fewer species in the Canadian forest.

White spruce, for example, is one of the most widespread and important species in the Canadian boreal forest. It is one of the very few dominants that lends character to the community. In the rain forest there are also dominants, but in terms of the total community their relative influence is much less because there are so many more species. Environmental resources as well as contributions to community function such as production and influence on the habitat are more equally divided. The degree of dominance is less, and no one species really "stands out" in a mixed rain forest. This is not to say that there is not "high dominance" expressed in certain tropical forests, but dominance in these latter cases results from factors other than climate such as adverse drainage and limited aeration of the soil. Under these situations, soil operates as an effective selection factor. It is another example of how environmental extremes generally encourage dominance and diminish diversity in terrestrial ecosystems. This aspect of the interrelationships of species composition and ecological factors will be discussed further in Chapter 8. The same process is occurring on an ever-widening scale through technological modification or elimination of natural habitats.

In looking at the dominance-diversity interaction, diversity must be qualified further. The simplest index of diversity is the total number of species found in a given area. This approach, however, gives equal weight to both rare and common species. Two communities thus may vary in *degree of diversity* even though they have an equal number of species because of the differing importance of their constituent species.

In the following example the figures in the array represent abundance values for each species, expressed as percentages of total population:

SPECIES

	A	B	C	D	E	F	G	H	I	J	
Stand I	27	2	1	3	5	50	5	4	2	1	100
Stand II	8	16	9	11	10	12	16	5	4	9	100

According to this example, stand I expresses more "dominance" than stand II because of the disproportionately large abundance values exhibited by the two species A and F, which together constitute 79% (27% + 52%) of the total community factor. In stand II there is relatively greater equitability of species, or less dominance; despite the fact that both stands contain the same number of species, stand II has a higher index of diversity. Before turning to a mathematical solution, it is possible to reason that stand I is less diverse because several species such as C and J hold tenuous positions. Assuming that both stands are at a maximum level of potential development, these so-called minor species might readily be eliminated if there were to be a slight change in the environment favoring the dominants.

Several indices have been derived for the purpose of quantifying relative degrees of dominance and diversity. The Simpson index that follows is frequently used for dominance evaluations:

$$c = (n_i/N)^2$$

where

c = Index of dominance
n_i = Value for each species
N = Total value for all species

Using the data presented for stand I, the dominance value is calculated in the following manner:

$$c = \left(\frac{27}{100}\right)^2 + \left(\frac{2}{100}\right)^2 + \cdots + \left(\frac{1}{100}\right)^2 = 0.356$$

Similarly the value for stand II is 0.114. The two quantities, then, are a basis for comparing the dominance trait in these two communities and support our reasoning that stand I has *less* species equitability, or *more* dominance, than stand II.

There is also a relatively simple formula that may used to determine diversity. This is the Shannon index. Using the same data, we would speculate that stand I, showing the larger dominance value, would indicate a smaller diversity value than stand II. The Shannon index is borrowed from information theory developed by communication systems engineering. Briefly it states that more data can be stored in complex systems than in systems that are simpler and less diverse. Another way of expressing this relationship is to say that a diverse system with many different components inherently has more information to be discovered (or studied) than a simple one with few parts. Basically we are looking at diversity in terms of entropy, or uncertainty. The analogy can be applied with regard to the number of species in a biological system. The more species diversity, the greater the entropy, or lack of predictable structure. A comparison of the "magnificent confusion" of a rain forest and the structured rows of a cornfield demonstrates more entropy in the former. Therefore studying a system with more species should increase the opportunities to derive more kinds of information, in contrast to the opportunities available in a system in which most of the individuals are the same. This conclusion may

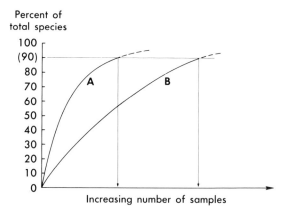

Fig. 4-6. Comparative differences in the number of samples required to ascertain a given percentage (in this example, nine tenths, although any fraction could be chosen) of the total species structure in simple as opposed to complex systems. The curve for the simple system with few species, designated as *A*, will rise more rapidly than the curve for the more diverse system with many species, indicated by *B*; *A* also requires fewer samples to yield the desired information.

be stated in another way by asking which system (or community) would require more samples for maximum information—one that is extremely diverse (seemingly confusing) or one with considerable homogeneity. Obviously the homogeneous community will yield its information in relatively fewer samples than the diverse community. In Fig. 4-6 we have idealized a set of curves showing this relationship between number of species in the community and sampling effort. In this example we have asked how many samples would be required to observe 90% of the total. Of course the curves will continue to rise as additional data are accumulated until all species are sampled. The Shannon index is as follows:

$$H = -\Sigma(P_i \log P_i)$$

where

H = Diversity index (or amount of entropy)
P_i = Relative importance of each species as a fraction of the total community (same as n_i/N in the Simpson equation)

We may determine the diversity index for stand I as follows:

$$H = -\Sigma\left(\frac{27}{100} \log \frac{27}{100}\right) + \left(\frac{2}{100} \log \frac{2}{100}\right) + \ldots + \left(\frac{1}{100} \log \frac{1}{100}\right) = 0.665$$

For stand II, H = 0.967. The comparative differences in dominance and diversity of the two communities are summarized as follows:

STAND	C	H	RELATIVE TRAITS
I	0.356	0.665	High dominance, low diversity
II	0.114	0.967	Low dominance, high diversity

Although the numbers themselves cannot tell us anything universally applicable to natural dominance and diversity relationships, they do allow us to compare communities and determine which has greater or lesser dominance and greater or lesser diversity. These examples have been presented to show clear-cut differences in community traits, but there are many situations in nature in which variations are more subtle and elucidation by inspection alone would be more difficult. Care must be taken by the investigator in adapting any model or formula, however, for its possible misuse, based on tenuous or inadequately derived assumptions about particular working situations. For example, studying part of a community or samples therein would most likely lead to erroneous estimates of community diversity.

NICHE

In every community each species occupies a particular niche, which in its *total aspect* is different from that of any other species. No two species have exactly the same niche, although a degree of replacement does occur. For example, when one species is eliminated from the habitat, resources that would have been used by that species may in part be utilized by another species or group of species. This phenomenon is designated as niche overlap. Generally we might say that the more species in the community, the greater the probability of some degree of overlap. In a sense this is a form of redundancy, which affords a measure of protection against disruptive factors that may develop from time to time. It could be considered analogous to the backup systems built into a spacecraft to ensure its safety. Similarly there is greater assurance of functional constancy in a diverse ecosystem because the extirpation of one species would not make as much difference as would the loss of a dominant species in an ecosystem that was less complex.

In nature, species with the physiological capacity to tolerate a broad range of environmental conditions often play a disproportionately large role in community function in terms of such factors as food production. They are the dominants and the principal components. Such organisms are said to occupy a "wide" niche. Thus one property of the niche is its variability. In those environments with comparatively sharp selection influences, we might expect some community expression or manifestation of species dominance and corresponding low diversity. Wide fluctuations in such factors as temperature, salinity, and water levels or a perennial stress such as cold or drought are naturally occurring factors. Human pollution in its many forms also constitutes environmental stress and works in the same direction, lessening species equitability.

Conversely, where there is greater uniformity and a more constant environment, as inside the canopy of a tropical rain forest, for example, external or physical selectivity is lessened. These circumstances offer a greater number of biological options, allowing more species to vie for niche position. Yet there are limits to the number of species even at the minimal densities necessary to maintain viable populations. In any finite system there must be accommodation to the competition factor. The answer, at least in part, is specialization. Competition is minimized through niche differentiation. Niches are accordingly numerous but "narrow" under these conditions, and specialization takes the form of different requirements for food, space, nutrients, and other critical factors. Niche separation may be of a temporal nature; also, closely allied forms that share *similar* requirements differ in that their peak demands on resources occur at different times of the year (Table 4-6).

Despite the specialization of one type or another that creates a number of narrow niches, most natural communities nonetheless display some measure of dominance. Generalized relationships have been devised to depict the niche aspect of the community as a whole. Expressed mathematically, niche width, compounded as a quantitative assessment of the community, shows a negative trend as the number of species increases. An interesting example of this inverse relation is shown in Fig. 4-7. Note that for a given number of species the calculated niche value is greater in the forest than in the grassland. In other words, the functional opportunities would appear to be greater in forest areas, a situation that is probably linked to greater biomass and complexity of structure, as compared to the grassland system. Such an augmentation of functional opportunities, if carried far enough, conceivably would lead to further niche differentiation. What is presently one species enjoying a relatively broad niche would then become two species, each with a narrower niche. In a gradually changing environment, we can also visualize realignments in different parts of the total niche structure. With different sets of variables being produced, there may be active selection for diverging populations and ultimately new species formation.

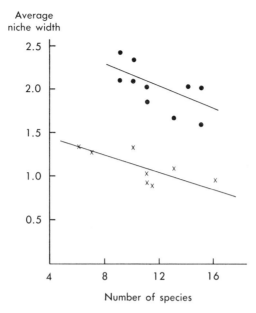

Fig. 4-7. Relationship between the average niche width (for the community) and the number of species in grassland and forest communities, indicated by crosses and dots, respectively. As the number of species increases, the average niche width becomes narrower. (From McNaughton, S. J. and L. L. Wolf. 1970. Science **167**:131-139.)

An interesting example of niche specialization is found in the finches that inhabit the Galapagos Islands, which are located on the equator approximately 600 miles west of Ecuador in the Pacific Ocean. There are 13 species of the so-called Darwin finches, and they are found nowhere else. Each species occupies a particularly distinctive food niche, a fact evident in eating habits and differences in bill structure. Some of the finches are ground feeders utilizing seeds and various plant materials, whereas others subsist on insects of different sizes. Some of these insect feeders are forest inhabitants in the higher elevations, whereas still others live in the desert and scrub vegetation of the lowlands. Despite these differences, it is theorized that all have a common ancestry in a single species. It is a sad commentary on our human exploitation of resources that one of these 13 species of finches first discovered by Charles Darwin more than a century ago is now extinct. The interested reader should consult Lack's work on this fascinating and instructive story of niche evolution in an environment that is both isolated and limiting. The two species of Galapagos finches shown in Fig. 4-8 illustrate differences in bill size and structure. *Geospiza* is a seedeater; *Certhidia* lives on a diet of insects.

Specialization is observed in the grazing habits of several savannah species in East Africa, where there is, on the whole, a spectacular diversity of animal life. In this region wildebeest, zebra, and gazelle demonstrate selective preferences not only for different plant groups but also for different parts of the same plant species (Table 4-8). When the available resources are divided in this manner, numerous niches can be maintained, and

Table 4-8. Variations in grazing habits of three savannah species in East Africa*

HERBIVORE	PART OF TOTAL DIET (%)		GRASS USE (%)		
	GRASSES	DICOTYLEDONS	LEAF	SHEATH	STALK
Zebra	100	0	—	49	51
Wildebeest	100	0	17	53	30
Thomson's gazelle	61	39	3	37	21

*From Gwynne, M. D. and R. H. V. Bell. 1968. Nature **220**:390-393.

A   B

Fig. 4-8. Two species of Galapagos finches, *Geospiza,* **A,** and *Certhidia,* **B.** The former forages for seeds on the ground; the latter is a tree dweller living on insects.

the intensity of competition is thus minimized. Many species are able to live together, each utilizing a designated part of the food resource and each contributing a measure of stability to the community. When a community comprises a variety of species, each having different feeding habits, the ecosystem is being utilized at a high level of efficiency. Food energy can flow through many channels of the system. If one channel (species) is eliminated, the functional efficiency is not impaired as severely as it would be if a particularly dominant species had been eliminated. When domestic cattle are substituted for these numerous wild species, it is shown that the available forage is not used effectively because feeding preferences become more nearly similar. The species "dominance" factor becomes more important; grazing pressures may be increased on certain plant species and decreased on others. Some species that were utilized previously are now ignored, so that in time there will be a noticeable shift in plant composition toward an increasing population of the more unpalatable types. The vegetation becomes

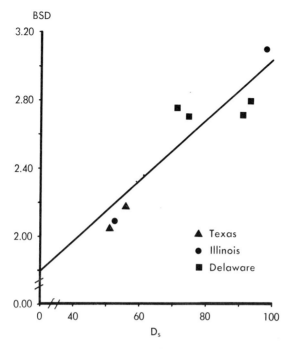

Fig. 4-9. Bird species diversity, *BSD,* and habitat heterogeneity, D_s, for shrub lands and forest vegetation in Texas, Illinois, and Delaware. (Redrawn from Roth, R. R. 1976. Ecology **57**:773-782; Copyright by the Ecological Society of America.)

more simplified and less protective of the soil, and the result is a general deterioration of the site and greater fluctuation of environmental factors.

Another example of specialization in the use of habitat resources may be cited using plant species. Earlier we noted the more or less even flow of production in tropical environments, attributable to nearly uniform year-round growing conditions. Using that same example (Table 4-6), we can demonstrate another kind of specialization that involves the way in which numerous species exhibit growth and reproduction that are temporally spaced on a seasonal basis. This strategy is also significant in reducing competitive interactions between closely allied life forms that presumably make similar demands on the ecosystem in terms of water and nutrients. In this way birds and other consumers that feed on fruits are assured a continuous supply of food. Plant diversity invites animal diversity, and it follows that the destruction of either impairs the function of the other as stable, self-perpetuating populations.

The degree of patchiness in grass-shrub vegetation shows a positive correlation with bird diversity, as illustrated in Fig. 4-9. Thus in addition to a vertical component such as the number of plant strata, there is also a horizontal factor concerning spatial arrangement between the dominant species of plants.

By visualizing a monoculture such as a crop species, it is possible to point out numerous differences between it and a native plant system. Crop species are developed for uniformity of structure and growth response. Life cycles are very nearly identical, and competition is maximized, since demands for light, nutrients, water, and space occur simultaneously. Crop plants are, of course, selected for high yield, and agronomists have enjoyed remarkable success in this regard. Unfortunately, however, adaptive traits that permit a species to meet and survive unknown contingencies are often lost. Indeed crop varieties might be called "specialists" with regard to their physiological requirements and limited adaptability. Because crop species are bred selectively to grow best within given climatic zones, local field environments within such zones are modified to suit the species' narrow ecological tolerance by fertilization, chemical control of diseases and insects, and irrigation.

The dominance of crop plants is artificially contrived and lacks the selective base characteristic of dominant species in a truly competitive system. If a crop species is to be successfully perpetuated, there must be continuous manipulation of the habitat, which in the process becomes unsuitable for many native species. Such manipulation in effect constitutes a choice between simplified systems consisting of sensitive species with a high food production potential and more diverse systems capable of self-regulation but not usually as productive of available food. With an increasing dependence on monocultures for food supplies and artificial dominance expressed by a single species, native plant and animal diversity must of necessity decrease in a finite system. Even so it is predictable that such trends cannot go on indefinitely. Eventually the environment will begin to deteriorate from its continuous use as a one-species system.

THE VALUE OF DIVERSITY

All stable communities require some kind of working system of checks and balances. Native species that are freed from competition or predation may often assume sufficient dominance to begin monopolizing available resources. As this monopoly continues to grow, further simplification is effected through the decimation of other species. Diversity prevents the development of excessive pressures on any part of the ecosystem (food, cover, space, and so on), and the threat of unstable conditions is thus minimized. When niches are vacated, unfettered dominance by species introduced from other regions or even other continents is a common occurrence. Weeds and pest animals are examples of opportunistic monopolies. In the New World, crabgrass, Russian thistle, the European starling, and the mongoose are but a few of such unwanted dominants whose ecological requirements are broadly suited to their newfound habitats.

A much-publicized illustration of the importance of predators was provided by the systematic program of exterminating the cougar and wolf in northern Arizona in the early part of this century. The far-reaching results were unexpected. The deer population, which normally would have been kept in check by these predators, increased from three- to fourfold, reaching population levels that far exceeded the carrying capacity of the habitat. Starving deer denuded the vegetation. Without an effective plant cover to protect the soil, there was excessive erosion; loss of soil, water, and fertility; and a general deterioration of the total environment. Yet other factors may also be involved, and the ecologist should be careful not to seek easy solutions to complex problems. In an interesting report on the history of the Kaibab deer populations, Burk presents other causal relationships, involving in addition to predator decimation, the removal of competing herbivores such as domestic livestock. Nonetheless human intervention and its impact on key species carries with it the potential to effect large-scale modifications in the biota and the environment.

The case of lake trout in the upper Great Lakes is another example of the removal of a predator, although in this case not deliberately. Marine lampreys were introduced into the Great Lakes through the Welland Canal. The lampreys, parasitic fish, reduced lake trout populations to low levels with far-reaching effects on commercial fishing. In addition, the alewife, which had also been unintentionally introduced from the eastern seaboard, was a principal prey of the lake trout. The reduced number of trout allowed the alewife population to multiply. They became a source of pollution as they died in tremendous numbers.

The introduction of the European rabbit to Australia also illustrates dominance problems associated with an adventive species. This animal responded to its newfound niche with a rapid population expansion. As a result, there was a large-scale destruction of the plant life on which the species subsisted. To combat a worsening condition, the myxomatosis virus was brought in to infect the rabbit population, and for a time the disease successfully checked further growth of the rabbit population. At present, however, the rabbits have developed a resistance to the disease, and this form of control is less effective. This is still another example of the results of a disruption of ecological balance. The countermeasures employed by people to correct mistakes are often inadequate and frequently create more problems than they solve.

Islands are particularly vulnerable to newly introduced plant and animal species that have the capacity for dominance. The island of Macquarie in the Antarctic Ocean is one of the most remote of all oceanic islands. This, coupled with a cold climate typical of the higher latitudes, accounts for its predictably sparse flora. The total plant life includes 38 species; several grasses constitute the principal ground cover. The introduction of the European rabbit in the late nineteenth century produced the classic pattern of selective utilization by a dominant species, followed by denudation and soil erosion. The same story of exotic species destroying island resources has been repeated around the world. In Hawaii, for example, the process began with the first goat set ashore by Captain Cook. The smaller and more remote the island, the less diversity there is. This fact emphasizes the fragility of island ecosystems and the care that should be taken in introducing plant and animal species.

Where dominance and host uniformity are present, conditions are often favorable for the outbreak of disease. The widespread occurrence of the southern leaf blight of corn in 1970 was made possible when the right combination of weather conditions and an abundant host created a situation that was favorable to the spread of the disease-causing fungus. Other situations are comparable. The development of high densities of the European rabbit in the British Isles caused epidemics of myxomatosis.

When individuals of the same species are crowded and numerous, the rapid spread of pathogens and pests through the population is made easier. The opportunity for pro-

Fig. 4-10. The killing of hawks and other raptor birds endangers the species and reflects public ignorance concerning their ecological importance as predators. (Courtesy Leland Payton, Columbia, Mo.)

liferation is lessened, however, when other nonhost species are present to break up the momentum of the attacking organism.

The loss of diversity through the expanding development of monocultures, killing of predators, introduction of aggressive exotics into fragile ecosystems, or simple absence of native species to an appreciable degree in harsh or fluctuating climates almost invariably means less regulation within the system. If we as custodians and managers of the land can keep diversity at the highest possible level, this most important element of self-regulation is retained. Without biological control, dominance by undesirable species creates the necessity for other forms of control, primarily through the use of synthetic chemicals such as DDT and 2-4,D. This substitution of artifacts for diversity and its natural checks is frequently harmful to nontarget species and contributes to the ever-growing instability of our environment.

There is another facet of species preservation that is perhaps not realized so readily when an extinction occurs. Every plant and animal species and every microorganism is a reservoir of biological characteristics not found in any other species, an irreplaceable reservoir. Each is an intrinsic part of the total "gene pool" of the biosphere. Yet we continue to offer bounties on valuable predators, and everywhere there is still a rank disdain for wild species (Fig. 4-10). It is this irreplaceable genetic wealth on which the biosphere depends for its future adaptations. We do not know when these genetic stocks might be required for our own survival within a changing environment. Maintaining the richest possible natural communities of both animals and plants simply improves our chances of meeting such contingencies.

SUMMARY

There are important reasons to preserve native diversity. Each living species has earned its functional place or niche in the environmental complex. To remove a species is to threaten the balance of a community, and the severity of the threat is commensurate with the relative significance of that species within the total structure. Stability refers to the resilience a community is able to exhibit in the face of detrimental external effects. Diverse systems undergo less change than relatively simple ones. The buildup of disease and pests, for example, is more common in the latter. There are fluctuations that do occur as part of the natural phenomena even in stable systems. These naturally occurring fluctuations tend to be cyclic and repetitive. Wild populations undergo periodic declines, but they can recover.

However, when innovative and frequent impacts of great magnitude occur, the trends that may be established as a result may diverge from predictable patterns. The biotic community is in a state of flux. Take, for example, the effect of thermal pollution on aquatic systems. The community, a viable unit with feedback mechanisms, begins to lose its capacity for self-regulation under the changes created by pollution.

The genetic composition of each native species has been distilled through and by competition in the struggle for survival that has been going on for millions of years. To weaken this genetic reserve is to narrow the potential range of biotic tolerance and reduce the functional capacities of the biosphere as a whole. Besides, if the world consisted only of cornfields, how dull it would be!

DISCUSSION QUESTIONS

1. Discuss the principal reasons for saving species from extinction and, in general, for maintaining the highest possible biotic diversity.
2. What are the primary factors underlying decimation of species? Provide several examples of declining species.
3. What is the dominance-diversity interaction? Give several ecosystem examples. What is the human effect on this interaction?

REFERENCES

Adams, M. W., A. H. Ellingboe, and E. C. Rossman. 1971. Biological uniformity and disease epidemics. Bioscience 21:1067-1070.

Bertram, C. R. 1975. The social system of lions. Sci. Am. 232:54-60, 65.

Burk, J. 1973. The Kaibab deer incident: a long-persisting myth. Bioscience 23:113-114.

Cockrum, E. L. 1962. Introduction to mammalogy. The Ronald Press Co. New York.

Cooke, H. B. S. 1963. Pleistocene mammal faunas of Africa, with particular reference to southern Africa. In F. C. Howell and F. Bourliere (eds.). African ecology and human evolution. Aldine-Atherton, Inc. Chicago.

Costin, A. B. and D. M. Moore. 1960. The effects of rabbit grazing on the grasslands of Macquarie Island. J. Ecol. 48:729-732.

Darlington, P. J., Jr. 1965. Biogeography of the

Southern end of the world. Harvard University Press. Cambridge, Mass.

Dobzhansky, T. 1950. Evolution in the tropics. Am. Sci. **38:**209-221.

Dorn, H. F. 1968. World population growth: an international dilemma. In J. B. Bresler (ed.). Environments of man. Addison-Wesley Publishing Co., Inc. Reading, Mass.

Ehrlich, P. R. and P. H. Raven. 1964. Butterflies and plants: a study in coevolution. Evolution **18:**586-608.

Fischer, A. G. 1960. Latitudinal variations in organic diversity. Evolution **14:**64-81.

Guilday, J. E. 1967. Differential extinction during late Pleistocene and recent times. In P. S. Martin and H. E. Wright, Jr. (eds.). Pleistocene extinctions: the search for a cause. Yale University Press. New Haven, Conn.

Gwynne, M. D. and R. H. V. Bell. 1968. Selection of vegetation components by grazing ungulates in the Serengeti National Park. Nature **220:**390-393.

Holmes, A. 1965. Principles of physical geology. The Ronald Press Co. New York.

International Union for Conservation of Nature and Natural Resources. 1966. Red data book. Vols. 1 and 2. Morges, Switzerland.

James, F. C. 1971. Ordinations of habitat relationships among breeding birds. Wilson Bull. **83:**215-236.

Janzen, D. H. 1966. Coevolution of mutualism between ants and acacias in Central America. Evolution **20:**249-275.

Kendrew, W. G. 1961. The climates of the continents. The Clarendon Press, Oxford, England.

Lack, D. 1969. Tit niches in two worlds: or homage to Evelyn Hutchinson. Am. Nat. **103:**43-49.

Lowe-McConnell, R. H. 1969. Speciation in tropical freshwater fishes. Biol. J. Linn. Soc. London. **1:** 51-76.

MacArthur, R. H. 1955. Fluctuations of animal populations, and a measure of community stability. Ecology **36:**533-536.

Mayfield, H. 1960. The Kirtland's warbler. Cranbrook Institute of Science Bulletin No. 40. pp. 9-33. Bloomfield Hills, Mich.

McNaughton, S. J. and L. L. Wolf. 1970. Dominance and the niche in ecological systems. Science **167:**131-139.

McVay, S. 1966. The last of the great whales. Sci. Am. **215:**13-21.

Mehringer, P. J. Jr., J. E. King, and E. H. Lindsey. 1970. A record of Wisconsin-age vegetation and fauna from the Ozarks of Western Missouri. In D. Wakefield, Jr. and J. K. Jones, Jr. (eds.). Pleistocene and recent environments of the Central

Great Plains. Special Publication No. 3. University Press of Kansas. Lawrence, Kan.

Miller, R. S. 1969. Competition and species diversity. In Diversity and stability in ecological systems. Brookhaven Symposium No. 22. Brookhaven National Laboratory. Upton, N.Y.

Moss, C. 1975. That ugly hyaena turns out to be a superb predator. Smithsonian **6:**38-45.

Paine, R. T. 1966. Food web complexity and species diversity. Am. Nat. **100:**65-75.

Pielou, E. C. 1966. Shannon's formula as a measure of species diversity: its use and misuse. Am. Nat. **100:**463-465.

Porsild, A. E. 1951. Plant life in the Arctic. Can. Geog. J. **42:**120-145.

Raven, P. H. 1967. Introduction to the Costa Rica flora. Organization for tropical studies. San José, Costa Rica.

Rodin, L. E. and N. I. Basilevic. 1968. World distribution of biomass production. In Proceedings of the UNESCO symposium on functioning of terrestrial ecosystems at the primary production level. Copenhagen.

Roth, R. R. 1976. Spatial heterogeneity and bird species diversity. Ecology **57:**773-782.

Russo, J. P. 1964. The Kaibab north deer herd—its history, problems and management. Wildlife Bull. No. 7. Arizona Game and Fish Department. Phoenix, Ariz.

Servicio Metereologico Nacional. 1962. Boletinmetereologico. San José, Costa Rica, pp. 1-42.

Seward, A. C. 1929. Greenland: as it is and as it was. Nature **123:**455-462.

Simpson, E. H. 1949. Measurement of diversity. Nature **163:**688.

Smith, S. H. 1968. Species succession and fishery exploitation in the Great Lakes. J. Can. Fish. Res. Bd. **25:**667-693.

Snow, D. W. 1964. A possible selective factor in the evolution of fruiting seasons in tropical forest. Oikos **15:**274-281.

Swan, L. W. 1961. The ecology of the high Himalayas. Sci. Am. **205:**68-78.

Terborgh, J. 1974. Preservation of natural diversity: the problem of extinction prone species. Bioscience **24:**715-722.

Thomas, A. S. 1960. Changes in vegetation since the advent of myxomatosis. J. Ecol. **48:**287-306.

Tramer, E. J. 1974. On latitudinal gradients in avian diversity. Condor **76:**123-130.

Wiens, J. A. 1976. Population responses to patchy environments. In R. F. Johnston (ed.). Ann. Rev. Ecol. Systematics **7:**81-120.

Ziswiler, V. 1967. Extinct and vanishing animals; a biology of extinction and survival. Springer-Verlag New York, Inc. New York.

ADDITIONAL READINGS

Allen, D. L. 1966. The preservation of endangered habitats and vertebrates of North America. In F. F. Darling and J. P. Milton (eds.). Future environments of North America. Natural History Press. Garden City, N.Y.

Andrewartha, H. G. and L. C. Birch. 1954. The distribution and abundance of animals. The University of Chicago Press. Chicago.

Banker, S. 1977. And now, good news: endangered species is saved. Smithsonian 7:60-64.

Caras, R. A. 1966. Last chance on earth; a requiem for wildlife. Chilton Book Co. Philadelphia.

Chase, A. and C. D. Niles. 1962. Index to grass species. G. K. Hall & Co. Boston.

Cott, H. B. 1968. Nile crocodile faces extinction in Uganda. Oryx 9:330-332.

Cox, J. A. 1975. The endangered ones. Crown Publishers, Inc. New York.

Dorst, J. 1970. Before nature dies. Houghton Mifflin Co. Boston.

Douglas-Hamilton, I. and O. Douglas-Hamilton. 1975. Among the elephants. The Viking Press, Inc. New York.

Eaton, T. H., Jr. 1970. Evolution. W. W. Norton & Co., Inc. New York.

Errington, P. L. 1967. Of predation and life. Iowa State University Press. Ames, Iowa.

Gause, G. F. 1932. Ecology of populations. Q. Rev. Biol. 7:27-46.

Guggisberg, C. A. W. 1970. Man and wildlife. Arco Publishing Co., Inc. New York.

Hayes, H. T. P. 1976. The last place. New Yorker, December 6:52.

Kruuk, H. 1972. The spotted hyena. The University of Chicago Press. Chicago.

Lack, D. L. 1945. The Galapagos finches, a study in variation. California Academy of Science. Occasional Papers. No. 21. San Francisco.

Leopold, A. S. 1966. Adaptability of animals to habitat change. In F. F. Darling and J. P. Milton (eds.). Future environments of North America. Natural History Press. Garden City, N.Y.

Lloyd, M. and R. J. Ghelardi. 1964. A table for calculating the "equitability" component of species diversity. J. Ann. Ecol. 33:217-226.

Maddox, D. M., et al. 1971. Insects to control alligatorweed. Bioscience 21:985-991.

McVay, S. 1966. The last of the great whales. Sci. Am. 215:13-21.

Mertz, D. B. 1971. The mathematical demography of the California condor population. Am. Nat. 105:437-453.

Orr, R. T. 1970. Animals in migration. Macmillan Inc. New York.

Pianka, E. 1974. Evolutionary ecology. Harper & Row, Publishers. New York.

Pooley, A. C. and C. Gans. 1976. The Nile crocodile. Sci. Am. 234:114.

Schorger, A. W. 1955. The passenger pigeon, its natural history and extinction. University of Wisconsin Press. Madison, Wis.

Simon, N. and P. Geroudet. 1970. Last survivors; the natural history of animals in danger of extinction. World Publishing Co. New York.

Van Gelder, R. G. 1969. Biology of mammals. Charles Scribner's Sons. New York.

Zimmerman, D. R. 1975. Panic in the pines. Audubon, May, 77:88.

5 Energy relationships

ENERGY FROM THE SUN

Life on our planet would shortly cease to exist without the steady flow of energy from the sun. Let us briefly consider the steps or transmutations that make the energy from this distant source available to the earth's ecosystems. In the high temperatures of the sun, a fusion reaction converts hydrogen to helium, releasing radiant energy of different wavelengths ranging from about 150 to more than 4000 millimicrons. This is the familiar solar light spectrum, whose energy is transmitted through space. At the earth's outer atmosphere (about 60 kilometers or 35 miles in altitude), the average rate of energy transfer is about 2 gram calories per square centimeter of surface per minute (g-cal/cm²/min).* This value is called the *solar constant*. However, not all this energy is actually available for distribution throughout the biosphere. About one third is readily returned into space as a function of the reflective properties of the earth's atmosphere and surface. This property, the ability to reflect the sun's radiation, is called the earth's albedo. The radiant energy entering the atmosphere is diminished further by about 20% through absorption by gases and particulates, so that a little less than one half of the radiation found in the outer atmosphere reaches the biosphere itself.

The amount of solar energy actually received in the biosphere varies widely; not all regions receive the same amount, and a global pattern is apparent. Regions in the tropics receive the most sunshine because of relatively more direct radiation throughout most of the year. With increasing distance from the equator, the angle at which the sun's rays enter the atmosphere results in a longer distance of passage and a greater screening effect. As a consequence, energy values per unit of surface area at the ground are less. The latitudinal implications of seasonal variations in the number of hours of sunlight per day and the shifting angle of incidence thus are key factors in the allocation of energy to different parts of the biosphere. Average energy values (summing all seasons) received at the earth's surface in several latitudinal zones are shown in Table 5-1. The northernmost climates with their characteristic biomes of tundra and boreal conifer forest receive only about 40% as much radiation as the tropics.

Such broadly based variations help to establish patterns of relative heat and cold.

*One gram calorie (g-cal) of energy will raise the temperature of 1 gram of water by 1° C. A kilocalorie (kcal) equals 1000 small calories.

Table 5-1. Total annual solar energy received on square centimeter of horizontal surface under average sky conditions for latitudinal zones in Northern Hemisphere*

LATITUDE	ENERGY (kcal)	RATIO
0-20	173	1.00
20-40	163	0.94
40-60	114	0.66
60-80	73	0.42

*Extrapolated from Sellers, W. D. 1965. Physical climatology. The University of Chicago Press. Chicago.

Solar radiation supplies the energy for such meteorological processes as wind, evaporation, and precipitation. The sun's energy, in conjunction with the spin of the earth as it
rotates around the sun, influences major atmospheric circulatory systems, the Trade
Winds and the prevailing westerlies, and even the ocean currents. (see Chapter 2). These
phenomena may, of course, be modified by such factors of physical geography as elevation, size of land masses, and proximity to large bodies of water. Through photosynthesis
by green plants, chemical or food energy is provided for essentially all life in the biosphere. Obviously the distribution and utilization of solar energy on so vast a scale are of
primary significance to the function of ecosystems and their land and water environments
over the face of the earth.

TECHNOLOGICAL FACTORS AND HEAT BALANCE

Human activities, through modifications on the surface of the earth and in the atmosphere itself, could conceivably alter the capacity of the biosphere to receive and retain
solar energy. The rapid expansion of urban areas, highway systems, and other developments usually associated with destruction of vegetation may increase the average surface
reflection of the earth. Deserts, some of which are the result of intensive exploitation of
other types of vegetation, have a higher albedo than grasslands; in some cases their albedo
is three times as great.

The 1970 report of the Council on Environmental Quality indicates that such modifications, if extensive enough, might substantially affect climatic conditions by decreasing
the heat budget of the earth. Airborne particulates from industries, automobiles, and
municipalities, particularly those in the 1.0 micron or smaller range, act as screening
agents, partially absorbing and thus blocking out incoming radiation. Soot, dust, smoke,
and aerosols emitted into the air in high concentrations on a sustained basis may produce
an effect that is sufficient to lower the earth's average annual temperature. Statistics for
the last 30 years show that the temperature has in fact dropped about 0.2° C, but it is not
known whether this drop is a result of these artifacts or some cyclic phenomenon of
weather that is not yet fully understood. Such warnings and concerns are perhaps speculative at present; yet consider the tremendous amount of material released to the atmosphere. Some estimates run as high as 800 million tons a year on a global basis. The rate
of emission of smoke particles alone is calculated at 20 million tons, with approximately
one half originating in the United States. The atmospheric component of the biosphere
through which sunlight must pass to sustain life is indeed subject to the large-scale and
heretofore unprecedented influences of human activities.

It has been argued that natural catastrophes such as volcanic eruptions also cause
atmospheric contamination through the release of dust, gases, and smoke particles. This is
true, but it should be pointed out that human pollution in various parts of the world is
steadily growing worse. According to some estimates, the total particulates from human
sources are just now beginning to approach the amount of long-term loading attributed to
volcanism. We can see this growing contamination of the air all around us. In the desert
southwest, heretofore clear skies show signs of increasing pollution. The Four Corners
Area of Arizona, New Mexico, Colorado, and Utah is a prime example. Large power

plant facilities using fossil fuels for producing electricity have been set up and are being expanded.

Since Krakatoa erupted to send its ejecta around the world in 1883, only two other eruptions of similar magnitude have been recorded. There is a long list of other eruptions, of course, and they all have contributed to the contamination of the atmosphere. According to Mitchell, many of these particulates are in the 0.1 to 1.0 micron ($\frac{1}{1000}$ millimeter) range, the same size as smoke particles introduced by human activities (see Chapter 11). These sizes have the greatest influence on the absorption of solar radiation. An objective analysis thus reveals that, to date, volcanism probably has had more influence on global cooling than man-made contamination of the atmosphere. Yet our human impact is represented by a steadily rising curve. The question is when will our influence become greater than that of natural causes.

There is also an opposing effect to be considered—the prospect of large-scale warming as a result of the addition of carbon dioxide (CO_2) to the atmosphere. One of the principal sources of CO_2 is the combustion of fossil fuels such as coal, oil, and natural gas. The annual rate of consumption of these fuels is accelerating rapidly. Although coal has been used for about 800 years and petroleum products for almost a century, more than one half of the world's utilization of energy from these sources has occurred in the last 25 years. In the past 100 years man has added 360 billion tons of CO_2 to the atmosphere, increasing its concentration by 13%. This rising trend is shown in Fig. 5-1. No natural catastrophes have been known to equal this increase during a comparable period of time. According to Rohrman et al., by the year 2000 the annual emission of CO_2 from fossil fuels in the United States alone will exceed 9 billion tons, a figure twice the 1969 rate.

Another source of CO_2, although certainly less extensive than the burning of fossil fuels, is the clearing of native vegetation for open cropland. One of the by-products of

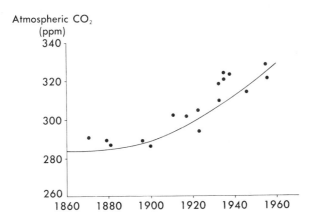

Fig. 5-1. Increasing amounts of CO_2 in the atmosphere during the last century, attributed largely to combustion of fossil fuels. (Redrawn from Plass, G. N. Carbon dioxide and climate. Copyright © 1959 by Scientific American, Inc. All rights reserved.)

ecosystem function and development is the deposition of residual organic matter and humus in the soil. The time required to reach an equilibrium between organic matter that is being stored and matter that is constantly being broken down by microorganisms may be many decades or even centuries. In the steady state there is a balance between the two processes. The CO_2 equivalent in the soil under these conditions would remain the same. The decomposition process is accelerated by clearing and cultivation, which effect increased aeration and mixing of the soil and higher soil temperatures caused by removal of the native plant cover. The CO_2 stored as carbon in organic matter in communities such as tundras, grasslands, and forests is thus a vast reservoir. When this CO_2 is released through oxidation, it cannot readily be replaced by photosynthesis.

In unbroken Missouri prairie, for example, the reserve of organic matter in the topsoil, if decomposed completely, is equivalent to 75 tons of CO_2 for each acre of ground. This equals more than 20 tons of carbon stored in the soil, or about 40 tons of organic matter. It is estimated that more than 100 years would be required to accumulate this amount in an equilibrium state. In colder climates the amount of time required to reach an equilibrium would be even longer. (See Chapter 6 for additional discussion of these time-rate relationships.) Yet we know that much of this original organic wealth is lost as CO_2 in a relatively short time when native ecosystems are changed to cultivated crops. By keeping some of the plant material in place, however, carbon losses from the soil could be slowed down. CO_2 is also being released into the atmosphere from other sources such as the annual burning of grasslands around the world. This process does not cause a net increase in atmospheric CO_2 because the grasslands are a renewable, self-perpetuating system. An amount of CO_2 equivalent to that released through the burning of standing crops is reincorporated by photosynthesis the following season. Under these conditions there is no sustained net increase in atmospheric CO_2, as there is when fossil fuels are burned and soil organic matter is lost and not replaced.

CO_2 in the atmosphere absorbs infrared radiations emitted from the earth. When the release of these long-wave or heat radiations is inhibited, the heat energy is trapped and may cause higher average temperatures at the earth's surface. This is popularly referred to as the "greenhouse effect," although strictly speaking this is an inaccurate designation, as there is very little difference between the screening effect of glass on short-wave (incoming) radiation and long-wave (outgoing) radiation. The real effect of glass is that it prevents the mixing of air of different temperatures. However, there is a "trapping" effect on these long-wave radiations because they are absorbed by CO_2. The question is whether increased atmospheric levels of CO_2 will significantly raise temperatures in the biosphere. With the ever-increasing consumption of fossil fuels (coal, for example, is 90% carbon) and the clearing of native ecosystems (organic matter in the soil is 55% to 60% carbon), will the buildup of CO_2 in the atmosphere have the climatic repercussions about which some scientists are warning us?

Fossil fuels are presently being consumed more than 1000 times faster than they are being formed as fresh deposits resulting from current photosynthesis. This means that in our highly complex, energy-demanding society we not only harvest almost all the immediate products of photosynthesis (or their animal by-products) but that we also must use

Table 5-2. Global relationships of photosynthesis, combustion of fossil fuels, and current organic deposition (fossilization) expressed in tons of CO_2 per year*†

PROCESS	CO_2	RATIO TO PHOTOSYNTHESIS
Photosynthesis	1.2×10^{11}	1.00
Combustion	1.6×10^{10}	0.13
Fossilization	1.1×10^{7}	0.0001

*Modified from Johnson, F. S. 1970. In S. F. Singer (ed.). Global effects of environmental pollution. Springer-Verlag New York, Inc. New York.
†Values were converted from grams to ton equivalents by the author.

the tremendous supplies of stored energy from ancient photosynthesis (coal and other fossil fuels). This might be considered a kind of "deficit spending," since we are not living within the budget of current photosynthesis. The energy relationships of the biosphere are not in balance. In a recent symposium on the human impact on the global environment, these significant energy problems were pointed out (Table 5-2). In weighing the relative cooling effects of atmospheric particulates such as dust and smoke versus the warming influence of CO_2 from fossil fuels, the best information suggests that the latter will win out. It has been projected that by the year 2000 temperatures on the earth will have risen 0.5° C above the 1969 values.

Water vapor is also effective in absorbing heat. Since water vapor is another by-product of fuel combustion, concern has been voiced by some scientists about supersonic transport planes (the SST) that would fly in the stratosphere at heights of 50,000 to 60,000 feet or more. Problems could conceivably arise in the stratosphere, where little moisture is present. Even small additions of water vapor to this dry air might have long-range effects by increasing the heat-absorption capacity of the upper atmosphere. Furthermore there is very little weather activity at these high levels. The residence time of water vapor, or the time required for a complete changeover, is considerably longer in the stratosphere than in the weather-making troposphere, as it takes about 4 years for a turnover of moisture in the former, compared to only 6 months, or less, in the latter. Thus the chances that water would accumulate in the stratosphere conceivably could be increased (see Fig. 2-1 for a vertical profile of atmospheric structure). It has also been hypothesized that water vapor is periodically emitted into the stratosphere from high-altitude clouds that reach into the tropopause. Therefore it is suggested in some quarters that the quantity and absorptive influences of water vapor emitted by airplanes may be insignificant in comparison to these natural sources of high-altitude moisture.

At present, we are only able to make projections concerning the human impact on general energy relationships; however, even if the long-range consequences are only remote possibilities, the implications are serious. These should be weighed most carefully against the immediate or short-term benefits that are so often used as a justification for continued economic growth and development.

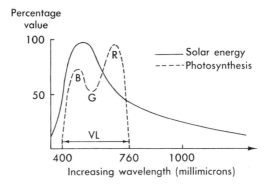

Fig. 5-2. Distribution of energy in solar radiation spectrum and relative photosynthesis in the visible portion, *VL*, or photosynthetically active radiation, sometimes designated as PAR. The two peaks occur in the blue, *B*, and red, *R*, regions, with less activity in the green, *G*.

THE PLANT EQUATION

The solar radiation distributed throughout the biosphere consists of a wide range of energy wavelengths, although most of the harsh, ultraviolet rays (shorter than 300 millimicrons) are screened out by the ozone layer of the atmosphere. Photosynthesis, the process in which chlorophyll in plants absorbs the energy necessary to produce chemical or food energy, occurs in the presence of light in the visible portion of the spectrum. The approximate limits of this energy range fall between 400 and 760 millimicrons. This part of the electromagnetic scale constitutes about 45% of all solar energy. Furthermore only about four fifths of this amount is potentially useful in photosynthesis, for the green portion of the visible light spectrum is only partially absorbed by the chlorophyll, so that utilization of energy for photosynthesis takes place mainly in the blue and red wavelengths. These two peaks of relative photosynthetic activity are shown in Fig. 5-2.

Photosynthesis is the vital link between solar energy and the life requirements of the biosphere. There are no practical substitutes for green plants. They are important not only as sources of food or chemical energy but also for the oxygen they produce as a byproduct of photosynthesis. For each unit (molecule) of carbohydrate produced, there are six of oxygen, as shown in the following shorthand for this important process:

$$6CO_2 + 12H_2O \xrightarrow[\text{(400-760 m}\mu\text{)}]{\text{Light energy}} C_6H_{12}O_6 \text{ (carbohydrate)} + 6H_2O + 6O_2$$

Since oxygen, O_2 in the equation, is produced only when molecules of water are split by the energy supplied in light, the process requires two units of water to produce one unit of molecular oxygen. Maintaining a viable plant environment is a critical factor in oxygen replenishment, although we know there is a tremendous reservoir of free oxygen in the atmosphere, about 20% by volume. Pollution in any form that destroys or impairs the photosynthetic process on land or in the ocean would interfere with this renewal. Such interference may be of particular significance in the oceans, where a deficiency of dissolved oxygen would readily develop without the photosynthetic activities of marine phytoplankton. The oxygen available to seawater from the atmosphere could not adequately overcome such a deficiency, for atmospheric oxygen does not penetrate the oceans to any appreciable depth.

It may come as a surprise that, on a worldwide basis, native plant communities and crop systems convert less than 1% of all the potential energy available for photosynthesis to organic energy. Essentially all life depends on the use of this very small percentage of available solar energy, a value sometimes called the photosynthetic efficiency of the plant community. This particular efficiency determination is based on a direct comparison between the available radiation energy and the amount of chemical energy produced in photosynthesis. Let us say that the measured solar energy (between 400 and 760 millimicrons) falling on a square meter of temperate grassland amounted to 47.0×10^5 kilocalories for the total growing season. Harvest of foliage and roots that grew during the same period in unit areas averaged 1300 grams per square meter. Using calorimetry, the energy content of this plant production is found to be equivalent to 5.7×10^3 kilocalories. An efficiency value is calculated as follows:

$$\text{Photosynthetic efficiency} = \frac{5.7 \times 10^3 \text{ kcal}}{47.0 \times 10^5 \text{ kcal}} \times 100 = 1.21\%$$

Certain communities have higher efficiency values than others. Forests, with their multilayered canopies, are generally the most efficient of native vegetations in converting solar to chemical energy. On a worldwide basis, forest communities of all types occupy about one third of the land area but produce 65% to 75% of the annual growth of terrestrial ecosystems. Grasslands are less productive per available unit of sunlight energy, and deserts are even less efficient. Chlorophyll is essential to photosynthesis, and certainly the amount of chlorophyll present in forests as opposed to grasslands accounts in part for this difference in efficiency. The amount of chlorophyll present is related to the amount of functioning leaf area in the plants that constitute a community. Thus it is possible to relate the productive capacity of the plant community to a simple measurement called the leaf area index (LAI), calculated as follows:

$$\text{LAI} = \frac{\text{Total leaf area (one surface)}}{\text{Ground area}}$$

In order to determine the productive capacity of needle-bearing trees such as pines, the total surface area is used, and the measurement is called the needle surface area index (NSAI). The more leaf area superimposed as layers over a given unit of ground area in any community, the greater the chlorophyll content and generally the more productive the system (Table 5-3).

Obviously within a given ecosystem there is an upper limit to the number of leaf layers that can be effectively arranged or structured to receive light from the sun. The filtering effect on light as it passes through one layer of leaves to the next decreases its intensity and changes its relative quality or wavelength, until a point is reached at which the light is inadequate for growth. There should be more light in the green area of the spectrum at the lower levels. Why?

The level of light intensity at which "gross photosynthesis" is balanced by respiration

Table 5-3. Comparison of leaf area index (LAI), chlorophyll content, and growth rates as exhibited by several crop species*

SPECIES	LAI	TOTAL CHLOROPHYLL (kg/ha)†	GROWTH RATE (kg/ha/day)
White clover	3.0	14.1	121
Kale	3.1	14.5	127
Red clover	4.8	20.0	188
Perennial rye grass	6.0	22.9	156
Maize (corn)	7.4	30.7	261

*Modified from Brougham, R. W. 1960. Ann. Bot. **24:**463-474.

†1 kg = 1000 grams = 2.2 pounds; 1 hectare (ha) = 10,000 square meters = 2.5 acres; the conversion to pounds per acre = kg/ha $\times \dfrac{2.2}{2.5}$ = kg/ha $\times$ 0.88.

is called the light compensation point. The plant is "breaking even" when all its photosynthate is used for maintenance or metabolic requirements, and there is none left for growth or the development of more biomass. The term "net photosynthesis" refers to the biomass energy in such forms as wood, forage, grains, and fruits. These plant materials are available for consumption by other organisms as a source of energy or for fossilization in lake and bog habitats. Net photosynthesis is thus the difference between what the plant produces and what is used to meet its immediate requirements. This relationship can be expressed as follows:

$$\text{Gross photosynthesis} - \text{Plant respiration} = \underbrace{\text{Net photosynthesis}}$$
$$\text{Food-chain energy} + \text{Current fossil deposition}$$

However, practically all the net plant production goes into immediate food chains, including that of people. As shown in Table 5-2, only a tiny fraction of photosynthate is preserved beyond what might be considered a normal turnover time.

Compensation points vary among species: some plants can continue to grow at light intensities that would be entirely inadequate for others. Plants that flourish in deep shade are called *sciophytes;* those adapted to more open conditions and higher light intensities are *heliophytes.* Communities comprised of many species use a greater range of the light energy that filters through the leaf canopies than do simple communities. In monocultures such as crop systems, in which the objective is to maximize production per unit of ground area, the effective LAI frequently reaches its upper limit through dense crowding of individual plants. Such limits will vary according to leaf size, shape, and angle of attachment, all of which depend on the crop in question. In addition to mutual shading, other environmental factors such as water and CO_2 may limit excessive densities. Generally the effective LAI of tropical forests is greater than that of other communities. One reason for this is the great diversity of species, each of which has a slightly different tolerance of light attenuation and variations in light quality. Leaf area values ranging from 10 to 12 have been determined for rain forests. The LAI is less for temperate deciduous forests, in which leaf area values equal to about six times the ground area are common.

In recent years plant physiologists have made interesting discoveries concerning the different mechanisms of carbon fixation, or photosynthesis. These mechanisms provide a basis for evaluating efficiency at the cellular level. Since the details of this field are beyond the scope and intent of this text, let us say only that productivity potentials in terms of ecological distribution, leaf anatomy, and chlorophyll structure are inherent in these differences. In general, if we exclude plants with so-called crassulacean acid metabolism, the CAM plants,* the green plants studied to date may be divided into two main groups. In the first group the process of *photorespiration* is characteristic, as evidenced by the fact that plants receiving light intensities adequate for photosynthesis produce CO_2. These are called C_3 plants. Plants in the second group do not release CO_2 to the air under similar conditions. Rather CO_2 is trapped within the chlorophyll-bearing tissue of the leaf and reused in photosynthesis. These are called C_4 plants. Based on these findings it is possible to generalize that species in the second group, those not demonstrating photorespiration, have warm-climate origins. Examples of C_4 plants are corn and sugar cane. These are more efficient photosynthesizers than certain other grasses, which are included in the first group, those demonstrating photorespiration. Oats and wheat are examples of C_3 grasses. Other examples from still other families might be cited. Several explanations that have been advanced to account for these differences in photosynthetic potential involve plant distribution, ecology, and evolutionary relationships. For the interested reader, three references on photorespiration and its implications in plant productivity are Hatch et al. (1971), Loomis et al. (1971), and Zelitch (1971).

PRIMARY PRODUCTION

Estimates of primary productivity for major ecosystems on land and in the ocean show wide variations. Only certain limited parts of the sea, for example, can be classified as highly productive, and about 90% of the sea is considered a kind of biological desert. The lack of radiant energy at deeper levels and deficiencies of phosphates and nitrates are important growth-limiting factors. The more productive sections of the ocean are the shallower areas on continental shelves and areas of upwelling, where there is a more plentiful supply of nutrients and light can be used more effectively by phytoplankton (see Chapter 9 for additional discussion).

On land the amount of energy produced by photosynthesis differs strikingly from one region to another. Varying degrees of moisture, seasonal temperatures, length of growing season, number of daylight hours, and availability of nutrients are reflected in varied community structures, biomass accumulations, chlorophyll concentrations, photosynthetic systems, and so on; they are important considerations in the assessment of primary production. Desert communities have smaller yields than neighboring grasslands that receive more moisture. Annual growth in the desert generally averages less than 100 grams per square meter, an amount equivalent to less than ½ ton per acre. Similarly,

---

*Most of us are acquainted with the great family of succulents, the Crassulaceae, many of whose species are grown as house plants. Unlike many succulents, or plants with fleshy stems and leaves that are tolerant of high salt concentrations in the soil, CAM plants are probably all salt intolerant, or nonhalophytic.

Table 5-4. Comparison of dry-matter production for marine phytoplankton and major terrestrial communities, expressed as a range of values*

BIOME TYPE	NET PRODUCTION $(g/m^{-2}/yr^{-1})$
Open ocean	50-400
Arctic tundra	50-300
Deserts	25-200
Steppe grasslands	150-300
Prairies and savannahs	250-2000
Coniferous forests	250-3000
Temperate deciduous forests	500-3000
Tropical rain forests	2000-5000

*Data from Chew, R. and A. E. Chew. 1965. Ecol. Monogr. **35**:355-375; Johnson, P. L. 1969. Arctic **22**:341-355; Kira, T. et al. 1964. Bot. Mag. Tokyo **77**:428-429; Kucera, C. L. et al. Ecology **48**:536-541; Ovington, J. D. 1965. Biol. Rev. **40**:295-366; Westake, D. F. 1963. Biol. Rev. **38**:385-425; and Whittaker, R. H. 1970. Communities and ecosystems. Macmillan Inc. New York.

Fig. 5-3. Wide and generally regular spacing of plants in the desert community. Plants in the foreground are creosote bush. Limited supplies of water necessitate sparse structural development as an adaptation for survival. (Courtesy E. M. Schmutz, University of Arizona, Tucson, Ariz.)

tundras are low producers. As climates moderate, plant yield on a unit-area basis increases. This trend is illustrated for several biome types in Table 5-4. The open ocean is included for purposes of comparison with production values similar to cold or dry climates on land.

Desert communities lack the structural mass and complexity necessary for greater production. In addition to typically low LAIs and wide spacing of desert perennials such as that seen in Fig. 5-3, the sparse canopies often shed their leaves in the dry season. Water losses are reduced as leaves are dropped, but photosynthesis is also curtailed, if not stopped completely. The demands of a stringent physical environment are reflected in the structural and functional features of desert plants. These features strike a balance between drought avoidance such as shedding leaves on the one hand and the minimum photosynthetic activity necessary to ensure a sustained flow of energy through the system on the other. There are other mechanisms for survival, of course, and these are discussed in Chapter 8.

The point here is that the productivity of some natural systems is of necessity compromised in the interest of biotic survival. Desert vegetation is characterized to a high degree by this type of compromise. Such a strategy maximizes diversity and adaptive processes for self-perpetuation within the limits imposed by the physical environment. It is executed at the expense of high-population densities and the large energy outputs that would be required to support them. Stability of the system thus is weighed against production. However, when people manipulate ecosystems, their strategy frequently is one of maximizing the energy yield while ignoring the integrity of the environment from which that energy is derived. The relationship between net plant production and actual water loss from soil and plants, or evapotranspiration, for a large number of ecosystem types is shown in Fig. 5-4. We see that the availability of water is a key predictor of plant growth and structural development in the community, with low-producing deserts at one end of the spectrum and complex rain forest types at the other. There is a varied array of plant communities in which water availability is intermediate between these extremes.

In Table 5-4 some average rates of production on a unit-area basis were presented to illustrate the wide range of differences between ecosystems as well as within each ecosystem type. On a worldwide basis, what does each of these ecosystem types contribute to total primary production? How does grassland compare with forest, for example? More importantly, what percentage of the total do crop plants produce? Estimates of total net photosynthesis are presented in Table 5-5, together with the amount of area occupied by each type of ecosystem. Here we see that the open ocean (not including offshore waters and areas of upwelling) constitutes a little over 60% of the earth's surface and provides about 25% of the photosynthesis. Can the ocean, which occupies such a large percentage of the earth's surface, be made more productive? The potential as well as the limitations are discussed in Chapter 9.

Usable energy flows along one-way paths, decreasing as it goes, always in the direction of nonusable energy, or heat conversion, which is radiated back into space. Ultimately all photosynthetic energy in the biosphere is changed to heat. We might visualize the energy as a ball of string unwinding as it rolls down an incline. Energy thus is non-

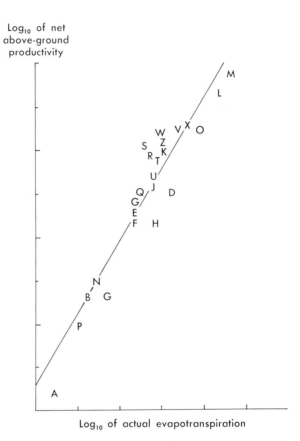

Log₁₀ of net
above-ground
productivity

Log₁₀ of actual evapotranspiration

Fig. 5-4. Net above-ground productivity in grams of dry matter per square meter in relation to actual evapotranspiration in millimeters. The letters are coded as follows: *A*, creosote-bush desert; *B*, Arctic moist tundra; *C*, alpine moist tundra; *D*, tall grass prairie; *E*, heath bald; *F*, heath bald; *G*, heath bald; *H*, mixed heath; *J*, mixed heath; *K*, beech-maple forest; *L*, secondary tropical forest; *M*, tropical forest; *N*, cool desert sand dunes; *O*, oak-hickory forest; *P*, cheatgrass; *Q*, Fraser-fir forest; *R*, spruce-fir forest; *S*, spruce-fir forest; *T*, Gray-beech forest; *U*, Gray-beech forest; *V*, hemlock-mixed forest; *W*, upper-core forest; *X*, de-ciduous-cove forest; *Z*, hemlock-rhododendron forest. (Modified from Rosenzweig, M. L. 1968. Am. Nat. **102**:67-80.)

Table 5-5. Estimates of net photosynthesis expressed as dry-matter production for oceanic zones and terrestrial communities*

SYSTEM	TOTAL AREA (mi² × 10⁶)	TOTAL PRODUCTION (tons × 10⁹ yr⁻¹)	TOTAL AREA (%)	TOTAL PRODUCTION (%)
Forests and woodlands	23	77.2	11.1	48.6
Deserts	17	1.4	8.2	0.8
Grasslands and savannahs	10	15.0	4.7	9.5
Agricultural crops	6	9.1	2.7	5.7
Tundra and alpine regions	4	1.1	1.6	0.7
Marshes, lakes, streams	1	5.0	0.8	3.1
Subtotal	61	108.8	29.1	68.4
Open ocean	130	40.8†	63.3	25.8
Coastal zone	14	9.0	6.9	5.7
Areas of upwelling	1	0.1	0.7	0.1
Subtotal	145	49.9	70.9	31.6
World total	206	158.7	100.0	100.0

*Data from Ryther, J. H. 1969. Science **166**:72-76; and Whittaker, R. H. 1970. Communities and ecosystems. Macmillan Inc. New York.
†Original data expressed as carbon converted here to total dry matter using a factor of 2.5x.

returnable. Its replenishment is possible only through the additions of fresh energy via photosynthesis. The universal tendency of energy to flow from available to nonavailable forms is described by the second law of thermodynamics as the increase in entropy of the system, that is, ultimate randomness and lack of structure. Viable communities are organized along lines of varying structural complexity and degree of functional stability. To keep the biosphere in running condition, there must be a constant supply of usable energy from photosynthesis flowing into it.

THE ENERGY PYRAMID

All green plants are autotrophs, for since they contain chlorophyll, they can engage in photosynthesis. Hence they are literally self-nourishing, that is, have the ability to feed themselves. Certain bacteria such as the photosynthetic green and purple sulfur bacteria are also autotrophic. They can grow under low light intensities such as those found in lake muds. There are also chemosynthetic bacteria that obtain their energy for growth through the oxidation of inorganic salts. All other organisms in the ecosystem are obligatory heterotrophs—in order to live, they must derive their energy from outside sources. A simple scheme for classifying the producers and consumers is outlined as follows:

AUTOTROPHS
Green plants
Photosynthetic and chemosynthetic bacteria

HETEROTROPHS
Herbivores   Scavengers
Carnivores   Decomposers

The energy initiated by the autotrophs is distributed to the ecosystem through a series of consumer levels. We may visualize a "pyramid of energy" consisting of several horizontal layers. The lowermost layer is, of course, the plant community. The layer above is composed of primary consumers, or herbivores, whose diet consists of plants. The third level is occupied by secondary consumers who prey on the herbivores, and so on, up to the top of the pyramid. This is the familiar *grazing food chain,* although it is not always a one-step, sequential process. Some consumers may select food from more than one level in the food chain. The fox, for example, preys on the herbivorous rabbit and on certain predators such as snakes in addition to utilizing several plant foods. Some herbivores are coprophagous, that is, they reingest some portions of their fecal waste, a characteristic of certain animals including the rabbit.

Scavengers also play important roles, as they derive the energy to meet their needs from plant and animal debris. Microorganisms, or decomposers, accomplish the final reduction and decay of all organic matter. These residues come from all levels in the energy pyramid and are the basis of the *detritus food chain.* The CO_2 that was once fixed in or-

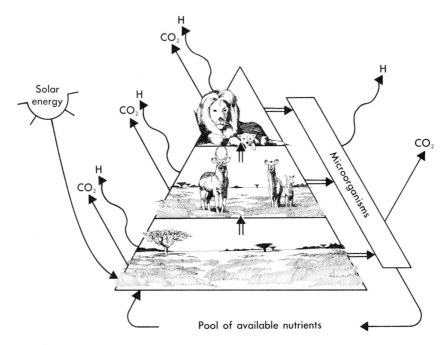

Fig. 5-5. Schematic diagram of energy pyramid with its several trophic (food) levels, representing grazing and detritus food chains. In a balanced system the total energy of photosynthesis is accounted for by heat losses at all levels, from green plant to top consumer. CO_2 is released to the atmosphere, and nutrients are made available through decay by microorganisms for reuse by plants. Double arrows represent flow of chemical energy through living components of the ecosystem. (Courtesy Fred Landa, Virginia Commonwealth University, Richmond, Va.)

ganic compounds through photosynthesis is released to become a part of three reservoirs: the air, water, and soil. In this way nutrients that were bound within an organic framework are released back into the soil to be recycled by new generations of plants. Heat, the final form of energy, is emitted into space. Fig. 5-5 is a simple diagram illustrating the relationships of both the grazing and detritus food chains, the two basic food chains in the biosphere. The word "trophic" means nourishment. The word "atrophy," meaning ill-fed, is commonly used to refer to arrested development or to withering or loss of a body part or organ. When destruction of a species disrupts the energy flow in the ecosystem, the result might also be interpreted as a kind of atrophy affecting the biotic community.

Comparison of food chains

Does a comparison of the two basic types of food chains reveal any consistent differences or patterns in the amount of energy that flows through each in any given ecosystem? In general we can say that grazing food chains receive their greatest allocations of energy in herbaceous communities such as grasslands, savannahs, and tundra vegetations. In young or developing ecosystems (see discussion of biotic succession in Chapter 7) this type of utilization is also relatively important. In all these communities the herbivore is a focal point, as illustrated by the lemming population in the coastal tundra of Alaska on which a number of predators rely for food. These relationships are diagrammed in Fig. 5-6. Grazing food chains are also a significant channel of energy flow in some aquatic systems such as the following sequence found in a farm pond:

Algae → Zooplankton → Small invertebrates →
Forage fish (bluegill) → Predator fish (large-mouth bass)

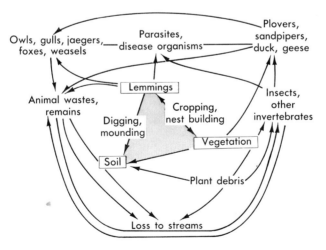

Fig. 5-6. Diagram of food-chain relationships in Arctic tundra, illustrating focal point of the lemming, dominant herbivore in a relatively simple system. (Redrawn from Pitelka, F. A. 1967. In H. P. Hansen [ed.]. Arctic biology. Oregon State University Press. Corvallis, Ore.)

In other aquatic environments such as an estuary, in which there is a large input of decaying organic material from outside the system, or allochthonous matter, detritus feeders, including scavengers, may account for a very high percentage of the total energy flow. In these energy-rich environments, important benthic (bottom-dwelling) faunas such as clams and shrimp derive their energy from the organic particulates that are transported into the water from other ecosystems and settle to the bottom.

Detritus organisms are important in the utilization of energy synthesized in forests. Woody material and fallen leaves come under attack by microbial decomposers and a host of soil organisms. As much as 90% of the net photosynthesis, or energy, available to the heterotrophs in a forest system may be channeled through decomposers. From time to time, however, swarms of insects may utilize much of the foliage, thus depriving detritus feeders of this energy source. Even in sustained-yield forests in which people make periodic harvests, decomposers may account for most of the total energy produced in photosynthesis. Within these decomposer groups, trophic levels are intricately related. Mites feed on bacteria that attack fungi, which in turn generally initiate the breakdown of plant residues. The reduction of detritus materials in ecosystems is more complex and perhaps less understood than the use of energy in grazing food chains, because the grazing food chains are a more conspicuous phenomenon compared to the activities of microorganisms in a field soil. As more is learned about these food pathways, it becomes

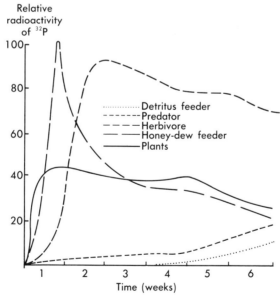

Fig. 5-7. Sequential conversion and utilization of biomass through several trophic positions of the ecosystem, analyzed by the use of ^{32}P, the radioactive isotope of normal phosphorus that occurs in all plant and animal cells. (Modified from Weigert, R. G. et al. 1967. Ecology **48**:75-83.)

increasingly apparent that much of the energy travels through detritus feeders, even in grasslands and other herbaceous systems. Recent studies also show that the oceans have a significant decay food chain comprising the most minute microorganisms. Both types of food chains are critical, however, to the success of any ecosystem. The elimination of large herbivores from a prairie may cause stagnation of the grasses. They and the detritus organisms that assist in the breakdown and decay of leaf litter on the floor of a forest are of equal functional importance.

In addition to the considerable fraction of energy flow attributed to the decomposers, other differences between grazing and detritus pathways occur in most systems. Obviously the use of energy is delayed in the detritus chain. Decomposers that depend on organic matter in the soil exhibit a more uniform pattern of energy intake than grazers that rely on the seasonal availability of green grass, for example. Weigert investigated the transfer of biomass from one trophic level to another through the use of phosphorus, which is a component of all plant and animal cells. Its radioisotope, ^{32}P, can be applied to a plant that then incorporates the substance, making it possible to trace the radioactivity through several consumers (Fig. 5-7). These data demonstrate that radioactivity does not appear in the detritus feeders until several weeks after its application to living plants and subsequent transmission through the grazing components of the food chain.

A conspicuous feature of grazing food chains of terrestrial systems is that the biomass of vegetation exceeds by far that of the animal consumers. For example, the total weight of all herbivores in a typical section of the East African savannah, despite their great diversity and abundance, may not exceed an average of 100,000 pounds per square mile (256 hectares). In this same square mile there may be 30 to 50 times as much plant material.

In some aquatic systems, however, the reverse may be true, so that the biomass or weight of the grazing organisms may *at any one time* exceed that of the plant community. This may occur because the standing crop of phytoplankton or algae has a more rapid turnover rate, and there are therefore more generations of new crops whose life cycles are completed in a shorter period than those of the herbivore feeders at the next level in the food chain. Nonetheless, as in any system, each successive level of the food chain or energy pyramid produces less energy than the level immediately below over a given period of time. This general rule does not apply in cases in which energy enters from a foreign source. An example would be stream pollution caused by the dumping of a municipality's organic wastes, which provide an excess of food energy for detritus organisms in the ecosystem and cause temporary dominance and species changes. When this happens, physical changes may also occur. Oxygen is required for the breakdown and assimilation, or utilization, of this excess energy. Therefore the supply of oxygen in these streams may temporarily become depleted, reaching levels that are insufficient for other organisms such as fish. This increased use of oxygen is called the *biological oxygen demand* (BOD). Although energy is an absolute necessity, either too much or too little will cause changes in the composition of biota throughout the food pyramid as well as varying degrees of change in the physical environment.

Trophic structure

A study of the trophic structure tells us how energy is exploited by different species and provides an insight into the amount and rate of energy flow through the ecosystem. Just as importantly, we get an idea of the degree of efficiency with which these transfers are made from one food level to the next. This topic is discussed in the next section. It is obvious that the greater the number of transfers between a consumer in the food chain and the photosynthetic base, the less energy there is to which the consumer has access. Generally the number of food transfers seldom exceeds four or five because of the diminishing availability of energy at these levels in the pyramid. The reason for this attenuation is that maintenance, or upkeep, involves the dissipation of energy at each level in the form of heat. The situation is analogous to the water pressure in a pipe, which drops as the distance from the source increases and water is drawn off through shunts or openings along the length of the pipe.

Because there is less food available in the upper levels of the food chain, the predators expend more energy in such food-gathering efforts as stalking and running down prey. A proportionately larger share of the total food intake is required to sustain progressively smaller but more active populations of predators. In these predator-prey relationships the carnivore is normally outnumbered by its herbivore prey. Such interactions suggest that viable populations of carnivores require more space per animal than their prey at lower levels in the food chain. The energy needs of each individual can thus be met only in a territory larger than that required by an individual from a population lower in the food chain. This is one factor involved in the preservation of large predators such as the mountain lion in our western states, for large tracts of land are required to maintain balanced predator-prey relationships. As an area is encroached on by various factors that diminish its viability as an ecological unit, both predator and prey can make some adjustment to change. With continued depreciation of the ecosystem, however, it is reasonable to assume that eventually it becomes impossible to maintain viable populations of the predator. Prey densities may actually increase as a result of these imposed shifts in predator-prey relationships. This is an application of the Volterra principle, which states that when both prey and predator are subject to the same impact factor, populations of the predator are more adversely affected, whereas populations of the prey benefit and increase (see Chapter 10).

In some biotic relationships the numbers of organisms do not decrease at the higher trophic levels. Parasites and plant-eating insects, for example, are found in higher densities than their much larger hosts. The size of the organism and the time required for the completion of a life cycle are important factors in the composition of the energy pyramid.

ECOLOGICAL EFFICIENCIES

Energy represented by different forms of biomass diminishes as it moves through the ecosystem, thus limiting the number of transfers that are normally possible. We may look at the fate of this energy in two ways with regard to the efficiency of its use by the several consumer types. The first question deals with intratrophic relationships, those that exist within a particular level of the food chain, that is, how effective is a given species in converting ingested food to new biomass or net growth? This is an important consideration, for it is the net growth that becomes the food available to the next level in the balanced system. The second question deals with intertrophic efficiency, or the efficiency of food transfer across trophic levels. Information of this type provides a means of assessing the biotic community as an energy-processing organization. As our native resources dwindle and human demands on ecosystems increase, these become important factors if we are to manage the environment in a safe, protective, and yet productive way.

Valid measurements of efficiency require that all calculations be made on a caloric basis. Measurements based solely on biomass or weight would be particularly misleading because the energy content of biological materials from one end of the trophic structure to the other varies rather widely between the approximate limits of 3500 and 7500 calories per gram on a whole organism basis. Therefore a pound of grass and a pound of herbivore tissue do not represent similar energy values. Following are some average data for selected biological materials* expressed as gram calories of energy per gram of substance (dry weight):

MATERIAL	WEIGHT (g-cal)
Cereal grain	3600-4300
Root crops	3400-3800
Forest litter	4800
Mammalian flesh	6400-7400
Bird flesh	5900-7000
Fish	4000-4400

The higher values for birds and mammals reflect a greater fat content as compared to fish with their characteristically high protein levels.

The following equation may be used to examine efficiency relationships within a trophic level:

$$I = E + R + P$$

where

I = Energy ingested
E = Energy egested
R = Respiration or heat loss
P = Growth or energy stored
A = Assimilation ($R + P$)

The biomass energy that is ingested is divided initially into two compartments, E and A. A is the sum of P and R (Fig. 5-8). The following intratrophic efficiency formulas can be derived from these allotments of energy:

(1) Assimilation efficiency (%) = $\frac{A}{I} \times 100$

(2) Net growth efficiency (%) = $\frac{P}{A} \times 100$

*Data from Petrusewicz, K. and A. Macfayden. 1970. Productivity of terrestrial animals. F. A. Davis Co. Philadelphia.

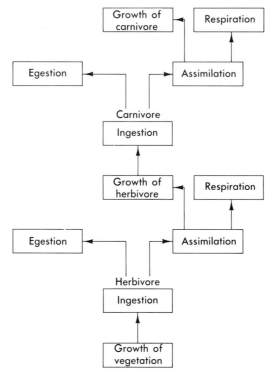

Fig. 5-8. Diagrammatic scheme of energy flow and distribution in the ecosystem. See text.

These two equations allow us to develop a third formula, which is used to determine growth efficiency. It is derived as the product of 1 and 2:

(3) Gross growth efficiency (%) $= \dfrac{A}{I} \times \dfrac{P}{A} \times 100 = \dfrac{P}{I} \times 100$

Can differences among the several trophic levels be seen in these efficiency values? Although oversimplification of these relationships is hazardous, it is possible to generalize that assimilation efficiencies at the herbivore level tend to be lower than at the carnivore level. There are several reasons for this. On the whole, the plants of which herbivore diets consist are less digestible than the diets of carnivores, so that in the case of herbivores a greater proportion of food intake appears as egested material. On the other hand, a predator that has a higher assimilation rate may have a lower net growth efficiency because more energy is expended in acquiring food and engaging in other necessary activities. Respiration or heat losses therefore are greater. From these relationships, then, we might expect an inverse relationship between A/I and P/A such as that shown for aquatic organisms in Fig. 5-9.

There are exceptions to or modifications of these trends, accounted for in part by the size and age of organisms and the supply of food. Invertebrates, for example, will pass

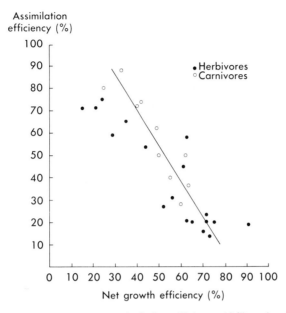

Fig. 5-9. Inverse relationship between assimilation efficiency (A/I) and net growth efficiency (P/A) for several aquatic consumers ranging from zooplankton to fish. (Modified from Welch, H. E. 1968. Ecology **49**:755-759.)

Table 5-6. Efficiency values for herbivore, carnivore, and detritus feeders*

CONSUMER	A/I (%)	P/A (%)	P/I (%)
Grasshopper No. 1 (H)	37	53	19.5
Grasshopper No. 2 (H)	36	37	13.3
Mouse (H)	70	3	2.1
Deer (H)	78	2	1.6
Elephant (H)	33	2	0.6
Weasel (C)	95	2	1.9
Snail (D)	45	13	5.9

*Data from Golley, F. B. 1960. Ecol. Monogr. **30**:187-206; Odum, E. P. et al. 1962. Ecology **43**:88-96; Petrides, G. A., and W. G. Swank. 1969. Proc. 9th Int. Grassland Congr. **9**:831-842; Smalley, A. E. 1960. Ecology **41**:672-677; and Weigert, R. G. 1965. Oikos **16**:161-176.

more food as waste when it is available to them in excessive quantities. This, of course, lowers the assimilation efficiency. Animals in confinement have higher growth efficiencies than those allowed to forage in the open. Differences between poikilotherms and homoiotherms probably also occur. Data on efficiency values from several sources are shown in Table 5-6. The P/I values for the mouse, deer, elephant (herbivore), and weasel (carnivore) are generally low in comparison to those of the invertebrates (grasshoppers and snails), suggesting that trophic position may not be as important in regulating gross growth efficiency as other factors such as body temperature in relation to the ambient environment. The P/I is an important value, however, in assessing intertrophic efficiencies. The efficiency with which an organism converts its food to net production determines the availability of energy for consumers at the next level in the food chain.

Assuming an average P/I value of 10% (too high for some animals and probably too low for others), we can generalize that for every 10 units of energy entering a particular level, only 1 unit is available at the next. Through three levels, then, the remaining energy is reduced to 1% of the initial input, as follows:

$$n^{+2} = 1$$
$$n^{+1} = 10$$
$$n \ \ = 100$$

Ecosystem (or intertrophic) efficiency as shown here is 10%. This is a convenient, much-quoted value, but it is probably too high for most ecosystems. When a given consumer utilizes all the food available to it, the system is functioning at maximum efficiency, indicating balanced levels.

The attrition of energy as it is filtered through the ecosystem explains the decrease in carrying capacity at each consumer level. When a consumer becomes too efficient (eating more than just the net growth), there is the possibility that its energy source in the food chain will be exterminated. If this should happen, the predator might become extinct. However, if the predator is potentially adaptable, it may utilize the next lower link (n^{+2} resorting to n, for example), thus shortening the food chain. The baleen whales are a good example. These are the largest animals on earth, and yet they feed on small crustaceans, including the well-known krill of the Antarctic seas.

Our own species is rapidly assuming a dominant position in the trophic structure of the biosphere. Because our food demands are increasing, we are placing greater reliance

Table 5-7. Daily caloric intake per capita and percentage distribution of calories by food groups for selected countries or regions of the world*

REGION	CALORIC LEVEL (kcal)	DAILY INTAKE (%)			
		CARBOHY-DRATE FOOD	FATS, OILS, NUTS	MEATS, EGGS, FISH, MILK	VEGETABLES AND FRUITS
Northern Europe	3260	48	20	28	4
United States	3190	40	24	30	6
Canada	3100	42	17	36	5
Southern Europe	2720	60	20	13	7
South America	2485	71	13	13	3
West Central Africa	2460	81	15	3	1
Central America and Mexico	2412	70	14	12	4
Japan	2360	78	11	7	4
India	2060	74	17	7	2

*Modified from Economic Research Service, Foreign Regional Analysis Division, United States Department of Agriculture. 1970. The world food budget. Foreign Agricultural Economic Report No. 19. United States Government Printing Office. Washington, D.C.

on plants as a source of energy. Where populations have expanded rapidly, the cost of food-chain energy obtained from other animals is too great, forcing a reliance on other sources for our energy needs. The variety of sources used by populations in different parts of the world is shown in Table 5-7. As part of the cost of maintaining these human populations at even subsistence levels in some parts of the world, the energy pyramid in effect is being flattened by removal of the herbivore. Heat losses associated with the transfer of energy from one level to another are avoided to make the ecosystem more efficient, but at the cost of animal protein in the diet.

HIDDEN COSTS

In view of the ever-increasing demands on environmental resources, it would be well to consider the energy required to produce plant foods for human consumption. Each year a population one third that of the United States is added to the world community. Agriculture has been notably successful in raising crop yields, so that in the United States a single farm worker feeds approximately 50 other people. Mechanization, fertilization, and application of herbicides and other chemicals involving the increasing use of fossil fuels are among the factors that have made these remarkable achievements possible. Over 90% of the population of California lives in urban concentrations, the balance in rural areas; in Tanzania the ratio is nearly reversed! Without innovations, the inadequate diets and malnutrition that daily afflict millions of people would affect even millions more. Today, in a world of over 4 billion, between 500 million and 1 billion people suffer from malnutrition, and many millions live on the brink of starvation.

In India and other less industrialized countries where human labor still is the main source of power, great progress has been made with the aid of genetically superior plant varieties and improved agronomic methods. Although it is relatively easy to determine the efficiency of wild or uncultivated plant communities solely on the basis of solar input, figuring total energy costs of producing a cultivated plant community such as an acre of wheat is more difficult. The costs of producing the fertilizer, pesticides, and herbicides, their transportation, the use of fossil fuels in running mechanized equipment, even the work hours expressed as heat expenditures to accomplish management, harvest, and storage are only some of the inputs.

It would be interesting to discover the extent to which the food energy of a given crop corresponded to these added inputs of artifact energy. Most of the time the farmer or grower must make some margin of profit to stay in business, but is the profit real in an ecological sense or with regard to amounts of accountable energy, regardless of the form it takes? Our biological environment is run on energy; it is organized and structured by constant inputs of usable energy. What then does the balance sheet show?

The problem of the relative efficiencies of managed ecosystems measured on a strict energy basis should be a rich field of investigation, yet there are still little data and information on this subject. A comparison of energy production against the solar input alone (for which no charge is made) still reveals efficiency values on a worldwide basis that are on the order of 1%, although photosynthetic efficiencies for certain crops may be considerably higher during periods of peak activity. The problem, whatever the effectiveness of solar capture, becomes one of determining the extent to which higher yields balance the cost of the artificially introduced or artifact energy required to achieve those yields. Reasoning tells us, on thermodynamic grounds, that taking the cost of these additional sources of energy into account can only decrease the ultimate value calculated for crop efficiency. The question is how nearly the efficiency value approaches zero, or whether it is in fact a negative value. The question may be highly academic, but its implications are nonetheless deserving of attention.

People do not often realize that the "green revolution," which on the surface appears to be a bonanza, a solution to the world food problem, also implies an accelerated use of finite energy resources. It is certainly true that improved crop strains are selectively bred for maximum suitability to the environmental requirements of broad geographical areas such as adaptability to different growing seasons and day lengths and to frost and other such factors. Beyond this, however, environmental manipulation such as cultivation, chemical spraying, and irrigation as extra energy expenditures is necessary if the greater yields residing in those genetic potentials are to be realized. Without these additional inputs, the green revolution could not possibly be the success that it is at the present time. Something must be given for something, and supplying additional energy to the food chain, like any other ecological process, is a *quid pro quo,* not a magic something for nothing.

This point can be illustrated by reviewing data discussed by van Overbeek concerning increased corn yields as a result of additional inputs of energy, including fertilizer and herbicides. Yields were increased 138% during the period from 1945 to 1970, but this gain was achieved with an increase of 222% in materials and techniques that have as a basis the expenditure of fossil-fuel energy. There are even greater disparities between input and output if we consider food *actually consumed* under conditions of a modern, industrialized society. When we add the energy required for processing, packaging, refrigerating, and finally cooking and serving, the American public is paying for a caloric, or energy, input approximately 7 to 10 times the actual food value. As fuel prices from the Oil Producing and Exporting Countries (OPEC) continue to rise, the differential will presumably become even greater. The lesser developed countries, who are attempting to mechanize their agriculture to increase food production for growing masses of people, will also be severely affected.

The fact that our food production and consumption are based on such complex elements poses serious questions for the future. There are two important aspects to this type of deficit or near-deficit spending. One is that we are caught on a treadmill from which we cannot escape as long as the world's population continues to increase at the present rate. We find ourselves in a situation that requires us to continually increase our pace, that is, expend more energy, just to maintain present standards of living. In some parts of the world we know this is not being achieved, although great strides have been made in certain countries.

Norman Borlaug, awarded the Nobel prize for his work on improved strains of wheat, says that the greatly enhanced production of food for human use is a device for buying time that can be used to develop programs for population control. However,

William Paddock has stated that such a hope is premature at best. There are economic, political, and religious obstacles to achieving even these minimal goals for improving the human condition. Diminishing resources and a growing energy deficit are receiving insufficient attention in policymaking at national and international levels. In his book H. T. Odum states as follows:

> The citizen in the industrialized country thinks he can look down upon the system of man, animals, and subsistence agriculture that provides some living from an acre or two in India when the monsoon rains are favorable. Yet if fossil and nuclear fuels were cut off, we would have to recruit farmers from India and other underdeveloped countries to show the now-affluent citizens how to survive on the land while the population was being reduced a hundredfold to make it possible.*

It should be realized, too, that weather conditions for the past 20 years have been, with few exceptions, rather favorable for large-scale crop production. Now in the mid-70s it is being stated by some scientists that cyclic changes in weather patterns may occur, causing droughts of major proportions. If these forecasts are realized, serious food shortages may develop even in those countries with advanced technology. A reduction of 50% in the corn crop of Iowa alone, which produces one tenth of the world's supply, would have a serious impact on the food problem. Kansas and North Dakota together produce more wheat than Australia, the world's fourth largest producer. Extended drought even in limited regions such as these might well affect the world situation. There is a food crisis today, as we recall that up to one fourth of the world's population may be affected by malnutrition. Yet despite our most advanced technology, the agricultural enterprise still largely depends on favorable weather for its ultimate success in feeding the world's populations.

Furthermore as these populations continue to proliferate, the arable land per capita decreases accordingly, lessening the options for developing future food reserves. The United Nations Food and Agricultural Organization provides the data shown in Table 5-8. To the naive and uninformed, the large stretches of our continent with sparse habitation offer false comfort in addressing population and food problems, but when reduced to an arable, food-producing base, our individual share suddenly seems very limiting to future well-being. The following three sets of variables emerge in the food-energy production picture: (1) arable land adaptable for the kinds of crop plants used in human consumption, (2) weather conditions in the context of an extended period of time, and (3) the type of agricultural technology and genetic selection employed. We have some control over only the third set of variables at present. Even here we depend on unfavorably large expenditures of fossil-fuel energy to develop and maintain our modernized techniques. Discussing social systems and the energy crisis, Boulding states that a collapse of developed agriculture threatens major disaster for the human race.

The second aspect of the food-energy problem involves the spiralling use of these fossil fuels. Accumulations of organic matter in fossil form are the unused products of

---

*Odum, H. T. 1971. Environment, power and society. John Wiley & Sons, Inc. New York. p. 120.

Table 5-8. Arable land of the world*

REGION	ARABLE LAND × 10⁶ (ha)	ARABLE LAND PER CAPITA (ha)
Europe	139	0.31
U.S.S.R.	230	0.94
North America	233	1.02
Latin America	118	0.40
Asia	458	0.22
Africa	212	0.60
Oceania	46	2.35
World	1440	0.39

*Data from United Nations Food and Agriculture Organization. 1972. FAO Yearbook. Rome.

Table 5-9. Unoxidized organic carbon from photosynthesis over geological time and molecular oxygen equivalent required for its combustion*†

CARBON SOURCE	CARBON (tons)	OXYGEN (tons)
Fossil fuels	3.3×10^{12}	8.7×10^{12}
Living biomass and current turnover	1.7×10^{12}	4.5×10^{12}
Sedimentary rocks	5.5×10^{15}	15.7×10^{15}

*Carbon data from Johnson, F. S. 1970. In S. F. Singer (ed.). Global effects of environmental pollution. Springer-Verlag New York, Inc. New York.
†Oxygen equivalents were calculated by the author.

Table 5-10. Total energy consumption for 1974 by sources in the United States*

SOURCE OF ENERGY	EXPENDITURE (×10¹⁵ kcal)	PERCENT OF TOTAL
Petroleum	8.45	46
Natural gas	5.57	30
Coal	3.25	18
Hydroelectricity	0.73	4
Nuclear power	0.30	2
Total	18.30	100

*Modified from Morgan, J. D., Jr. 1975. The mineral position of the United States. In R. W. Marsden (ed.). Politics, minerals and survival. University of Wisconsin Press. Madison, Wis. The original values were expressed in British thermal units (BTUs), here converted to kilocalories (kcal); 1 BTU = 0.251 kcal.

ancient photosynthesis. Because the oxygen liberated during photosynthesis was not used in combustion, it accumulated in the atmosphere and reached a level of about 20% by volume. Prior to the evolution of green plant cells, essentially no free oxygen existed on earth. Would utilization of the total reservoir of *usable* fossil fuels affect that balance? The answer is supplied in Table 5-9; it seems that there would be very little effect, because the free oxygen represented by organic matter stored today in the earth's sediments is at least three orders of magnitude (1000×) greater than that represented by fossil fuels. Clearly oxygen in the atmosphere would suffer little depletion. Even the burning of all our forests would have little impact. However, as fossil energy is rapidly depleted, conservation measures as well as alternate sources of energy must be found to sustain the so-called green revolution.

ISSUES AND ALTERNATIVES

To feed people and turn the "wheels of progress," what will be the principal energy sources of the future? We have been and continue to be a spendthrift society exhibiting a remarkable lack of foresight concerning the conservation of depletable resources. An oft-quoted statistic bears repeating: we represent 6% of the world's population, but use up to one third of its energy resources. Our consumption has doubled in 20 years. Fossil fuel is still our main "out" and will continue to be for some years in the future (Table 5-10). Over 90% of our energy needs are generated by coal, natural gas, and petroleum products. The second point to emphasize is the tremendous consumption on a per capita basis that bears out the disparity between percentage of population and energy use in the world context. Our consumption per capita (based on 220 million people) averages approximately 228,000 kilocalories per person per day. Based on daily caloric intake (Table 5-7) and up to 10 times this value attributed to modern agricultural and production methods of making food available (see the van Overbeek discussion in the preceding section), we find the balance of our daily consumption is approximately 193,000 kilocalories. This amount of energy is 5.5 times the total food factor. It is assigned to other expenditures of energy in our daily lives; that is, the use of electricity, private transportation, shelter, education, recreation, and other demands on the resource base including national programs such as military defense.

The rank of selected regions of the world based on per capita energy used each year is shown in Fig. 5-10. The units are equated to the 1977 world average of approximately 2000 kilograms of primary energy such as coal and natural gas, using one unit as a convenient basis of comparison. By this standard we use nearly six times as much as the average nation, with an even greater disparity when making comparisons with the less-than-average consumption of the poorer nations of Africa, Asia, and Latin America. The concept that increasing consumption indefinitely means an ever-rising standard of living should be brought into question as to its validity. Yet such a concept has always been the basis of economic prosperity and has helped to bring on the present energy crisis, posing serious problems for the future. Based on 1971 figures the energy consumption for Sweden is less than two thirds that of the United States. The Swedish standard of living, however, exceeds ours by several criteria. There are more teachers, more books pub-

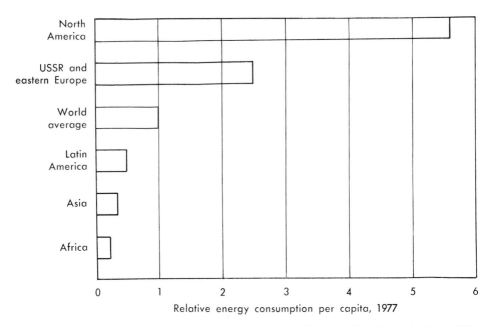

Relative energy consumption per capita, 1977

Fig. 5-10. Energy consumption by world regions, 1977. (Redrawn from Interchange. 1977. Population Ed. Newsletter **6:**1-3.)

lished, more newspapers, and twice the number of hospital beds *per 1000* people in Sweden. It does not necessarily follow that spending more provides more. There is a point of diminishing returns that we have already reached, if not exceeded. One factor in the difference between the two countries is that in Sweden passenger transportation is heavily concentrated on rail and bus transit, compared to motor cars and air travel in the United States. Furthermore, according to Schipper and Lichtenberg, even when similar modes of travel are compared, the United States has the lower passenger-vehicle ratio per mile traveled. This is especially true of the private automobile.

Nuclear power

There are two possible sources from which we can derive nuclear energy, one produced by fission of atomic nuclei, the other by fusion. The latter process discussed briefly at the beginning of this chapter is also the reaction taking place in the high temperatures of the sun's surface. Atomic fission reactors are currently in use on a limited basis (Table 5-10). Fusion reactors are years away from practical application in satisfying national power needs. At present, conventional reactors use fissionable uranium (U 235) as a principal fuel. It constitutes only about 0.7% of natural uranium ore. According to Hammond, if the United States relied entirely on this energy source at the 1972 rate of use, the known supply of U 235 available to us would be depleted in a little over 2 years. At current rates of use, this supply of U-235 would run out in even less time, hence the thrust in "breeder reactor" technology. This type of fission reactor, which will begin production in the 1980s, can utilize the more plentiful but nonfissionable uranium-238 (U 238). The technology involves the use of fast neutrons to bombard the uranium nucleus whereby fissionable plutonium (Pu 239) is the fuel product. Even in the conventional reactors now in use, some Pu 239 is produced. For each 1000 kilograms (1 metric ton) of fresh uranium consisting of 967 kilograms of U 238 and 33 kilograms of U 235, there are 8.9 kilograms of Pu 239 in the spent fuel. It is estimated that breeder reactors could supply our energy needs for 100 years, using U 238 to produce plutonium.

Despite its tremendous potential, nuclear power has its drawbacks. In addition to "reactor safety" problems, that is, the hazards associated with the reactor operation it-self, there is a more important issue for which no long-range solutions have been devised. This is the disposal of very dangerous and long-lived radioactive wastes and the re-processing of spent fuel cores. The radioactive half-life ($T_{\frac{1}{2}}$) of plutonium is 24,000 years. Even after 5 $T_{\frac{1}{2}}$s, or 120,000 years, the environment would still carry about 3% of the original contamination. Plutonium, essentially a synthetic element, is one of the most dangerous known. Not only is it a source of destructive, ionizing radiation (alpha par-ticles), but it is also of itself a dangerous chemical and a serious hazard to human health. In addition to plutonium, there are other radioactive by-products from the reactor opera-tion that must be dealt with. Besides the problems associated with reactor safety and waste disposal, vast quantities of water are required in the cooling process during opera-tion of the facility. According to Novick nuclear power plants release approximately 50% more hot water to the environment (streams, lakes, estuaries, and so on) per unit of electricity produced than do fossil-burning plants. Consequently they have been identified as a serious potential in thermal pollution of the aquatic environment. If projections for the number of nuclear plants going into operation hold true, it is estimated that one sixth of the river systems of eastern North America will be affected by heated water at the turn of the century. In Chapter 2 thermal pollution was initially presented as a hazard to aquatic life based on its extent today. In contemplating the future, heating of natural waters poses severe threats to biological diversity and community composition, adversely affecting productivity, aesthetics, and ecosystem stability.

Nonnuclear power

Thus far we have discussed resources that sooner or later will be exhausted. Another option in this category is the development of geothermal power, which is produced from energy stored in hot water and rock within the earth itself. Its advantage is that no hazards to health and environment are raised, but it, too, is in finite supply. There is a growing awareness that eventually we must turn to those energy sources that are renewable. These include solar radiation, wind power, and tidal energy among others. Of these, solar radia-tion provides the most opportunity for practical utilization in terms of direct solar heating of homes and use of solar cells to generate electrical power. Congressional fund-ing for research, however, is comparatively low in the solar energy field in relation to the support given to nuclear development and to the exploitation of fossil fuels such as the extraction of usable forms of energy from high-sulfur coal or oil shales. More attention should be given to the possibilities of solar energy, although lesser options should not be overlooked. It has been suggested that perhaps there should be a decentralization of energy sources rather than an emphasis on the development of massive electric grids for ever-larger power stations.

Organic waste

Another source of renewable energy is the utilization of organic refuse such as wood, paper, and kitchen garbage. The heating value varies with the composition, but for most urban refuse, the ranges are from 1.5 to 3.0 kilocalories per gram (3000 to 6000 BTUs per pound) dry weight. Such materials are being used in some cities as a fuel source in power plants. A problem associated with direct combustion is the emission of certain pollutants such as chlorine gas when polyvinyl chloride plastic containers and bottles are burned. An alternate method to direct combustion is microbial digestion under anaerobic conditions to produce methane (CH_4). Methane is a principal constituent of natural gas. It has the advantage of being essentially pollution free. Research on an ever-widening basis is being conducted on this form of solid-waste conversion.

A rich field for CH_4 production is in the treatment of agricultural and animal wastes. The amount of material from feedlots, for example, is considerable. According to the American Association for the Advancement of Science* animal wastes alone make up nearly 25% of the total amount of organic refuse generated in the United States. This amount exceeds urban wastes plus municipal sewage. With the trend toward larger concentrations of livestock in feedlot areas, the use of such materials as an energy source is expanding in significance. The reuse of both urban and agricultural waste products is a conservation measure we should not ignore. It is an important factor in helping to answer the nation's energy needs and in simultaneously minimizing health hazards and the deterioration of the environment.

SUMMARY

Photosynthesis is the basic source of essentially all food energy in the biosphere. This energy is distributed throughout the ecosystem by grazing and detritus food chains. It may be visualized as a pyramid diminished as heat of respiration passes from one level of consumers to the next higher level. This energy attrition limits the potential number of consumer transfers. In balanced systems all the energy made available by green plants is ultimately converted to heat. In order to survive as a viable structure, the biosphere must constantly receive fresh inputs from photosynthesis. Changes in the earth's albedo, atmospheric pollution, and other factors affecting the ability of the biosphere to receive solar energy are important considerations that could affect the earth's heat balance and photosynthetic potential. Our species is a dominant one whose numbers are continuously expanding; we are effecting a simplification of energy transfer in the biosphere by destroying native habitats, extirpating animal consumers, and becoming increasingly dependent on the direct utilization of plants as a source of food. In some nations, rapidly

---

*Hammond, A. L., W. D. Metz, and T. H. Maugh II. 1973. Energy and the future. American Association for the Advancement of Science. Washington, D.C.

growing populations already experience protein deficiencies and even caloric shortages, and these are becoming more prevalent. To meet the growing demands for more and better foods, more space will be needed for monoculture conversions, which in turn invite an ever-growing dependence on artifact pollutants (chemical fertilizers, pesticides, and herbicides) to ensure high yields and sustained production.

Other alternatives to fossil fuels and nuclear energy should be explored. These efforts are urgent on the grounds that fossil fuels such as natural gas and petroleum are depletable in the near future and that nuclear energy is a recognized health hazard, primarily in terms of the storage of dangerous and long-lived radioactive wastes such as plutonium. Only coal as a fossil fuel is plentiful and readily usable. But here, too, there are obstacles to environmental protection and the safeguarding of human health when we consider present mining procedures and the high sulphur content of our coal reserves. Oil shales are another source of energy in vast supply, but these constitute an energy problem in their processing, requiring large quantities of water. Most of these deposits are in regions in which rainfall is limited. The alternatives to fossil fuels and nuclear power include solar energy, wind, geothermal power, and a national effort to convert organic waste such as manures to synthetic methane or oil products. In addition, it is paramount that energy and population policies be established around the world and that every nation have the means and motivation in a common cause to coordinate and implement such policies. Otherwise major catastrophes may subdue both developed and lesser-developed societies alike—ecology draws no fine lines between them. Witness the growing water shortages of the 1970s for agriculture in the United States, where techniques are the most advanced in the world or the lack of water for nomadic grazing in parts of Africa. All of us, whether we drive automobiles or ride camels, "suck from the same straw."

DISCUSSION QUESTIONS

1. Photosynthesis "feeds the biosphere." Explain.
2. Discuss the limitations on the number of trophic levels in the food pyramid. How does an expanding human population ultimately affect the "shape" (number of trophic or energy transfers) of the pyramid?
3. What are the disadvantages of using nuclear power? List the threats to the environment where such power plants are located.

REFERENCES

Abercrombie, K. and A. McCormack. 1976. Population growth and food supplies in different time perspectives. Population Dev. Rev. 2:479-498.

Army, T. J. and F. A. Greer. 1967. Photosynthetic limits on crop yields. In A. S. Pietro, F. A. Greer, and T. J. Army (eds.). Harvesting the sun. Academic Press, Inc. New York.

Berg, G. G. 1973. Hot wastes from nuclear power. Environment 15:36-44.

Bethe, H. A. 1976. The necessity of fission power. Sci. Am. 234:21-31.

Borgstrom, G. 1973. World food supplies. Intext Educational Publishers. New York.

Borlaug, N. E. 1971. The green revolution, peace and humanity. World Sci. News 8:9-16.

Boulding, K. E. 1974. The social system and the energy crisis. Science 184:255-257.

Bray, J. R. 1960. The chlorophyll content of some native plant and managed plant communities in central Minnesota. Can. J. Bot. 38:313-333.

Brougham, R. W. 1960. The relationship between the critical leaf area, total chlorophyll content, and maximum growth-rate of some pasture and crop plants. Ann. Bot. 24:463-474.

Carter, L. J. 1977. Failure seen for big-scale, high technology energy plans. Science 195:764.

Chew, R. and A. E. Chew. 1965. The primary productivity of a desert-shrub (Larrea tridentata) community. Ecol. Monogr. 35:355-375.

Cochran, N. P. 1976. Oil and gas from coal. Sci. Am. 234:24-29.

Council on Environmental Quality. 1970. The first annual report to the President. United States Government Printing Office. Washington, D.C.

Coupland, R. T. 1975. Productivity of grassland eco-systems. In D. E. Reichle et al. (eds.). Productivity of world ecosystems. Proceedings of the International Biological Programme Symposium, Seattle, Washington. 1972. National Academy of Sciences. Washington, D.C.

Dahlman, R. C. and C. L. Kucera. 1965. Root productivity and turnover in native prairie. Ecology **46:**84-89.

Economic Research Service, Foreign Regional Analysis Division, United States Department of Agriculture. 1970. The world food budget. Foreign Agricultural Economic Report No. 19. United States Government Printing Office. Washington, D.C.

Gates, D. 1970. Weather modification in the service of mankind: promise or peril. In H. W. Helfrick, Jr. (ed.). The environmental crisis. Yale University Press. New Haven, Conn.

Golley, F. B. 1960. Energy dynamics of a food chain of an old field community. Ecol. Monogr. **30:** 187-206.

Hammond, A. L. 1972. Energy options: challenge for the future. Science **177:**875-876.

Hammond, A. L., W. D. Metz, and T. H. Maugh II. 1973. Energy and the future. American Association for the Advancement of Science. Washington, D.C.

Hardy, R. W. F. and U. D. Havelka. 1975. Nitrogen fixation research: a key to world food? Science **188:**633-643.

Hellmers, H. 1964. An evaluation of the photosynthetic efficiency of forests. Q. Rev. Biol. **39:**249-257.

Hinckley, A. D. (project manager). 1976. Impact of climatic fluctuation on major North American food crops. Funded by the C. F. Kettering Foundation. The Institute of Ecology. Washington, D.C.

Horsfall, J. G. 1970. The green revolution: agriculture in the face of the population explosion. In H. W. Helfrick, Jr. (ed.). The environmental crisis. Yale University Press. New Haven.

Hubbert, M. K. 1969. Energy resources. In Resources and man. W. H. Freeman & Co., Publishers. San Francisco.

Interchange. 1977. Energy: increasing consumption, decreasing resources. Population Ed. Newsletter **6:**1-3.

Johnson, F. S. 1970. The oxygen and carbon dioxide balance in the earth's atmosphere. In S. F. Singer (ed.). Global effects of environmental pollution. Springer-Verlag New York, Inc. New York.

Johnson, P. L. 1969. Arctic plants, ecosystems and strategies. Arctic **22:**341-355.

Johnston, H. 1971. Reduction of stratospheric ozone by nitrogen oxide catalysts from supersonic transport exhaust. Science **173:**517-522.

Kira, T. et al. 1964. Primary production by a tropical rain forest of southern Thailand. Bot. Mag. Tokyo **77:**428-429.

Kucera, C. L., R. C. Dahlman, and M. R. Koelling. 1967. Total net productivity and turnover on an energy basis for tallgrass prairie. Ecology **48:**536-541.

Macfadyen, A. 1963. Heterotrophic productivity in the detritus food chain in soil. Proc. Int. Congr. Zool. **16:**318-323.

Mayer, J. 1976. The dimensions of hunger. Sci. Am. **235:**40-49.

Meadows, D. L. 1972. The limits to growth. Universe Books. New York.

Mitchell, J. M., Jr. 1968. A preliminary evaluation of atmospheric pollution as a cause of the global temperature fluctuation of the past century. In S. F. Singer (ed.). Global effects of environmental pollution. Springer-Verlag New York, Inc. New York.

Morgan, J. D., Jr. 1975. The mineral position of the United States. In R. W. Marsden (ed.). Politics, minerals and survival. University of Wisconsin Press. Madison, Wis.

Newman, J. E. and R. C. Pickett. 1974. World climates and food supply variations. Science **186:**877-881.

Novick, S. 1973. Toward a nuclear power precipice. Environment **15:**32-40.

Odum, E. P., C. E. Connell, and L. B. Davenport. 1962. Population energy flow of three primary consumer components of old field ecosystems. Ecology **43:**88-96.

Odum, H. T. 1971. Environment, power, and society. John Wiley & Sons, Inc. New York.

Ovington, J. D. 1962. Quantitative ecology and the woodland ecosystem concept. Adv. Ecol. Res. **1:**103-192.

Ovington, J. D. 1965. Organic production, turnover and mineral cycling in woodlands. Biol. Rev. **40:**295-366.

Paddock, W. C. 1970. How green is the green revolution? Bioscience **20:**897-902.

Peterson, E. K. 1970. The atmosphere, a clouded horizon. Environment **12:**32-39.

Petrides, G. A. and W. G. Swank. 1969. Estimating the productivity and energy relations of an African elephant population. Proc. 9th Int. Grassland Congr. **9:**831-842.

Petrusewicz, K. and A. Macfadyen. 1970. Productivity of terrestrial animals. F. A. Davis Co. Philadelphia.

Pimental, D. et al. 1976. Land degradation: effects on food and energy resources. Science **194:**149-155.

Pitelka, F. A. 1967. Some characteristics of microtine cycles in the Arctic. In H. P. Hansen (ed.). Arctic biology. Oregon State University Press. Corvallis, Ore.

Plass, G. N. 1959. Carbon dioxide and climate. Sci. Am. **201**:41-47.

Rieber, M. 1975. Low sulphur coal. J. Environ. Economics Management **2**:40-59.

Rohrman, F. A., B. J. Steigerwald, and J. H. Ludwig. 1971. Industrial emissions of carbon dioxide in the United States: a projection. Science **156**:931-932.

Rose, D. J. 1974. Nuclear eclectic power. Science **184**:351-359.

Rosenzweig, M. L. 1968. Net primary productivity of terrestrial communities: prediction from climatological data. Am. Nat. **102**:67-80.

Ryther, J. H. 1969. Photosynthesis and fish production in the sea. Science **166**:72-76.

Schipper, L. and A. Lichtenberg. 1976. Efficiency energy use and well-being: the Swedish example. Science **194**:1001-1013.

Sellers, W. D. 1965. Physical climatology. The University of Chicago Press. Chicago.

Shapley, D. 1976. Crops and climatic change: USDA's forecasts criticized. Science **193**:1222-1224.

Smalley, A. E. 1960. Energy flow of a salt marsh grasshopper population. Ecology **41**:672-677.

Tamplen, A. R. 1975. Solar energy. In B. Commoner et al. (eds.). Energy and human welfare. Alternative technologies for power production. Macmillan Inc. New York.

Turner, F. B. The ecological efficiency of consumer populations. Ecology **51**:741-742.

United Nations Food and Agriculture Organization. 1972. FAO Yearbook. Rome.

van Overbeek, J. 1976. Plant physiology and the human ecosystem. In W. R. Briggs et al. (eds.). Ann. Rev. Plant Physiol. **27**:1-17.

Wade, N. 1972. Gasification: a rediscovered source of clean fuel. Science **178**:44-45.

Watson, D. J. 1958. The dependence of net assimilation rate on leaf area index. Ann. Bot. **22**:37-54.

Weigert, R. G. 1965. Energy dynamics of the grasshopper populations in old field and alfalfa field ecosystems. Oikos **16**:161-176.

Weigert, R. G. et al. 1967. Forb-arthropod food chains in a one-year experimental field. Ecology **48**:75-83.

Welch, H. E. 1968. Relationship between assimilation efficiencies and growth efficiencies for aquatic consumers. Ecology **49**:755-759.

Whittaker, R. H. 1970. Communities and ecosystems. Macmillan Inc. New York.

Whittaker, R. H. and G. E. Likens. 1973. Carbon in the biota. In G. M. Woodwell and E. V. Pecan (eds.). Carbon and the biosphere. Proceedings of the 24th Brookhaven Symposium. 1972. Upton, N.Y.

Whittaker, R. H. and G. M. Woodwell. 1967. Surface area relations of woody plants and forest communities. Am. J. Bot. **54**:931-939.

Wiens, J. A. 1974. Climatic instability and "ecological situation" of bird communities in North American grasslands. Condor **76**:385-408.

Wilson, J. W. 1965. Stand structure and light penetration. J. Appl. Ecol. **2**:383-390.

Wittwer, S. H. 1975. Food production: technology and the resource base. Science **188**:579-584.

Wolf, M. 1974. Solar energy utilization by physical methods. Science **184**:382-386.

ADDITIONAL READINGS

Bach, W. 1972. Atmospheric pollution. McGraw-Hill Book Co. New York.

Brown, L. R. 1970. Seeds of change. Praeger Publishers, Inc. New York.

Center for the Biology of Natural Systems. 1970. Hidden cost of economic growth. Washington University, St. Louis. CBNS Notes **4**:1-18.

Chandler, T. J. 1969. The air around us. Natural History Press. Garden City, N.Y.

Cook, E. 1971. The flow of energy in an industrial society. Sci. Am. **224**:134-144.

Freedman, R. and B. Berelson. 1974. The human population. Sci. Am. **231**:31-39.

Gates, D. M. 1971. The flow of energy in the biosphere. Sci. Am. **224**:88-100.

Gilmore, C. P. 1976. Sunpower: Saturday Rev. October 30, pp. 20-22.

Hatch, M. D., C. B. Osmond, and R. O. Slayter (eds.). 1971. Photosynthesis and photorespiration. John Wiley & Sons, Inc. New York.

Leith, H. and R. H. Whittaker (eds.). 1975. Primary production of the biosphere. Springer-Verlag New York, Inc. New York.

Loomis, R. S., W. A. Williams, and A. E. Hall. 1971. Agricultural productivity. In L. Machlis, W. R. Briggs, and R. B. Park (eds.). Ann. Rev. Plant Physiol. **22**:431-468.

Lovins, A. B. 1975. World energy strategies. Facts, issues and options. Friends of the Earth International. San Francisco.

Lowry, W. P. 1967. Weather and life; an introduction to biometerology. Academic Press, Inc. New York.

Morgan, D. 1976. American agripower and the future of a hungry world. Saturday Rev. November 13, pp. 7-12.

Morowitz, H. J. 1968. Energy flow in biology. Academic Press, Inc. New York.

Rappaport, R. A. 1971. The flow of energy in an agricultural society. Sci. Am. **224**:116-132.

Spedding, C. R. W. 1975. The biology of agricultural systems. Academic Press, Inc. New York.

Stacks, J. F. 1971. Stripping: the surface mining of America. Sierra Club. San Francisco.

Steinhart, C. E. and J. S. Steinhart. 1974. Energy: sources, use and role in human affairs. Duxbury Press. North Scituste, Mass.

Tanzer, M. 1969. The political economy of international oil and the underdeveloped countries. Beacon Press. Boston.

UNESCO. 1965. Methodology of plant eco-physiology. In F. E. Eckardt (ed.). Proceedings of the Montpelier Symposium. Place de Fontenoy, Paris.

Wald, G. 1976. The nuclear-power-truth maze. Bioscience. 26:631-632.

Wallace, D. (ed.). 1976. Energy we can live with. Rodale Press. Emmaus, Pa.

Woodwell, G. M. 1970. The energy cycle of the biosphere. Sci. Am. 223:64-74.

Zelitch, I. 1971. Photosynthesis, photorespiration and plant productivity. Academic Press, Inc. New York.

6 Organic turnover and cycling processes

BIOLOGICAL TURNOVER

Walking through a forest with its ever-present carpet of leaves, we can appreciate the fact that decay and replenishment are recurring processes. How much biomass (measured in grams or pounds) or its energy equivalent (expressed in calories) is represented in just this one compartment or reservoir of the ecosystem? Under equilibrium conditions, how much time would be required for the energy represented here to be completely used and then replaced by a similar amount of energy? These are questions dealing with the "turn-over" process of systems in a state of energy balance. Here we are using only one part of the total biomass structure as an example. Obviously the turnover process and its attendant release of inorganic substances, the mineral nutrients, apply to all forms of energy accumulation in the ecosystem—the wood, leaves, animal biomass, and micro-organisms in environments that are above as well as below ground.

Under equilibrium conditions, all the gross primary production (GPP) from photo-synthesis is utilized in autotrophic and heterotrophic respiration (R_A and R_H). This rela-tionship can be expressed as follows:

$$GPP = R_A + R_H$$

Net plant growth, or net primary production (NPP), is equal to GPP minus R_A. This is the biomass available to the heterotrophs, and all of it is accounted for in the grazing and detritus food chains of the biotic community as per the following equation:

$$NPP = R_H$$

Under these conditions, net ecosystem production (NEP) then equals zero as follows:

$$NEP = (NPP - R_H) = 0$$

Although NEP is specified as zero, this does not mean there is no *new* growth in the plant and animal populations making up the community. Indeed there must be growth to re-place losses that result from such factors as herbivory, predation, disease, and old age in the various levels of the food web. When NEP is zero, all the energy in the form of new growth is exactly balanced by the amount of energy that the ecosystem uses to main-tain itself, so that there is no energy accumulation or storage. This topic is discussed in greater detail in Chapter 7.

Periods of growth and expansion are inherent in all biological forms, including eco-systems. After a time, however, net growth or increase in biomass begins to slow and eventually ceases as negative feedback processes appear in the system. When NEP is positive, the system is in the process of storing energy in the form of biomass structure. As long as net growth continues for the community as a whole, the system has not reached a state of equilibrium with the physical potential of the environment. An NEP that is negative, or less than zero, implies excessive use of available resources. Harvesting the trees in commercial forests at a rate faster than they are being replaced by new growth is an example. Another is the overgrazing of range lands that is reflected in decreasing yields of the forage grasses and their replacement by less productive species of plants.

The rate at which energy is used is a characteristic feature of balanced communities, and therefore the rate differs from system to system. Furthermore it differs from compartment to compartment within the same system. Energy flows readily through the leaf litter on the floor of a tropical rain forest, releasing with commensurate speed the mineral nutrients held within the organic framework of the litter. These processes occur much more slowly in a far northern forest where biological activity is greatly retarded by comparison. A knowledge of the different rates of energy flow and dispersal of nutrients is fundamental to an understanding of ecosystem function. A knowledge of how such processes work and an awareness of the distinctions that arise between ecosystems provide much-needed insights to wise land use and resource conservation. All too often the technology developed in one region is applied to another with less than satisfactory results. Can land-clearing practices used in cool-temperature zones, for example, be applied to the tropics where temperature and rainfall may make denuded landscapes particularly vulnerable to erosion and laterization?

The turnover rate is of special significance to the energy dynamics of a community, for it provides information on the amount of energy that passes through the system in a given period of time and on the energy storage characteristics of the system. We may take as an example the annual leaf fall in a forest. Leaves represent net productivity (P) for a particular energy compartment of the ecosystem. How does this value relate to the residual litter (R), which is a part of the forest that was built up over a period of years during the development of the ecosystem? In balanced systems we know that the amount of energy dissipated over time is equal to energy input; however, disappearance is not represented by the actual biomass produced during that particular period. Rather it is a sum of decay increments from the total biomass (B) that makes up the residual litter from a number of past production periods (R) plus the current net productivity (P). The turnover rate is expressed as follows:

$$TR\ (\%) = \frac{P}{P + R = B} \times 100$$

Using the example of annual leaf fall in relation to total biomass, we may calculate an annual turnover rate for leaves. The annual production is 5000 pounds per acre and the residual is 15,000 pounds per acre as per the following:

$$TR = \frac{5000}{5000 + 15,000} \times 100 = 25\%$$

Since production and loss are equal in balanced systems, we may also write as follows:

$$Loss = TR \times (P + R) = 5000 \text{ pounds per year}$$

These 5000 pounds represent a composite of decay losses from the total biomass (P + R), which is made up of residual materials of different ages. Turnover rate, then, is an estimate of the decay factor, often designated as k.

Another example for which data are available is the 50% annual turnover rate for a tall-grass prairie, which is based on an average foliage production of 500 grams per square meter and a total biomass (P + R) of 1000 grams per square meter.* Production

---

*There are 454 grams in 1 pound and 4000 square meters in 1 acre; therefore the conversion to pounds per acre = $g/454 \times 4000/m^2 = g/m^2 \times 8.8$.

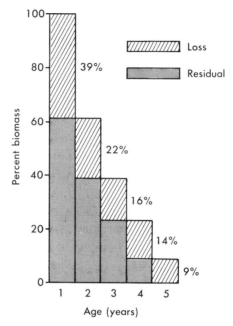

Fig. 6-1. Annual turnover in prairie litter as a composite of losses from several age classes in the total biomass. Of the annual loss, 9% is 5 years old or more. (Modified from Kucera, C. L., R. C. Dahlman, and M. R. Koelling. 1967. Ecology **48**:536-541.)

and decay are assumed to be in balance; therefore decay is also 500 grams per square meter. Simulated litter samples were used to study decay losses through a 4-year period. Findings show that the annual turnover consisted of 195 grams from the preceding crop, 110 grams that were 2 years old, 80 grams that were 3 years old, 70 grams that were 4 years old, and 45 grams that were at least 5 years old. In other words the prairie litter considered here comprises deposits from annual crops going back at least 5 years. This breakdown of annual turnover by age is shown in Fig. 6-1.

Some ecosystems accumulate large amounts of litter and others very little, depending on production and decay relationships characteristic of the system. When the accumulation (R) is large, that fraction of the total biomass attributable to current production is obviously small, suggesting slower turnover rates for the total biomass. Since the decay rate (k) is inversely related to turnover time (t), slow decay processes indicate that extended periods are required for the disappearance of a given biomass under equilibrium conditions. This relationship can be expressed as follows:

$$k \text{ or } TR = 1/t$$

If a given compartment has a storage value that is large in relation to its production, it will be less subject to fluctuation at any particular time. This relationship may be demonstrated in areas across a wide range of latitudes, showing the effect of temperature on the rate of decay processes (Table 6-1). Note that the rain forest has the greatest production rate and the lowest litter accumulation (R).

Fig. 6-2. Solid waste accumulation at Sagwon, a distribution center for oil operations in the Alaskan tundra. (Courtesy David R. Klein and International Union for Conservation of Nature and Natural Resources, Morges, Switzerland.)

In cold climates the total annual production of forests is less than in warmer regions, but despite this fact, organic storage in the form of litter tends to be greater in high-latitude forests because of retarded decomposition. If you have ever walked through either high montane or Canadian spruce forests, the thick, spongy quality of the forest floor is readily apparent. In these northern or high-altitude ecosystems the most aged fractions of this surface debris may be 100 years old or more! One can readily see how human intrusions under such conditions may leave junk and garbage heaps for years to come unless proper disposal procedures are adopted (Fig. 6-2). Although tropical forests have a greater productivity than those in higher latitudes, energy accumulation in the form of litter storage is disproportionately less because of the higher decomposition potential reflected in higher turnover rates.

These relationships (Table 6-1) demonstrate that the rate of disappearance or utilization of leaf biomass in a moist tropical forest is more than twice (0.60/0.25) the rate in a temperature forest, about 10 times (0.60/0.07) the rate in a northern conifer forest, and 30 times (0.60/0.02) the rate in the tundra. The rapid breakdown of materials in the tropics creates serious problems in the conservation of nutrients in the soil, since they are readily released and washed away in regions of great rainfall. Preservation of the plant cover that moderates soil temperatures, thus slowing down decay rates and protecting the soil against nutrient losses, becomes especially significant in the wet tropics. Yet it is in these areas that some of our most critical food shortages arise, prompting the clearing of rain forests on an ever-increasing scale for agricultural production.

Table 6-1. Estimates of annual leaf production (P), residual litter accumulation (R), derived turnover rate (P/B), and turnover time (t) for several ecosystems in a temperature gradient*

ECOSYSTEM	P (lb/acre)	R (lb/acre)	P/B (B = P + R)	t (yr)
Tropical rain forest	15,000	10,000	0.60	1.7
Temperate deciduous forest	5,000	15,000	0.25	4.0
Northern conifer forest	3,000	45,000	0.07	14.0
Shrubby tundra	1,000	50,000	0.02	50.0

*Data from Bray, J. R. and E. Gorham. 1964. Adv. Ecol. Res. **2**:101-157; Rodin, L. E. and N. I. Baslievic. 1968. In Proceedings of the UNESCO symposium on functions of terrestrial ecosystems at primary production levels. Copenhagen; and Jordan, C. F. 1971. Am. Sci. **59**:425-433.

ESSENTIAL ELEMENTS

Carbon, hydrogen, and oxygen are the basic building blocks of energy-rich compounds manufactured through photosynthesis. There are also a specified number of nutrient elements that are essential to plant growth and metabolism. The so-called macronutrients are used in concentrations up to several thousand parts per million (ppm), whereas the micronutrients, or trace elements, may be present in adequate amounts at levels of 1 ppm. For example, extremely small amounts of molybdenum are needed in the symbiotic fixation of nitrogen in leguminous plants. Boron is a trace element required in cell division. Often inhibited growth may be attributed to an insufficiency of a single element, an illustration of the well-known principle called the *law of the minimum*. This law states that a function or process is controlled by the factor that becomes deficient first. This principle, of course, applies to ecological factors other than nutrient supply, including soil moisture, light intensity, and carbon dioxide concentration. The basic premise of controlling factors can be expanded to the *law of tolerance,* which deals with such regulatory features of the environment as temperature extremes, salt concentrations, soil acidity (expressed as pH), and numerous types of pollution.

The essential nutrients are classified as follows:

MACRONUTRIENTS	MICRONUTRIENTS (trace elements)
Calcium	Boron
Magnesium	Cobalt
Nitrogen	Copper
Phosphorus	Iron
Potassium	Manganese
Sulfur	Molybdenum
	Zinc

Nitrogen, phosphorus, and sulfur, in particular, figure prominently in life processes—energy transfer in cell metabolism, amino acid production, and the building of proteins as the basis of protoplasmic growth could not occur without them. These three elements, plus carbon, hydrogen, and oxygen, make up more than 99% of all living matter.

Many other elements occur in plants. Silicon, for example, is found chiefly in grass plants, and sodium is found primarily in halophytes or salt plants. There may be as many as 40 different elements involved in plant uptake. However, the fact that a particular element is present in plant tissue does not always mean that it is essential, as its presence may result solely from its ready availability in the water or soil. Plants growing in soils in which cinnabar (mercuric sulfide) is found in commercial quantities present a good example. Under these conditions, the vegetation contains concentrates of mercury as great as 3500 parts per billion (ppb), whereas plants in normal soils may contain concentrations that do not exceed 500 ppb. Selenium-bearing plants are another example. Certain species of the pea (legume) family are called "selenium indicators" when this element occurs in high concentrations in plant tissue.

Radioactive elements that are chemically similar to elements normally taken up by the plant may also be absorbed and enter the food chain. Mixed fission products (MFP) of atomic testing enter the atmosphere and eventually rain down over wide areas far removed from their source. Strontium-90 and cesium-137 are among the important radionuclides that are the by-products of nuclear explosions. They have received wide attention during the years of atomic testing. Strontium, like calcium, becomes concentrated in bones and teeth. Cesium is deposited in muscle, and it moves through living systems much like potassium. The fact that plants fail to discriminate between those elements that are essential to their metabolism and those that are chemically similar but often deleterious is obviously a lesson in the importance of guarding the environment. If these elements are allowed entry into ecosystem processes, they can be carried into the food chain, eventually to man himself. The phenomenon of "biological concentration" manifests itself repeatedly when foreign substances that are persistent are released into the environment. DDT and other "hard" pesticides and radioactive elements are but a few examples of such substances.

With the principal exception of nitrogen, the essential nutrients listed previously are

Table 6-2. Composition of earth's crust*

ELEMENT	PERCENTAGE BY WEIGHT
Oxygen	46.6
Silicon	27.7
Aluminum	8.1
Iron	5.0
Calcium	3.6
Sodium	2.8
Potassium	2.6
Magnesium	2.1
All other elements combined	1.5
	100.0

*From Mason, B. H. 1966. Principles of geochemistry. John Wiley & Sons, Inc. New York.

derived from the weathering of parent materials. These are the so-called mineral elements. Although there is considerable variation in the abundance of these elements, they are far less prevalent than certain other elements that are naturally present in the soil (Table 6-2). Oxygen is the most plentiful element, followed by silicon, aluminum, and iron. Together these four elements make up almost 88% of the total mass. Oxygen is bound in this mineral framework, as in the iron and aluminum silicates. It is not free to evolve as a gas because of the ease with which it becomes bound in secondary products. The next most plentiful elements are calcium, magnesium, sodium, and potassium. All the other elements make up only about 1.5% of the total. *Phosphorus, among the macro-nutrients, is in shortest supply.*

NUTRIENT CYCLES

The circulation of plant nutrients is a process that is complementary to energy flow. When organic matter is not reduced heterotrophically and transformed into heat energy, carbon dioxide, and water, the only other means of achieving an appreciable release of nutrients is by physical leaching. This final breakdown process, called mineralization, is effected by microbiological activity. If green plants were deprived of a regular nutrient supply, they in turn would be unable to produce the steady supply of energy required by the ecosystem. There is a basic difference between the logistical patterns of energy and those of nutrients. Energy is depicted as having a one-way, irreversible flow process (Chapter 5). Nutrient elements such as calcium, magnesium, iron, or phosphorus, however, are nonreducible and therefore can be reused by plants. Nutrients are basic components of cyclical processes that vary in rate of movement, bulk of storage compartments, and origin. A knowledge of these characteristics is essential to the judicious management of both native and artificial (crop) ecosystems. Information concerning the importance and relative abundance of these elemental resources in a finite environment is also important. Furthermore, as we noted earlier, biological decay varies significantly with climatic conditions, so that environmental factors are important not only in regulating nutrient exchanges between the physical and biological components of ecosystems but also in the evolution of special mechanisms that are selected for adaptability to a given environment.

Tropical rain forests, for example, exhibit an adaptive process called direct nutrient cycling. This mode of cycling, described by Went in his studies of the Amazon forests, involves bypassing the soil in nutrient exchange under certain conditions. Leaf litter and humus accumulation are often only a few centimeters thick on the floors of such rain forests. The fine feeder roots of trees occur primarily in this organic layer of rotting leaves and other plant debris, and they seldom penetrate the underlying mineral soil. Mats of fungi are also present in the layer of surface debris. Their hyphal strands are closely intertwined with the delicate tree roots and decaying organic matter. Such physical associations between root and fungus are called *mycorrhizae*. The process is mutually beneficial to both host plant and the fungal population, since the green plant supplies usable energy to the fungus in organic form via photosynthesis, and the fungus in turn conserves nutrients that are available to the root system of the photosynthetic host. As nutrients are released through decay, they are retained in the surface layer by these mats of fungal growth. The mycorrhizae thus are a vital link, replacing soil, which is the common agent that holds nutrients against leaching and transfers them in a directly available form to living root systems.

This cooperative complex of vegetation and microorganisms can develop and sustain itself under the protective canopies of a dense forest. The system persists because it performs the vital functions of conserving nutrients and making them available in a medium otherwise unfavorable for such activity. Imagine the effect on nutrient storage and availability in areas in which these complex forests are leveled or partially cleared through cutting and burning. Such practices are widespread throughout the tropics where increased food production is a primary goal, despite the fact that delicately balanced communities are destroyed. One result is an impoverished system that has been depleted of essential nutrients and in a very short time is so depleted that it becomes inadequate for either closed forest or crops. Restoration can occur only through the process of biotic succession, if at all. The question here is not whether demands for food that require conversions from natural to man-made ecosystems should be met; the question involves the extent to which such destructive conversions can be made before environmental deterioration becomes irreversible.

The elements essential to life are interwoven through a series of intersecting cycles. Carbon, hydrogen, oxygen, nitrogen, and sulfur move through living matter, entering and leaving their respective sinks (storage compartments) of air, water, and soil in the process. These five elements appear in various chemical forms, but all have in common a volatile or gaseous phase. Note that of the six elements that constitute 99% of living matter, only phosphorus is limited to soil and water by the nature of its cycling process. At one time or another in the cycle, carbon dioxide, sulfur dioxide, hydrogen sulfide, water vapor, molecular oxygen, and nitrogen are the transported agents. The other requirements of plant nutrition are restricted primarily to soil and water systems. Dust may on occasion carry nutrients aloft into the atmosphere for varying periods before they settle again on land or water surfaces. Oxygen and nitrogen are the main constituents of the atmosphere. Their respective values are approximately 20% and 78% by volume. Carbon dioxide occurs as a minute percentage, equivalent to little more than 300 ppm (0.03%). All other gases make up the balance. The concentration of water vapor alone may vary widely, since it is affected by local climatic conditions on a day-to-day basis.

In the following pages four elements will be discussed. These are phosphorus, nitrogen, sulfur, and carbon. Phosphorus, nitrogen, and sulfur are critical plant nutrients that deserve our attention. Supplies of phosphorus may become scarce within a relatively short time; nitrogen is part of a complex cycle that may be interrupted by industrial synthesis from a vast atmospheric reservoir. Here the problem may be excess amounts of nitrogen, especially from agricultural runoff and animal wastes into streams and reservoirs. Both elements, however, are important in speeding up the process of biological succession, particularly in aquatic environments. These nutrients increase photosynthesis, often providing an oversupply of organic matter, or carbon, as an energy source. Such an abundance may have serious repercussions in the ecosystem. The microorganisms

Table 6-3. Amounts of those elements essential to plant growth dissolved in sea water *

ELEMENT	AMOUNT (tons/mile^3)
Magnesium	6,400,000
Sulfur	4,200,000
Calcium	1,900,000
Potassium	1,800,000
Boron	23,000
Nitrogen	2,400
Phosphorus	330
Iron	47
Zinc	47
Molybdenum	47
Copper	14
Manganese	9
Cobalt	2

*Modified from Goldberg, E. D. 1963. In M. N. Hill (ed.). The sea. John Wiley & Sons, Inc. New York.

that feed on this organic carbon may reduce the oxygen supply to levels that are inadequate to sustain other forms of life in the community. Sulfur is a third element of importance. It is becoming increasingly significant in biospheric cycling, not as an impending nutrient deficiency, but as an atmospheric contaminant from coal-burning power plants. The fourth element, carbon, is a main component of all organic material. By tracing the carbon cycle it is possible to plot the flow of energy throughout the biosphere.

Phosphorus

Phosphorus is the only plant nutrient utilized in relatively large quantities (a macronutrient) of which shortages in supply may well develop in the future. Although in the future other elements required for agriculture and industry may be extracted from seawater (as magnesium and sodium chloride, common table salt, are extracted now), it is unlikely that phosphorus can be obtained in this manner. Phosphorus is relatively scarce in the oceans (Table 6-3). Its average concentration is approximately ½ pound per million gallons, compared to about 45 tons of sodium and more than 85 tons of chlorine. This comparison emphasizes some serious obstacles to the human modification of the landscape. On the one hand, phosphorus is essential to plant growth, and the supply may become limited. On the other hand, seawater has such high concentrations of nonessential salt (for plant growth) that it cannot be used for irrigation in arid regions where expanded crop systems are contemplated.

In 1971 a symposium dealing with global ecological problems was sponsored by The Institute of Ecology, an international organization. At this symposium the estimates made of world population growth indicated that the use of phosphorus will continue to increase at rates that will deplete all known reserves of rock phosphate in about 60 years. At this point the population of the world will be approximately 11 billion. If the use of phosphorus could be stabilized at current levels, however, these reserves would last for 400 years. Once they are depleted, the earth's ecosystems would have to rely on the gradual release of phosphorus through geological weathering. It is estimated that with this limited quantity of phosphorus the earth could support a human population of only 1 to 2 billion people!

Aside from the increasing demands of a larger population, there are several other reasons for impending shortages. When phosphorus is applied to the soil as fertilizer, some of it becomes chemically immobilized as insoluble phosphates with several other elements that are relatively abundant in the soil: calcium, iron, and aluminum. When soils are alkaline, calcium phosphates are formed; under more acid conditions, phosphorus enters into chemical combinations with iron and aluminum. Since these phosphates are released very slowly for immediate plant use, sustained applications of phosphate fertilizers are necessary to supply the expanding needs of modern agriculture.

Phosphates also enter the drainage pattern in runoff, although not the same degree as nitrates, which have a greater solubility in the soil. Nonetheless it is estimated that over 10 million tons of phosphorus enter the ocean each year, with a large proportion of this amount deposited as insoluble precipitates of calcium phosphate. Phosphorus under these transfer conditions is essentially a noncycling resource. Some reclamation from the sea occurs in zones of upwelling such as the Humboldt Current, in which nutrients are carried upward, are utilized by phytoplankton in the euphotic (light) zone, and then enter a fish-bird food chain. Guano deposits that are rich in phosphates have been built up on offshore islands and coastal areas, and these are an important source of fertilizer. Harvest from the seas is another means of reclamation. These deep-sea deposits could only be replenished, however, through uplift of the earth's crust and exposure of new phosphate beds above sea level.

The phosphorus cycle is relatively simple (Fig. 6-3). Unlike nitrogen, phosphorus has

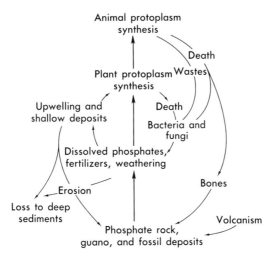

Fig. 6-3. Phosphorus cycle.

no atmospheric interface with land and water. When it is removed from the normal cyclic processes of the biosphere, the results are not easily overcome. Another factor to be considered in the conservation of phosphorus is the manner in which it is currently being used. In the United States about four fifths of the phosphate extracted from deposits is used in fertilizers. Approximately 10% to 15% is used in the manufacture of detergents, and the remainder is used for other industrial purposes. These nonfood uses are an additional drain on phosphorus reserves and are also an important factor in the accelerating eutrophication in aquatic systems.

If phosphorus shortages develop and no new deposits are discovered, will native ecosystems be able to maintain themselves as they have for millions of years without exhausting the supply of phosphorus? Some deficiencies certainly do exist in nature, but in general the answer lies in the characteristics intrinsic to native ecosystems. One is the "closed" nature of cycling in stable communities, which means that phosphorus and other mineral elements tend to remain within the system. They are not dispersed here as readily as they are in crop systems, which are more vulnerable to erosion and nutrient losses. Minerals gradually released through the decomposition of organic matter again become available to plants. The losses that do occur as a result of leaching are gradually made up by minerals released by weathering of the parent material. Large-scale monitoring studies are now underway in various parts of the country to assess the "nutrient budget" of whole ecosystems with regard to many other nutrients as well as phosphorus.

In addition, ecosystems develop through the activities of self-regulating mechanisms that create optimal conditions for survival rather than through a constantly expanding productivity. It is true that many native communities, particularly aquatic systems, respond to phosphorus applications with increased plant growth and subsequently more heterotrophic populations. Yet such increases in net production usually change community structure and composition. Although a balanced community tends to function as a nonexpanding system, the self-regulation and feedback processes inherent in native systems are lacking in artificial communities because of the external input of nutrients. In nature the feedback loop may be initiated by some stress or limitation that ultimately establishes a top level of production, thus preventing the development of excessive requirements and the exhaustion of a particular resource.

Nitrogen

Nitrogen is an inert, colorless, tasteless gas. Alone, it is incapable of supporting life, yet it is an essential component of all living organisms and is involved in the synthesis of proteins and the production of enzymes and nucleic acids. It is the main constituent of the earth's atmosphere, totaling almost four fifths by volume. In a sense atmospheric nitrogen is a lifeless envelope on which a viable biosphere depends. Certain steps, or transformations, are necessary to convert nitrogen from an unavailable state to one usable by green plants. In nature these steps are accomplished principally through the process of nitrogen fixation that is carried out by several types of soil microorganisms. These are certain bacteria and the blue-green algae.

Among the bacteria are two important and well-known genera, *Azotobacter* and

Table 6-4. Nitrogen content and dry matter comparisons between a stand of red alder 34 years of age and a Douglas stand 450 years of age*

FOREST TYPE	DRY MATTER (tons/ha)	NITROGEN (kg/ha)
Red alder stand	210.0	589
Understorey plants	9.5	103
Forest floor (litter, and so on)	66.4	877
Total	286.0	1569
Douglas fir stand	539.0	388
Understorey plants	9.9	58
Forest floor (litter, and so on)	98.6	566
Total	648.0	1012

*Modified from Turner, J. et al. 1976. J. Ecol. **64**:965-974.

Clostridium. These are free-living organisms. The first occurs in aerated soils, and the second is an anaerobe found in environments devoid of oxygen. Other bacteria living in the soil are also able to fix certain quantities of atmospheric nitrogen in forms that can be used by green plants. An important group of nitrogen-fixing bacteria belong to the genus *Rhizobium.* In contrast to the free-living types, these bacteria live in a symbiotic association on the roots of legumes such as peas, beans, and clover. Bacterial cells can invade the fine roots of these plants, producing nodules in which nitrogen is fixed. Since plants of the legume family are found around the world, their importance in the nitrogen enrichment of a wide variety of soils is apparent. Symbiotic relationships in nitrogen fixation also occur in other plants, including alder, podocarps, bayberry, and cycads, but involve microorganisms other than *Rhizobium.* The effectiveness of alder as a nitrogen-symbiont is especially marked, as expressed by the data in Table 6-4. This small tree species in the Pacific Northwest pioneers barren lands and in short periods of time develops relatively large amounts of nitrogen. The level of nitrogen in an alder stand exceeds that of old-growth Douglas fir occurring later in the succession.

In all the nonleguminous plants previously mentioned, nodule formation does occur. Bacterial associations or infections on the surface or in the cortex of roots have been found in several grasses, including corn. No nodules are produced, but these discoveries are creating wide genetic interest because grasses are among the most important plants in food and forage production. If microbial fixation in genetically altered plants could be substituted for the fossil fuel used in making nitrogen fertilizer, there would be large savings of depletable energy supplies. In addition to bacteria, blue-green algae can also fix nitrogen and are significant pioneer species in abandoned fields, where soils have been depleted by poor farming practices. Native ecosystems such as forests and grasslands depend on the sustained activity of the various nitrogen-fixing microorganisms occurring in the soil.

The nitrogen cycle includes two other significant processes, nitrification and denitrification, that involve soil microorganisms. In a sense these are two distinct and opposing processes, as nitrification increases the availability of nitrogen to plants and

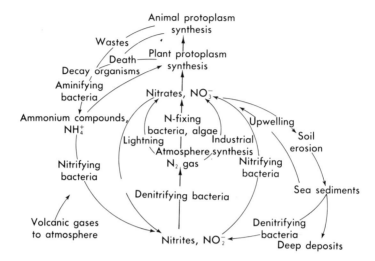

Fig. 6-4. Nitrogen cycle.

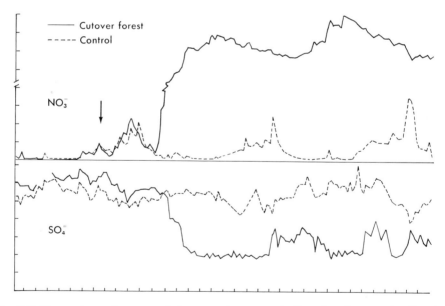

Fig. 6-5. Comparison of measured stream-water concentrations for nitrate and sulfate ions in cutover forest and control forest at Hubbard Brook, N.H. (Redrawn from Likens, G. F. et al. 1970. Ecol. Monogr. **40**:23-47. Copyright © 1970 by the Ecological Society of America.)

denitrification returns it to its original state as nitrogen gas (N_2) in the atmosphere. When plant and animal matter decay, the nitrogen locked up in complex molecules is progressively oxidized in a series of steps, each relying on a different group of soil microorganisms. Liberated ammonia is transformed by species of *Nitrosomonas* to nitrite salts (NO_2^-), which in turn are oxidized to the nitrate form (NO_3^-) by the genus *Nitrobacter*. The energy released through these oxidations can be used by the organisms in synthesizing organic energy from carbon dioxide and water. This sequence proceeds readily in well-aerated soils. Under waterlogged conditions in which the soil lacks oxygen, the process is reversed, and denitrification occurs. Nitrates are converted to nitrites, releasing the oxygen required for the metabolism of organic compounds in the soil. These processes are illustrated in Fig. 6-4.

As the result of studies at Hubbard Brook Experimental Forest in New Hampshire, Likens and co-workers presented data on nitrogen transformations adversely affected by clear-cutting heavily forested watersheds. Clearing the forest and exposing the soil caused a rapid increase in the nitrifying process, shown by accelerated losses of nitrate nitrogen from the ecosystem (Fig. 6-5). These findings suggested that in the undisturbed state conditions are maintained in which nitrification is minimized and that the climax vegetation draws on the ammonium form (NH_4^+) for its supply of nitrogen. The opposite effect was observed for sulfur (Fig. 6-5). Under conditions of accelerated nitrification with clearing of the watershed, there was a significant decrease in the production of sulfate ions (SO_4^{-2}). The toxicity of high nitrate concentrations to sulfur-oxidizing bacteria may be a controlling factor in sulfate formation. Thus cycles of the various elements, in this case nitrogen and sulfur, exhibit an interdependency that may be adversely affected by drastic changes in the ecosystem.

Nitrogen in nitrate form is very soluble and readily leached from the soil. Additionally in cropping systems where heavy applications are required for high yields, nitrate losses are considerable. Delwiche reports that losses to the sea amount to 30 million tons per year. Since nitrogen can be restored to the atmosphere in a gaseous form by denitrifying bacteria, this loss is not as serious as in the case of phosphorus, which remains bound in the sediments deposited on the sea floor.

In recent years nitrogen fixation by industrial methods has increased greatly, adding to the volume of transfer in the nitrogen cycle. If more nitrogen is being fixed by combined microbial and industrial activities than is being denitrified by microorganisms, accumulation obviously is occurring at particular points in the cycle. Two problems are implied. If the biosphere is indeed experiencing a gain in nitrogen, then enrichment, especially of aquatic systems, is a probable result. In a recent study it was shown that in the intensively farmed areas of Illinois, about 55% to 60% of the nitrogen (found as nitrate) in the outflow from drain tiles could be traced to fertilizer application. Obviously under these conditions the supply of nitrogen moving into the watershed is being increased. The second problem involves the nitrate contamination of underground water that is the source of some municipal water supplies. In the human body, nitrates are converted to nitrites, which in excess may lead to the development of deficiencies in oxygen metabolism, causing methemoglobinemia. This is the so-called blue baby condition that occurs in newborn infants. In some localities, including the San Joaquin Valley in California, nitrate poisoning is causing concern. Water in many of the valley's wells has been shown to exceed nitrate levels acceptable for human health.

The oxides of nitrogen emitted in automobile exhaust and by factories, electric power plants, and other consumers of fossil fuels are another source of pollution. The automobile contributes almost one half of the total. Oxides of nitrogen enter the atmosphere and, in a series of complex reactions in the presence of visible light, produce photochemical oxidants, including ozone (O_3). Certain plant species are very sensitive to ozone. Concentrations as small as 0.05 ppm can reduce radish crop yields by one-half. Increasing amounts of nitrogen oxides are entering the atmosphere; more than 20 million tons were emitted in 1968 in the United States alone. Still another industrial source of nitrogen is nitrilotriacetic acid (NTA), which has been suggested as a substitute for phosphates in detergents. However, other effects besides simple increases in nitrogen are being discovered. For example, there is some evidence that the breakdown products of NTA produce carcinogenic symptoms in test animals. Chapter 11 provides information on urban-industrial deterioration of the atmospheric environment, including "smog" formation and the rising incidence of "acid rain."

The nitrogen cycle depends on several specialized groups of soil organisms. Without the benefit of their diversified activities, native ecosystems would soon experience a disruption of the nitrogen cycle. Shortages would develop where nitrogen fixers and nitrifiers were lacking, and surpluses would appear in the absence of denitrifiers. Do chemicals used for weed and pest control affect these organisms? Researchers in several disciplines have been considering this question for the past 20 to 25 years. Generally herbicides such as 2,4-D are not a significant threat because under temperate conditions they are degraded and rendered harmless in a relatively short time, sometimes in a matter of weeks. Yet Pramer has recently suggested that when soil microorganisms break down these herbicides, molecular structures that are also toxic may be produced. Certain pesticides such as DDT, dieldrin, and BHC in the chlorinated hydrocarbon group are known to be relatively long lived in the soil. They do not decompose readily, and the secondary products resulting from dechlorination by certain bacteria are also a hazard. The dechlorination process is as follows:

It has been shown that residual amounts of DDT remain in certain soils for 10 years or more. Even after 14 years, 50% of the original application may still remain. This characteristic persistence of DDT is shown in Fig. 6-6. Fortunately rates of application in the field are not generally detrimental to the bacteria involved in the nitrogen cycle, for the rate of disappearance maintains an equilibrium at a level within the range of tolerance of these organisms. Yet studies show that the activities of some organisms are sharply re-

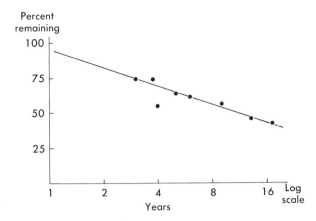

Fig. 6-6. Percentage of DDT remaining in test soil by years. Even after 15 years, approximately one half of the original application still remains. (Redrawn from Nash, R. G. and E. A. Woolson. 1967. Science **157**:924-927.)

Table 6-5. Effect of DDT on nitrate production by nitrifying bacteria, expressed as milligrams of nitrate per 2500 grams of soil*

STORAGE PERIOD (yr)	DDT LEVELS (% of soil by weight)						
	0	0.001	0.01	0.10	0.25	0.50	1.00
0	40.1	40.8	42.9	25.8	15.4	13.0	9.2
1	41.0	42.5	39.5	29.2	22.0	12.4	7.1
2	39.2	36.1	39.8	32.8	13.7	6.4	4.0
3	43.5	41.6	38.5	34.0	31.0	24.5	20.0
Average	41.0	40.2	40.2	30.5	20.5	14.1	10.1

*Modified from Jones, L. W. 1952. Soil Sci. **73**:237-241.

duced when certain levels are reached (Table 6-5). Based on these data, bacteria were curtailed when DDT was present at levels above 0.01% by weight of dry soil. This value is equivalent to 200 pounds per acre, far in excess of most practices. However, some cotton-growing areas have averaged 100 pounds of DDT per acre over a 10-year period because of repeated applications during each growing season. The long-range effects of these sustained applications on nutrient cycles and on soil organisms in general are not yet fully determined.

Sulfur

Sulfur is an essential constituent of protein and, therefore, of all life. The principal sources or reservoirs in the sulfur cycle are the earth's crust, including both igneous and sedimentary rocks, and the ocean and its sediments. Sulfur is the fourth most abundant element in seawater after chlorine, sodium, and magnesium. Its movement in the biosphere, like that of nitrogen, includes an atmospheric phase. Unlike nitrogen, how-

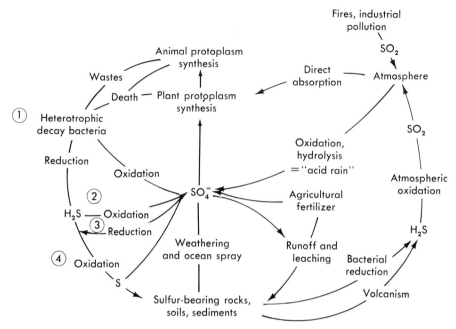

Fig. 6-7. Sulfur cycle. The circled numbers distinguish between heterotrophic bacteria, *1*, and chemosynthetic forms, *2, 3,* and *4*.

ever, sulfur occurs in relatively small quantities in the air. As mentioned in Chapter 2 sulfur enters into atmospheric pollution problems resulting from coal-burning facilities such as power plants. Sulfur occurs in several chemical forms in the ecological cycle, depending on the oxidative potential of the source material. Bacteriological metabolism is an important component in sulfur cycling in the ecosystem. The major pathways and forms of sulfur are shown in Fig. 6-7.

In Chapter 5 we mentioned briefly the autotrophic forms of bacteria—those with photosynthetic and chemosynthetic modes of self-nourishment. Both types are important in the movement of sulfur in the biosphere. The photosynthetic group includes the green and purple forms. As recipients of low-intensity light energy, these pigmented bacteria can split hydrogen sulfide (H_2S) and convert it to elemental sulfur (S) as shown in the following equation:

$$2H_2S + CO_2 \xrightarrow{\text{Light energy}} 2S + CH_2O + H_2O$$

In this simplified expression of a complex biochemical process, we can see the similarity of this equation to photosynthesis carried out by green plants (Chapter 5, p. 92). The product that differs is elemental sulfur instead of molecular oxygen. Other autotrophic, or chemosynthetic, forms can change oxidized sulfur to sulfides under conditions in which oxygen is lacking, as in lake muds. Hydrogen sulfide gas may be seen escaping as bubbles from poorly aerated water, with an odor conventionally associated with rotten

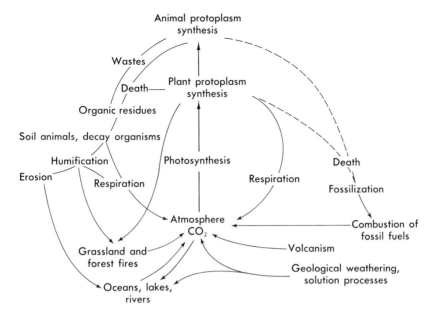

Animal protoplasm
synthesis

Wastes

Death

Plant protoplasm
synthesis

Organic residues

Soil animals, decay organisms

Humification

Photosynthesis

Death

Erosion

Respiration

Respiration

Fossilization

Atmosphere
CO_2

Combustion of
fossil fuels

Grassland and
forest fires

Volcanism

Geological weathering,
solution processes

Oceans, lakes,
rivers

Fig. 6-8. Carbon cycle. Broken lines indicate that relatively little primary and secondary production is being diverted to fossil deposits. Fossil fuels are presently being used at rates more than 1000 times greater than the rate at which they are being formed from current photosynthesis (see Table 5-2).

eggs. These bacteria are of the genus *Desulfovibrio*. Still other bacteria such as those of the genus *Thiobacillus* convert sulfides to the oxidized, or sulfate, form under aerobic conditions, or where oxygen is available. There are also heterotrophic bacteria that effect conversions of sulfur in organic matter such as decaying plants and wastes from animals, from which their energy is derived. Both anaerobic and aerobic processes can occur, depending on the availability of oxygen or its lack. The principal form of sulfur absorbed by green plants through the root system is the sulfate ion, (SO_4^{-2}). Sulfur dioxide (SO_2) from fires, industrial pollution, and volcanic activity can be absorbed directly from the air by foliage. According to Grey and Jensen, in studies of the Great Salt Lake, there are seasonal inputs of atmospheric sulfur (released by anaerobes) that are comparable to industrial sources. Overall the sulfur cycle significantly depends on the metabolic roles of several microorganisms in the soil and water environment of the biosphere.

Carbon

The entry of carbon into the cyclic process begins with atmospheric carbon dioxide that is reduced in photosynthesis involving hydrogen that is split from water molecules. This process initiates the dissemination of carbon-based energy through the food chains of the biosphere (Fig. 6-8). Carbon accounts for approximately 40% to 50% of the dry weight of all living matter.

Table 6-6. Carbon and nitrogen relationships in oven-dried prairie grass litter through 4 years of leaching and decay*

TIME (yr)	C/N	LOSSES (%) C	LOSSES (%) N
0 (fresh material)	90	0	0
2	44	63	24
4	15	94	64

*Modified from Koelling, M. R. and C. L. Kucera. 1965. Ecology **46**:529-532; Kucera, C. L. et al. 1967. Ecology **48**:536-541.

Table 6-7. Effect of chlorinated hydrocarbons on population densities of several soil invertebrates*

CHEMICAL COMPOUND	PREDATORY MITES	DETRITUS MITES	SOIL SPRINGTAILS	FLY LARVAE	BEETLE LARVAE	EARTHWORMS
Aldrin	0	– –	– – –	– – –	– – –	0
Dieldrin	0	– –	– – –	– – –	– – –	0
DDT	– – –	– – –	+ +	– –	– –	0
BHC	– –	– – –	+ +	– – –	– – –	0
Heptachlor	–	– – –	–	– – –	– – –	– –

*Modified from Edwards, C. A. 1964. In G. T. Goodman et al. (eds.). Ecology and the industrial society. Blackwell Scientific Publications, Ltd. Oxford, England.
0 = no effect; – = relative decrease in numbers; + = relative increase in numbers.

Table 6-8. Ratios of DDT in earthworms to DDT in soil in cotton region by soil type*

SOIL TYPE	RATIO
Silty clay	3
Silt loam	8
Fine sandy loam	10
Silt	11
Clay loam	15
Silty clay loam	18
Clay loam	40
Silt loam	44

*Modified from Gish, C. D. 1970. Pest. Monitor. J. **3**:241-252.

The ratio of carbon to nitrogen (C:N) provides some interesting insights regarding trophic position and organic turnover. The C:N ratio for certain plant products may be as great as 100:1. The ratio is less in the conversion of plants to animal tissue. However, in the breakdown and humification of these materials by soil detritus animals and micro-organisms, the C:N ratio always tends to become narrower because the nitrogen concentration increases, relative to total weight. The reason for this lies in the fact that at all trophic levels energy in the form of various carbon compounds is being extracted from the system and dispersed as heat. Carbon dioxide is released to the external environment as a by-product of metabolism.

When carbonaceous materials of plant origin such as oat straw or corn stalks are incorporated into the soil, the availability of nitrogen as a nutrient may be decreased for a time. This shortage occurs because the high carbon content of plant debris supplies the energy for an expanding population of microbes that release carbon dioxide in respiration but utilize nitrogen to make more cells. Until the bacteria themselves die and decompose, the nitrogen is tied up in their protoplasm and is not available for uptake by plants. Changing C:N ratios for plant material being decomposed over several years of field weathering are shown in Table 6-6. Although both elements are being lost, it is carbon, as a principal component of the total biomass, that is disappearing at the faster rate.

The carbon cycle is an intrinsic part of the world food picture. As greater demands are made for plant production, the interaction of carbon and the cycles of other elements such as phosphorus, nitrogen, and sulfur become sensitive areas of human concern. Sustained cycling processes determine the availability of nutrients throughout the biosphere. These cycling processes in turn depend on the turnover of plant and animal products that is accomplished by various populations of microorganisms and soil invertebrates such as earthworms and mites. Poisoning the soil environment adversely affects the ability of these key organisms to effect the turnover of carbon that leads to nutrient release. Dysfunction of the affected ecosystems may result.

Experimental evidence indicates that the decomposition processes associated with organic turnover and nutrient cycling can be modified through the application of soil chemicals. The results of one study conducted in the Soviet Union indicated that treated litter lost only one-sixth as much weight as untreated samples because the depleted populations of soil organisms were inadequate to achieve the natural breakdown of organic matter. The effects of pesticides on several groups of soil invertebrates are shown in Table 6-7. Decreasing populations indicate that a majority of these organisms are seriously affected. In cases in which no direct adverse effect is registered (the effect of DDT and several other chemicals on earthworms is an example), adverse results may be manifested in secondary consumers such as birds. Since earthworms are immune to relatively high concentrations of these chemicals, they store and transfer the deleterious effects of DDT to consumers farther up in the food chain. In several soils of the cotton region where heavy applications of DDT are still routine, the ratio of pesticide in the earthworm to pesticide in the soil is as high as 40:1 (Table 6-8).

BIOLOGICAL CONCENTRATION

As the productivity of our highly technological society continues to increase, wastes and by-products become more abundant in the mainstream of ecological cycles. One of the more spectacular increases in recent decades has been the amount of lead (Pb) being released into the environment. The Greenland ice- and snowfields are a biologically inactive reservoir and are therefore suitable for use as a monitoring system. By dating the

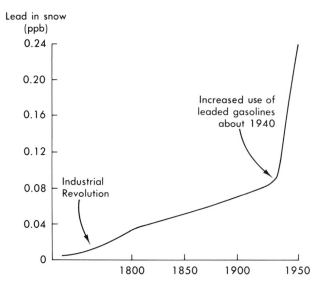

Fig. 6-9. Increase in lead in the Greenland snow and ice pack. Note the sharp rise since about 1940, attributed essentially to the combustion of leaded gasolines. The units are expressed as parts per billion. (Modified from Bryce-Smith, D. 1971. Chem. Br. **7:**54-56.)

ages of the snow layers and examining the lead content of each, it is possible to trace the pattern of rising pollution, as shown in Fig. 6-9.

Mercury (Hg) compounds, widely used for various industrial purposes, are also found in increasing amounts. Microbiological action converts this waste material to soluble methylmercury, which is then readily dispersed through aquatic systems, thus allowing for greater exposure and availability to consumers. Little is known about mercury turnover and the rates of transfer in plants and animals, with the exception of fish. However, one study found that the mercury waste from a paper mill sharply increased the mercury levels in a wide range of organisms sampled from locations downstream from the point of effluence at the mill. These concentrations, deposited in tissues and expressed as parts per million, were compared with samples taken from stream locations above the mills, where only natural mercury deposition would be present. Comparative data are given in Table 6-9. The fact that mercury can be accumulated over time is demonstrated in fish, as concentrations of mercury are higher in older organisms (Fig. 6-10).

The argument that such contaminants may not be present in concentrations harmful to consumers at the lower levels of the food chain may be valid, but is it true of consumers at higher trophic levels? Concentrations of contaminants tend to increase upward through the food pyramid because some substances are stored in certain tissues of the consumers, each in turn becoming the prey for a subsequent consumer. In addition, we know that total biomass representing usable food decreases upward through the pyramid. A particularly resistant material or a radioactive substance with a long half-life (the amount of time required for the initial quantity or effectiveness of a substance to be reduced by 50%)

Table 6-9. Effect of pulp and paper mill discharge on mercury levels in several aquatic plants and animals*

ORGANISM	SAMPLING LOCATION†	MERCURY (ppm)	ORGANISM	SAMPLING LOCATION	MERCURY (ppm)
Water moss	1	0.08	Stone-fly larva	1	0.07
	2	3.70		2	2.40
Water lily	1	0.02	Alder-fly larva	1	0.05
	2	0.52		2	5.15
Leech	1	0.02	Perch	1	0.37
	2	3.10		2	2.48
Isopod	1	0.06	Pike	1	0.55
	2	1.90		2	3.30
Caddis-fly larva	1	0.05			
	2	11.10			

*Data from Wallace, R. A. et al. 1971. Mercury in the environment. ORNL NSF-EP-1 Oak Ridge National Laboratory Oak Ridge, Tenn.
†1 indicates location above mills; 2 indicates location between mills.

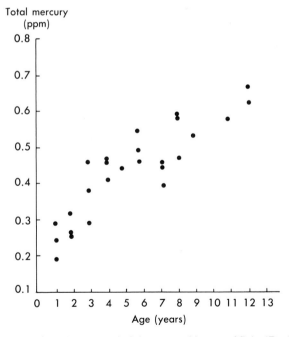

Fig. 6-10. Rising trend of total mercury in lake trout with age of fish. (Redrawn from Bache, C. A. et al. 1971. Science **172:**951-952.)

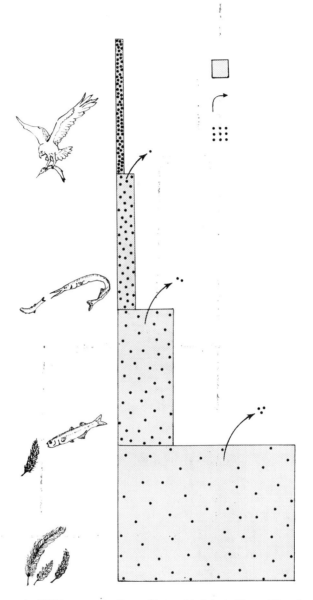

Fig. 6-11. Increase in DDT concentration with trophic level. (From Woodwell, G. M. Toxic substances and ecological cycles. Copyright © 1967 by Scientific American, Inc. All rights reserved.)

Table 6-10. Relative amounts of radioactive cesium-137 and strontium-90 in lichen-caribou-wolf food chain in northern Alaskan tundra*†

RADIOSOTOPE	LICHEN	CARIBOU	WOLF
Cesium-137	×	2.5× (flesh)	5.5× (flesh)
Strontium-90	×	6.7× (bone)	2.6× (bone)
		0.01× (flesh)	260× (bone to caribou flesh)

*Modified from Hanson, W. C. et al. 1967. In B. Aberg and F. P. Hungate (eds.). Radiological concentration processes. Pergamon Press, Ltd. Oxford, England.
†The original values expressed as picocuries of radioactivity are converted here to simple ratios.

could persist long enough to be transferred to the very top level in relatively high concentrations on a unit biomass basis. Of course, these substances are being lost through decay and excretion at each transfer level, but the relation of a substance to the biomass in which it is being stored can result in a buildup. Top carnivores are particularly vulnerable to this type of "trophic magnification," as illustrated in Fig. 6-11. The relative density of the dots depict differences in DDT concentration between trophic levels.

We have already noted that radioactive nuclides such as strontium-90 and cesium-137 enter food chains simply because green plants do not completely discriminate between these and the chemically similar elements that are metabolic requisites. Radioactive fallout particles can enter the food chain if they simply lodge on mats of vegetation and are consumed before they are washed into the soil by precipitation. In the tundra, lichens and mosses absorb radioactive fallout, which is then transferred to the caribou. The Eskimo in turn may ingest considerable quantities of cesium in caribou meat, which constitutes a large proportion of the diet of some high-arctic populations. Relative concentrations of radioactivity in the fallout from nuclear detonations are given in Table 6-10. Cesium is stored principally in edible tissue such as muscle. Each trophic transfer serves to approximately double the amount of cesium, with the result that muscle tissues of the wolf contain more than five times as much radioactivity as do tissues of vegetation on a unit biomass basis. Since cesium accumulates in edible tissues of the herbivore, it constitutes a more serious problem for top predators than strontium, which is deposited in bone, so that further magnification through this portion of the food chain is essentially halted. This is illustrated by the fact that the concentrations of strontium found in wolf bone are smaller than concentrations found in caribou bone, an indication of partial arrest. However, when the concentrations of strontium in wolf bone are compared with concentrations in caribou flesh, the ratio is greatly increased.

Another interesting aspect of radionuclide accumulation involves the dietary preferences of the herbivore. In the Arctic, lichens are a principal source of food for caribou and reindeer, particularly in winter. Lichens are relatively long-lived plants that persist for several years. This fact, coupled with their slow rate of growth, explains why such plants effectively concentrate fallout materials. On the other hand, grasses present a flush of growth during each summer and then die back and may therefore be relatively free of radioactivity. This difference is reflected in the fact that concentrations of radioactivity in herbivores vary on a seasonal basis; higher concentrations in winter than in summer are a result of preferential feeding and the seasonal availability of food.

Certain pesticides used in insect control have well-known magnification effects. Among these are the chlorinated hydrocarbons mentioned earlier in this chapter, although DDT is perhaps the best known. The value of DDT as an insecticide was discovered in 1939, and DDT initiated the age of organic pesticides. It and other hydrocarbon compounds have been widely applied in agriculture around the world. No one would deny that people have derived benefits from insecticides, but the question of how far can we go in using them at the risk of doing irreparable damage to ecosystems and communities of plants, animals, and microorganisms remains.

In recent yeras there has been increasing debate concerning the effects of these chemicals on ecological processes. They enter food chains and are transferred, amplified, and stored in the fatty tissue of the consumer. People are not immune to their effects. The average amount of DDT in tissues of Americans is 11 ppm, an acceptable level according to health officials. Nonetheless the presence of DDT is another indication of the degree to which all elements in the biosphere, including man, are interrelated. What are the effects of DDT and related chemicals on the total environment? The degree of stress is functionally related to several characteristics that affect the rate of turnover. These include selectivity, mobility, and resistance to decay processes. Often many nontarget organisms are adversely affected by large-scale applications such as aerial spraying. The ease with which chemicals can be transported throughout the ecosystem and the fact that they persist as effective compounds influence their concentration at any given point in the food chain. In addition, the nonselective nature of these compounds, together with their persistence in the food chain, extends the deleterious effects far beyond the intended target species.

Various studies have shown that the general effect of pesticides is to simplify biotic communities. Wildlife and hence food chains may be seriously affected. Fish are among the most sensitive of species, and even the most minute amounts of pesticide are lethal (Table 6-11). Note that the concentrations listed in Table 6-11 are expressed in parts per billion. It is fairly certain that DDT is accelerating the rate at which certain raptor birds

Table 6-11. Amounts of pesticides required to kill 50% of fish in 96 hours; concentrations are given in parts per billion*

PESTICIDE	BLUEGILL	MINNOWS	GOLDFISH	GUPPIES	CHINOOK SALMON	COHO SALMON	RAINBOW TROUT
Endrin	0.6	10.0	1.9	1.5	1.2	0.5	0.6
Dieldrin	7.9	16.0	37.0	23.0	6.1	10.8	10.0
Aldrin	13.0	28.0	28.0	33.0	7.5	45.9	17.7
DDT	16.0	32.0	27.0	43.0	11.5	44.0	42.0

*Modified from O'Brien, R. D. 1967. Insecticides, action and metabolism. Academic Press, Inc. New York.

Table 6-12. DDT-caused mortality in robins and other bird species recovered from sprayed and unsprayed localities*

LOCATION	NUMBER OF DEAD BIRDS	
	1963	1964
Hanover (sprayed)		
Robin	61	26
Other birds	90 (33 spp.)	46 (26 spp.)
Total	151	72
Norwich (unsprayed)		
Robin	4	2
Other birds	6 (6 spp.)	6 (6 spp.)
Total	10	8

*Modified from Wurster, D. H. et al. 1965. Ecology **46**:488-499.

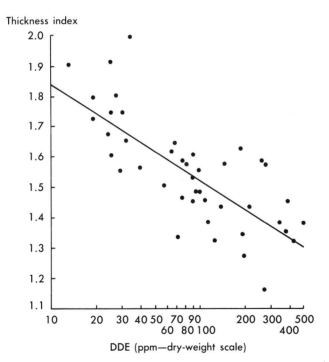

Fig. 6-12. Relationship of DDE residues in eggs of the Alaskan peregrine falcon and thickness of the shell. These birds breed in the tundra but migrate as far south as Argentina, a pattern that brings them into contact with heavily contaminated areas. DDE is a degradation product of DDT. (Redrawn from Cade, T. J. et al. 1971. Science **172**:955-957.)

such as the peregrine falcon, osprey, and bald eagle are becoming extinct. The eggshells of the California brown pelican, for example, contain high concentrations of pesticide and are so thin that it is impossible to pick them up without crushing them. Yet apologists for DDT say that the facts are not conclusive and that other factors may be involved. The effect of DDT in thinning the eggshells of the peregrine falcon is shown in Fig. 6-12. In California the bodies of dying grebes contained as much DDT as 1600 ppm and an early study by Wurster et al. established that DDT sprayed on elm trees in an effort to control bark beetles and the spread of Dutch elm disease also killed robins and other birds (Table 6-12).

Aerial spraying leaves DDT residues in the soil. These are ingested by earthworms in the detritus food chain, and the earthworms in turn are consumed by robins that accumu-

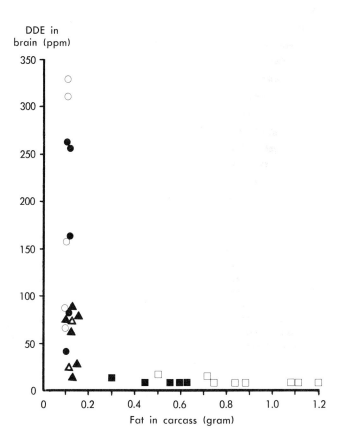

Fig. 6-13. Relationship between the concentration of DDE in the brain and the amount of fat in the Mexican free-tailed bat, a migratory species. Squares represent reference animals, triangles unexercised animals, and circles those animals exercised under experimental conditions. Closed symbols are older animals and open symbols younger individuals. (Redrawn from Geluso, K. N. et al. 1976. Science **194**[8 Oct.]:184-186. Copyright 1976 by the American Association for the Advancement of Science.)

late large quantities of DDT. The robin population today has recovered; yet the dramatic difference between sprayed and unsprayed areas is convincing evidence that DDT is lethal for some species in the environment. As an example, it has been shown recently that the bats of Carlsbad Caverns, New Mexico, are still dying from high body burdens of chlorinated hydrocarbons, causing a declining population that is now but a fraction of its former numbers. Why this decline if the use of DDT has been sharply curtailed in the United States? Again we see how vulnerable a migrant species may become, for this bat winters in Mexico whose government permits the continued use of DDT. When the animals are in flight, body fat is mobilized, releasing stored pesticide (as DDE) to the brain. Experimental animals not subject to stress but with similar amounts of storage in their bodies showed less mortality than those simulating migration stress. These interesting relationships are depicted in Fig. 6-13.

Can we always depend on the resurgence of nature to ameliorate the effects of these pollutants? The growing list of species whose survival is seriously jeopardized by chemical insecticides provides grounds for concern. Expediency often wins out over ecological considerations, but can we in the long run afford further reductions in native diversity and the stability of ecosystems? Borlaug defends the use of DDT because it cuts down on crop losses and thus furthers the expansion of food production. But are there other, less persistent chemicals to achieve the same effect? The following passage by O'Brien is appropriate:

> A great fraction of the wildlife problem is economic rather than technical. The use of methoxychlor in place of other chlorinated hydrocarbons would perhaps double the cost of chemicals, which presumably represents an ever-dwindling proportion of the total treatment costs, and enormously reduce hazards for wildlife. It is not unreasonable that to enjoy the advantages of cheap agricultural commodities and undamaged flowers and trees, along with full streams and a noisy Spring one should pay a little more.*

SUMMARY

Organic turnover is an intrinsic function of balanced ecosystems. A given input of biomass expressed as usable energy is balanced by a similar amount that is lost over the same period of time. Turnover rate may be defined as the fractional portion of the total biomass energy that enters and leaves the system on a periodic basis. It is inversely related to the turnover time, or the time required for the complete utilization of the equivalent energy stored in the total biomass. The more storage built into a system in relation to its increment characteristics, the more stability it is thought to have. The release and recycling of essential elements are associated with utilization and ultimate decay of ecosystem materials.

A knowledge of energy dynamics and nutrient dispersal provides the basis for better land use and conservation of biotic and abiotic resources. Ecosystems vary, and their management should be based on the traits unique to each. Human activities can threaten the steady flow of energy in the turnover process and in the transfer and reuse of nutrients.

---

*O'Brien, R. D. 1967. Insecticides, action and metabolism. Academic Press, Inc. New York.

Cycles may be interrupted by inadequate supplies of an essential element that results from waste, temporary displacement, or creation of permanent shunts. Nutrients may also be present in excessive amounts in given segments of the cycle, causing eutrophication or a speeding up of successional processes. Ecological cycles also imply the transport of undesirable and nonessential materials through food chains. Our own species is not immune to the ill effects of such substances. The entry of a toxic product into the ecosystem can cause great harm to its biotic constituents. Quite often the consumers at the upper trophic levels suffer most, as they are especially vulnerable to resistant substances that tend to become concentrated as they pass from one consumer level to the next. The artifacts we inject into the air, water, and soil are a threat to the normal functions and processes of turnover and cycling in the biosphere.

DISCUSSION QUESTIONS

1. Energy flows through a system, but nutrients are recyclable. Explain.
2. Compare imperfectly and perfectly balanced nutrient cycles. Give examples and explain the use of this terminology in biogeochemical cycling.
3. Discuss the impact of human pollution on the integrity of the cycling process. Give examples. What is "acid rain"?

REFERENCES

Ahmed, A. K. 1976. PCB's in the environment. Environment **18:**6-12.

Bache, C. A. et al. 1971. Residues of total mercury and methylmercuric salts in lake trout as a function of age. Science **172:**951-952.

Barrett, G. W. 1968. The effects of an acute insecticide stress on a semi-enclosed grassland ecosystem. Ecology **49:**1019-1035.

Beasley, T. M. and C. L. Osterberg. 1969. Natural and artificial radionuclides in seafoods and marine protein concentrates. Nature **221:**1207-1209.

Borlaug, N. E. 1971. The green revolution, peace and humanity. World Sci. News **8:**9-16.

Boykins, E. A. 1967. The effect of DDT contaminated earthworms in the diet of birds. Bioscience **17:**37-39.

Bray, J. R. and E. Gorham. 1964. Litter production in forests of the world. Adv. Ecol. Res. **2:**101-157.

Brill, W. J. 1977. Biological nitrogen fixation. Sci. Am. **236:**68-81.

Brooks, J. L. 1970. Eutrophication and changes in the composition of the zooplankton. In Planning Committee for the International Symposium on Eutrophication. Eutrophication: causes, consequences, corrections. National Academy of Sciences. Washington, D.C.

Bryce-Smith, D. 1971. Lead pollution—a growing hazard to public health. Chem. Br. **7:**54-56.

Cade, T. J. et al. 1971. DDE residues and eggshell changes in Alaskan falcons and hawks. Science **172:**955-957.

Carter, L. C. 1977. Radioactive wastes: some urgent unfinished business. Science **195:**661-666, 704.

Center for the Biology of Natural Systems. 1970. Nitrate sources in surface waters. CBNS Notes **3:**1-3.

Chacko, C. I., J. L. Lockwood, and M. Zabik. 1966. Chlorinated hydrocarbon pesticides: degradation by microbes. Science **154:**893-895.

Cloud, P. 1969. Mineral resources from the sea. In Resources and man. W. H. Freeman & Co., Publishers. San Francisco.

Commoner, B. 1970. Threats to the integrity of the nitrogen cycle: nitrogen compounds in soil, water, atmosphere, and precipitation. In S. F. Singer (ed.). Global effects of environmental pollution. Springer-Verlag New York, Inc. New York.

Crossley, D. A. and M. Witkamp. 1964. Effects of pesticides on biota and breakdown of forest litter. Trans. 8th Int. Cong. Soil Sci. **3:**887-892.

Davis, J. J. and R. F. Foster. 1958. Bioaccumulation of radioisotopes through aquatic food chains. Ecology **39:**530-535.

Delwiche, C. C. 1970. The nitrogen cycle. Sci. Am. **223:**136-146.

Duvigneaud, P. and S. Denaeyer-De Smet. 1970. Biological cycling of minerals in temperate deciduous forests. In D. E. Reichle (ed.). Analysis of temperate forest ecosystems. Springer-Verlag New York, Inc. New York.

Edwards, C. A. 1965. Effects of pesticide residues on soil invertebrates and plants. In G. T. Goodman, R. W. Edwards, and J. M. Lambert (eds.). Ecology and the industrial society. Blackwell Scientific Publications, Ltd. Oxford, England.

Edwards, C. A., D. E. Reichle, and D. A. Crossley, Jr. 1970. The role of soil invertebrates in turnover of organic matter and nutrients. In D. E. Reichle (ed.). Analysis of temperate forest ecosystems. Springer-Verlag New York, Inc. New York.

Geluso, K. N., J. S. Altenbach, and D. E. Wilson. 1976. Bat mortality: pesticide poisoning and migratory stress. Science **194:**184-186.

Gish, C. D. 1970. Organochlorine insecticide residues in soils and soil invertebrates from agricultural lands. Pest. Monitor. J. **3:**241-252.

Goldberg, E. D. 1963. The ocean as a chemical system. In M. N. Hill (ed.). The sea. John Wiley & Sons, Inc. New York.

Grey, D. C. and M. L. Jensen. 1972. Bacteriogenic sulfur in air pollution. Science **177:**1099-1100.

Grundy, R. D. 1971. Strategies for control of manmade eutrophication. Environ. Sci. Technol. **5:**1184-1190.

Hammond, A. L. 1971. Mercury in the environment: natural and human factors. Science **171:**788-789.

Hanson, W. C., D. G. Watson, and R. W. Perkins. 1967. Concentration and retention of fallout radionuclides in Alaskan arctic ecosystems. In B. Aberg and F. P. Hungate (eds.). Radiological concentration processes. Pergamon Press, Ltd. Oxford, England.

Hill, F. B. 1973. Atmospheric sulfur and its links to the biota. In G. M. Woodwell and E. V. Pecan (eds.). Carbon and the biosphere. Proceedings of the 24th Brookhaven Symposium. 1972. Upton, N.Y.

Hunt, E. G. and A. I. Bischoff. 1960. Inimical effects on wildlife of periodic DDT applications to Clear Lake. Calif. Fish Game **46:**91-109.

Jensen, S. and A. Jernelov. 1969. Biological methylation of mercury in aquatic organisms. Nature **223:**753-754.

Johnson, T. B., R. N. Goodman, and H. S. Goldberg. 1967. Conversion of DDT to DDD by pathogenic and saprophytic bacteria associated with plants. Science **157:**560-561.

Jones, L. W. 1952. Stability of DDT and its effects on microbial activities of soil. Soil Sci. **73:**237-241.

Jordan, C. F. 1971. A world pattern in plant energetics. Am. Sci. **59:**425-433.

Koelling, M. R. and C. L. Kucera. 1965. Dry matter losses and mineral leaching in bluestem standing crop and litter. Ecology **46:**529-532.

Kucera, C. L., R. C. Dahlman, and M. R. Koelling. 1967. Total net productivity and turnover on an energy basis for tallgrass prairie. Ecology **48:**536-541.

Lichtenstein, E. P. et al. 1960. Persistence of DDT, aldrin and lindane in some midwestern soils. J. Econ. Entomol. **53:**136-142.

Likens, G. F., F. H. Bormann, and N. M. Johnson. 1972. Acid rain. Environment **14:**33-40.

Likens, G. F. et al. 1970. Effects of forest cutting and herbicide on nutrient budgets in the Hubbard Brook watershed-ecosystem. Ecol. Monogr. **40:**23-47.

Lovering, T. S. 1969. Mineral resources from the land. In Resources and man. W. H. Freeman & Co., Publishers. San Francisco.

Mason, B. H. 1966. Principles of geochemistry. John Wiley & Sons, Inc. New York.

Menhinick, E. F. 1962. Comparison of invertebrate populations of soil and litter of mowed grasslands in areas treated and untreated with pesticides. Ecology **43:**556-561.

Mishustin, E. N. 1964. Effect of herbicides on microbiological processes in soils. Izv. Akad. Nauk. SSSR (Biol.) **2:**197-209.

Morgan, J. D., Jr. 1975. Mineral position of the United States. In R. W. Marsden (ed.). Politics, minerals, and survival. University of Wisconsin Press. Madison, Wis.

Nash, R. G. and E. A. Woolson. 1967. Persistence of chlorinated hydrocarbon insecticides in soils. Science **157:**924-927.

Noshkin, V. E. 1972. Ecological aspects of plutonium in aquatic environments. Health Phys. **22:**537-549.

O'Brien, R. D. 1967. Insecticides, action and metabolism. Academic Press, Inc. New York.

Perterle, T. J. 1969. DDT in Antarctic snow. Nature **224:**620.

Potter, L. D. and M. Barr. 1969. Cesium-137 concentrations in Alaskan Arctic tundra vegetation. Arctic Alpine Res. **1:**147-153.

Pramer, D. 1971. The soil transformers. Environment **13:**42-46.

Rickard, W. H. et al. 1965. Gamma-emitting radionuclides in Alaskan tundra vegetation. Ecology **46:**352-356.

Rochlin, G. I. 1977. Nuclear waste disposal: two social criteria. Science **195:**23-31.

Rodin, L. E. and N. I. Baslievic. 1968. World distribution of plant biomass. In Proceedings of the UNESCO symposium on functions of terrestrial ecosystems at primary production levels. Copenhagen.

Rodriguez-Barrueco, C. 1968. The occurrence of nitrogen-fixing nodules on non-leguminous plants. Bot. J. Linn. Soc. London **63:**77-84.

Shapley, D. 1977. Will fertilizers harm ozone as much as SST's.? Science **195:**658.

Turner, J., D. W. Cole, and S. P. Gessel. 1976. Mineral nutrient accumulations and cycling in a stand of red alder (Alnus rubra). J. Ecol. **64:**965-974.

United States Geological Survey. 1970. Mercury in the environment. Geol. Surv. Prof. Paper 713. United States Government Printing Office. Washington, D.C.

Viereck, L. and K. van Cline. 1972. Distribution of selected chemical elements in even-aged alder ecosystems near Fairbanks, Alaska. Arctic Alpine Res. **4:**239-255.

Wallace, R. A. et al. 1971. Mercury in the environment. ORNL NSF-EP-1 Oak Ridge National Laboratory. Oak Ridge, Tenn.

Wedemeyer, G. 1966. Dechlorination of DDT by Aerobacter aerogenes. Science **152:**647.

Wenk, E., Jr. 1969. The physical resources of the ocean. In D. Flanagan (ed.). The ocean. W. H. Freeman & Co., Publishers. San Francisco.

Went, F. W. and N. Stark. 1968. Mycorrhiza. Bioscience **18:**1035-1039.

Williams, L. G. and Q. Pickering. 1961. Direct and foodchain uptake of cesium[137] and strontium[85] in bluegill fingerlings. Ecology **42:**205-206.

Woodwell, G. M. 1967. Toxic substances and ecological cycles. Sci. Am. **216:**24-31.

Woodwell, G. M. and F. T. Martin. 1964. Persistence of DDT in soils of heavily sprayed forest stands. Science **145:**481-483.

Woodwell, G. M. and T. G. Marples. 1968. The influence of chronic gamma irradiation on production and decay of litter and humus in an oak-pine forest. Ecology **49:**456-465.

Wright, R. F. 1976. The impact of forest fires on nutrient influxes to small lakes in northeastern Minnesota. Ecology **57:**649-663.

Wurster, D. H., C. F. Wurster, Jr., and W. N. Strickland. 1965. Bird mortality following DDT spray for Dutch elm disease. Ecology **46:**488-499.

ADDITIONAL READINGS

Deevey, E. S., Jr. 1970. Mineral cycles. Sci. Am. **223:**148-158.

Dixon, B. 1976. Magnificent microbes. Atheneum Publishers. New York.

Flick, D. F., H. F. Kraybill, and J. M. Dimitroff. 1971. Toxic effects of cadmium: a review. Environ. Res. **4:**71-85.

Kellogg, W. W. et al. 1972. The sulfur cycle. Science **175:**587-596.

Montague, K. and P. Montague. 1971. Mercury-Sierra Club Battlebrook. Charles Curtis. New York.

Odum, E. P. 1967. Fundamentals of ecology. W. B. Saunders Co. Philadelphia.

Packard, V. 1960. The waste makers. David McKay Co., Inc. New York.

Rudd, R. L. 1964. Pesticides and the living landscape. University of Wisconsin Press. Madison, Wis.

7 Biotic succession

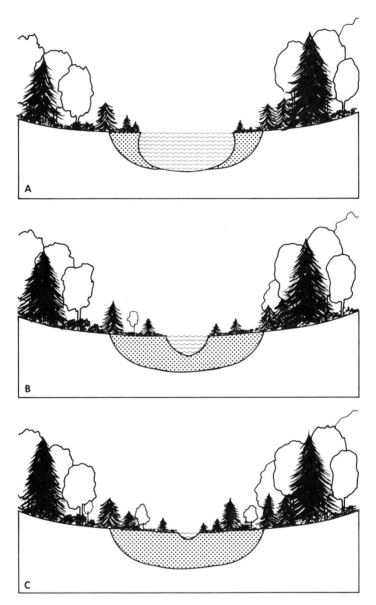

Fig. 7-1. Vertical profile illustrating development of bog from an open lake. Accumulation of organic materials decreases water depth, making possible the invasion of several stages or communities in a sequential pattern, **A** to **C.**

WHAT IS SUCCESSION?

Ten thousand years ago in the northern states of Minnesota, Wisconsin, and Michigan and in southern Canada eastward to New England, numerous lakes were formed by the meltwaters of retreating glaciers. Today some of these are swampy bogs and even areas of heavy vegetation. Where once there were open waters, dense ground cover or forests of coniferous and deciduous trees may occur (Fig. 7-1). Over shorter periods of time, abandoned pastures and croplands that have been subject to fire and erosion exhibit the sparse beginnings of a native cover (Fig. 7-2). How do these changes in habitat and community come about? By what process does a productive forest with its complement of plant and animal species supplant a sterile, or oligotrophic, lake? How is the land that was formerly a field exchanged for native trees and grasses? Nature is in a constant state of self-adjustment and repair. It is always changing and seldom static. Yet there is also the tendency toward equilibrium between the biota and its environment.

Biotic succession is a dynamic process and one of the most fascinating of all natural phenomena. It is a changing scene of adaptation, competition, establishment, and sur-

Fig. 7-2. Abandoned land showing regeneration of the native plant species in the Missouri Ozarks. (Courtesy Leland Payton, Columbia, Mo.)

Species
dominance

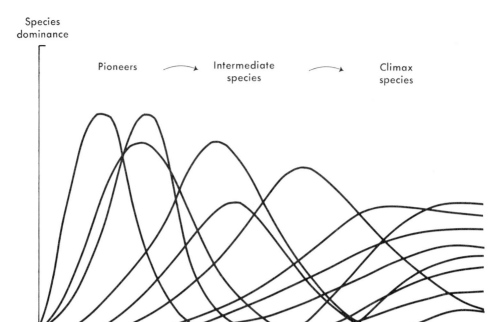

Fig. 7-3. Schematic diagram of successional development indicating a species continuum over a period of time. The time scale is relative.

vival as well as "weeding out" in a common mixing ground. With time there is a sequential replacement of one community by another that is better suited to the environment. However, these temporal patterns do not constitute an exchange of one discrete set of species for another. Generally the progression is gradual and subtle, particularly in terrestrial communities. This shift from one assemblage to the next is schematically shown in Fig. 7-3. Each curve represents a separate species in a hypothetical situation. Note the overlap of species at any point in time, showing that as some species fade out, others are beginning to invade. As succession progresses, the replacement rate of species slows down. With time an accumulation of species (more diversity) and a diminishing dominance on a relative basis usually occur.

In typically short-lived populations of phytoplankton, or free-floating microscopic algae, abrupt changes in density and composition can and do occur. Interesting seasonal patterns based on a comprehensive study of a small pond are depicted in Fig. 7-4. Populations of heterotrophic organisms with short life cycles such as soil bacteria also tend to demonstrate rapid sequences of change attributable to environmental factors such as food supply, temperature, oxygen supply, and moisture. The fungi are an interesting group of microorganisms that usually follow a general pattern of species succession as different compounds are made available from dead plant materials. Some fungi, including the

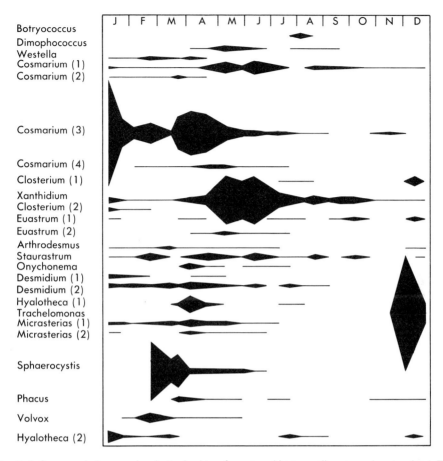

Botryococcus
Dimophococcus
Westella
Cosmarium (1)
Cosmarium (2)

Cosmarium (3)

Cosmarium (4)
Closterium (1)
Xanthidium
Closterium (2)
Euastrum (1)
Euastrum (2)
Arthrodesmus
Staurastrum
Onychonema
Desmidium (1)
Desmidium (2)
Hyalotheca (1)
Trachelomonas
Micrasterias (1)
Micrasterias (2)

Sphaerocystis

Phacus

Volvox

Hyalotheca (2)

Fig. 7-4. Seasonal changes in phytoplankton for a pond in a small, grassed watershed. The relative thickness of the lines is a measure of species density and seasonal dominance. Only the genus name is given. The numbers indicate different species of the given genus. (Modified from Proctor, V. W. 1951. Unpublished data.)

bread molds, use materials such as sugars and proteins that are more soluble and easily broken down; these fungi are among the first stages to appear in a succession. As the sugar and proteins are exhausted, the cellulose organisms become more prominent and are followed by fungi capable of breaking down substances such as lignin that are even more complex and resistant to decay. The fungi also demonstrate successional sequences on living plants. Sherwood and Carroll found a distinctive pattern of species replacement on Douglas fir needles and twigs. For example, there were some species of fungi on twigs of old-growth trees that did not occur on the younger specimens and vice versa. In general there were more kinds of fungi on the undersides of the twigs in both age categories, indicating sensitivity in response to microhabitat variation.

Succession eventually terminates with what is often called a *climax* community. Examples are a forest of beech and sugar maple as the dominant species on a Michigan till plain and the oak and hickory forests in the Ozark watershed of Missouri and Arkansas. Communities such as these are self-perpetuating and more or less stable as long as the ecological conditions under which they developed prevail. When such communities appear, an equilibrium between the biotic community and its environment is said to have been achieved. This equilibrium will persist for an indefinite period of time barring major site modifications such as severe soil losses through erosion or climatic changes.

The sequence from the first or pioneer to the terminal community is a *sere*. The time required for its completion varies, but generally many years or even centuries may be involved. Successional processes occur most rapidly in warm, moist climates. As climatic conditions become progressively colder or drier, biotic change is more retarded. Chronosequences for the Alaskan tundra (Table 7-1) demonstrate that even the development of a pioneer phase is a slow process. By comparison, a pioneer phase in the humid tropics may be completed in 3 to 5 years because of the rapid growth of invading species and such resultant modifications in site factors as shading and soil moisture. Yet it is estimated that even in the tropics 400 to 1000 years may be required for the completion of a sere that terminates in a mature rain forest.

The principles of succession were formally organized at the turn of the century and during the following two decades. H. C. Cowles, with his work on the vegetational development of sand dunes at the southern end of Lake Michigan, was a pioneer in this field. The pattern of communities, which was development toward a climax forest of beech, was demonstrated on successively older dunes at increasing distances from the present lake level. The old dunes were remnants of the former beaches of a receding glacial lake. This work and that of V. E. Shelford, who related animal populations to the same chronological sequences that pertained to plants, provided some of the earliest comprehensive studies of biotic succession. The first extensive treatise on plant succession, dealing with the salient features of initiation, continuance, and stabilization, was systematically prepared by F. E. Clements. Subsequent refinements in terms of energy flow, biomass accumulation, nutrient regulation, and homeostatic (stabilizing) mechanisms have been developed.

Table 7-1. Chronosequences of five stages in Alaskan tundra succession from pioneer communities to stable climax vegetation*

STAGE	TIME REQUIRED TO REACH GIVEN STAGE (yr)
I Pioneer species	25-30
II Meadow	100
III Early shrub	200
IV Late shrub	300
V Climax tundra	Less than 5000

*From Viereck, L. A. 1966. Ecol. Monogr. **36**:181-199.

THE CONCEPT OF CLIMAX

For many years two basic concepts of biotic stability have been advanced. The older of these is the *monoclimax theory*, whose principal proponent was Clements. It states that for each climatic region there is theoretically only one type of stable vegetation in which a sere could terminate. The oak-hickory forests of the Midwest would constitute such a climax vegetation, despite the fact that certain prairie communities persist under identical temperature and rainfall conditions. Even today, prairie relics that presumably are stable communities may be found on sites whose soils and topography differ sharply from those supporting the more generally distributed oak forests.

Supporters of the monoclimax theory call such prairie communities a *subclimax*. Subclimax types are examples of arrested succession, in which development is terminated

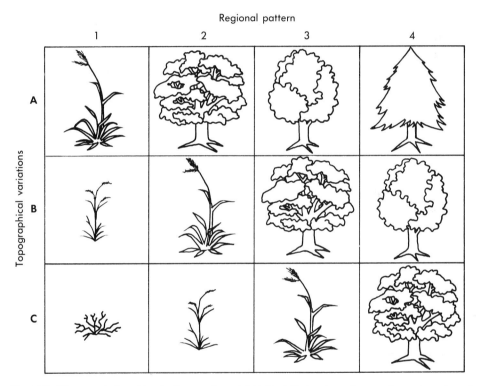

Fig. 7-5. Climax relationships in the continental United States from the southwestern desert to the northeastern conifer forests, representing a wide range of moisture and temperatures. For any given climatic region, the postclimax types, *A,* occupy sites that are cooler and/or have more favorable moisture conditions. Preclimax types, *C,* occupy warmer and usually drier sites than climax types, *B. A-1* = tall grasses, *A-2* = oak woods, *A-3* = beech and maple forest, *A-4* = northeastern conifer type, *B-1* = short grasses, and *C-1* = desert type. (Modified from Dansereau, P. 1957. Biogeography: an ecological perspective. The Ronald Press Co. New York.)

at a stage just short of the theoretical climax. They demonstrate ecological stability based principally on such local physical factors as soils and drainage. Climate, of course, is still an important influence. Many examples of this type of curtailed climax are found in native ecosystems. Other arresting factors are fire and grazing, which prevent a succession from attaining full development. Subclimax communities are also called disclimaxes (disturbance climaxes) under certain conditions.

The second basic concept of biotic stability is called the *polyclimax* theory. This approach views all biotic communities that represent stability and functional permanence as climaxes. Thus the poorly drained soils that are largely responsible for the persistence of prairies in a forest climate provide the basis for an edaphic, or soil, climax. Other factors, including fire and grazing, that impart stability to a given community type are similarly applied in numerous examples. The pine woods of the southeastern United States are considered in some situations to be fire climax communities, since succession results in a forest consisting of oaks, hickories, and other deciduous genera when burning does not occur.

Certain communities are classified as *postclimax* because they occur only under conditions superior to those that characterize the regional climax. Postclimax communities usually occur in localized habitats whose microclimate is sufficiently modified by terrain to support a community that is characteristic of the climax of another region. Within the broad oak-hickory climax of central Missouri, for example, there are localized stands of sugar maple in protected, covelike situations. Such communities have their more extensive counterparts farther north in Wisconsin and adjacent states where the environment is cooler and usually more humid. Exposed sites such as high bluffs and steep slopes that are warmer and drier than normal would produce a preclimax community similar to the vegetation one would expect to find in yet another climatic region. These regional relationships for a broad moisture-temperature gradient in the midcontinental United States are depicted in Fig. 7-5.

Both of these theories hold that climaxes, whether one or several, are discrete units of vegetation. A third interpretation of climax is the *population pattern* concept. Under this interpretation of stable vegetation, climax is a continuous array of species assemblages corresponding to changing intensity or supply factors along an environmental gradient. Such an approach is perhaps more realistic, for it more accurately describes and accounts for a multiplicity of stable yet merging community patterns in the landscape.

THE VARIED CAUSES OF SUCCESSION

The progression from pioneer communities through various developmental stages to a steady-state condition is universal. It is initiated by environmental change. Volcanism; floods; windthrow; weather changes; and pollution, fire, grazing, strip mining, and other human activities disrupt an existing biotic equilibrium or produce new areas open to successional development. When instability is created in any form, biological processes begin to effect a balance. The dispersal of plant propagules such as spores, seeds, burrs, and tumbleweeds is a first step in the development of a new sere. Visualize a barren till plain left by glaciers, the devastated hillsides after a forest fire, or even a tiny crevice on a boulder surface where a single seed or spore might lodge. All these areas, large or small, are potential testing grounds for biotic dissemination and adaptation. In modified environments such as abandoned croplands or cutover forests, the seeds of potential pioneers may remain dormant for prolonged periods, germinating only when conditions become favorable. Selective establishment of those species able to germinate, grow, and propagate leads to increasing densities and competition among community members. Not all would-be pioneers are successful. Temperature, moisture conditions, and nutrient supply may be unfavorable, as may the availability of oxygen, nutrients, acidity levels, or light intensity values.

After biotic succession is initiated, why does subsequent species change occur through time? What factors drive this progression toward a climax equilibrium? Even though their tenure is temporary, pioneer plants begin the biological process of habitat modification or amelioration. Humus from decaying plants is added to the soil, and the trend in improving water and fertility relations is begun. As these ecological conditions are changed by a given set of organisms, the competitive balance is shifted in favor of new functioning combinations of plants and consumers that can exploit the modified site more effectively than their predecessors. This process of species replacement continues until no further biotic changes of the environment are possible, and a steady-state system, or climax community, of plants and consumers has been reached. The sequential pattern of species has a physiological basis. In general those plants appearing later in succession are characterized by increasing tolerance of shade. In Chapter 5 we mentioned that the light compensation is the level of intensity at which the energy produced in photosynthesis is used up in plant respiration. When light intensity for seedlings on the forest floor is lowered through increasing shade to this point, no growth can occur, and such species are vulnerable to being crowded out by those with greater shade tolerance. The following list provides compensation values (expressed as percentage of full sunlight measured outside the forest canopy) for several eastern trees in the seedling stage:

SPECIES	COMPENSATION VALUES
Eastern larch	18
Northern red oak	14
Hackberry	12
Eastern white pine	10
Hemlock	8
Sugar maple	3

The values indicate that certain species are more tolerant of shade than others; these would be capable of succeeding the species requiring greater light intensities for growth. White pine would follow larch, and then would eventually be replaced by sugar maple. In making these comparisons it is implied that other factors such as soil moisture and nutrient availability are constant. Changes in compensation points might well and, in fact, do occur as these factors and others become more or less favorable.* Successions

---

*For further discussion see Horn, H. S. 1971. The adaptive geometry of trees. Monograph in Population Biology No. 3. Princeton University Press. Princeton, N.J.

are sometimes arrested through certain management practices that favor a commercially important tree that could not otherwise compete successfully with the climax or more shade-tolerant species.

When new areas devoid of soil and organic matter become open to biotic invasion, the sequence of events that follows is called a *primary succession*. Such areas have no previous plant history in which modifications such as soil development might have occurred. Examples of primary communities are the lichens and mosses that develop on bare rocks. On substrates such as these the chemical process of carbonation is effected by these early plants. Carbon dioxide, a product of plant respiration, combines with available water to form carbonic acid, a chemical that assists in the weathering of the rock substrate. This process is illustrated as follows:

$$CO_2 + H_2O \; \rightleftarrows \; H_2CO_3 \; \rightleftarrows \; HCO_3^- + H^+ \; \rightleftarrows \; CO_3^{-2} + H^+$$

The rate of breakdown would depend on the nature of the parent material and its resistance to chemical action and physical disintegration and also on the climatic factors of temperature and rainfall. High temperatures and adequate moisture would enhance the general weathering effect. Mixing mineral material with the products of organic decay forms a soil layer between plants and the underlying rock.

A large-scale primary succession was initiated in the island group of Krakatoa in Indonesia when some areas were completely devastated by volcanic activity in 1883. All plant and animal life was destroyed in one locale that was 25 miles from the nearest source of life. The first plants were propagated by wind-borne spores or seeds. An examination 50 years after the eruption revealed that 271 plant species had become established. A more recent example of primary succession that also involves volcanism is the newly born island of Surtsey in the North Atlantic, where the first plant species appeared a year and a half after the island was formed. This island will provide biologists with an opportunity to observe changes in primary succession for many years to come.

Weed communities that appear in abandoned fields or overgrazed range lands are common examples of *secondary succession* and have the potential to end in a climax vegetation. Plant changes that are a result of intensive grazing in tall grass prairie are shown in Fig. 7-6. When we refer to "weeds," the connotation is often one of exotic species. Russian thistle, brought to this country in flax seed before the turn of the century, is a good example of a species that invaded range lands of the West; cheat grass from Europe is another. When native ecosystems are modified, species from other locales have an opportunity to fill vacated niches. When the land is released from such domestic uses as grazing and cultivation, these species may dominate the early phases of succession. In time the original community of native species may be restored, and the exotics relegated to minor roles in the community. The success of the so-called weed species as pioneers is attributed to several factors. These include short reproductive cycles and life histories, abundant seed production, and efficient seed dispersal that enables the plant to move rapidly into barren or abandoned areas. Burrs, stickers, tumbleweeds, and light, wind-borne seeds are good examples of seed dispersal mechanisms for such plants. Annual species such as Russian thistle are thus well suited for the role of pioneer species.

Fig. 7-6. Overgrazed prairie (right) compared to exclosure plot. Intensive use reduced the number of native plants, resulting in a conspicuous increase in exotic or "weedy" species and a decrease in ground cover.

There are two general classes used to compare the reproductive strategies of pioneers versus those appearing in late succession, or climax: the r-selected group and the K-selected group. The pioneer types are classified in the so-called r-selected group and the climax species in the K-selected group. In the latter we usually observe lower seed production and less mobility, with a relatively greater energy allocation for developing stable plant structures of extended longevity. These two types of selection will be discussed in more detail in Chapter 10.

The effects of fire on succession may be widely observed. In temperate grasslands and tropical savannahs it is traditional to burn dead vegetation. Land clearing of forests, particularly in tropical regions, and the practice of rotating crops also make extensive use of fire as a cultural tool. What are the effects of these practices? In some cases they are detrimental, particularly in the case of large-scale forest destruction. Because forests are associated with the humid regions of the earth, their removal by any means contributes to accelerated erosion, loss of nutrients, and lowered productivity. In the case of grasslands, however, potential litter accumulations that are eliminated by fires may otherwise be a factor in redirecting successions toward other, less desirable systems. In the drier, eastern parts of the Serengeti Plains, for example, the grasses are naturally short, and litter accumulation that could suppress their growth is not a critical factor. In this general region, fire is not an important agent in maintaining the grasslands, and rainfall is inadequate to support any appreciable invasion of woody plants. In other, moister areas of this vast savannah complex, obvious increases in populations of woody plants at the expense of the grass vegetation occur when fire is not used to control trees and remove litter. Here is an example of fire as an important factor in a stable grass ecosystem; without it the spectacular herds of ungulates and associated predators could not be supported. Yet even here the benefits of fire are qualified in terms of other factors such as soils, topography, and drainage patterns, all of which combine to effect a very intricate ecological relationship.

On the range lands of the southwestern United States there has been a noticeable decline in native grass species during the past 50 years. What are the factors involved in this depletion? Why should a native ecosystem, presumably stable and self-perpetuating prior to settlement, undergo successional changes? Studies show that there is an interplay of several factors. The causes of successional changes from a grassland climax to other vegetation types are complex but in sum are attributable to misuse of the range. Overgrazing and selective grazing by domestic livestock are significant factors, and less fuel in the form of unused standing crops and grass litter has been the result. As fires become less frequent, there was an increase in the densities of such species as cholla cactus, creosote bush, and mesquite, the relative importance of which depended on local soil and topographical factors. The spread of these plants through the grassland promoted shading and a depletion of soil moisture, resulting in further decline of grasses. In addition, widespread control of such predator species as the coyote has led to an increase in rodent populations, and rodents are effective disseminators of the seeds and propagules of these invading plant species. Prior to these changes, mesquite was largely confined to the region's river flats, but it is estimated to have invaded 75 million acres of grassland. Attempts now are being made to control the mesquite and eliminate it from these areas through large-scale spraying and land-clearing operations. Here is an example of technological remedies used in an attempt to recover from a biological imbalance initiated and sustained by expanding human activities.

There are numerous examples of human intervention in the succesional process. We see these disruptive influences in strip mining, smelter operations, clear-cutting of steep-sloped forests, marsh drainage, filling of estuaries, aerial spraying of herbicides and pesticides—the list is long and continues to grow. When the impact is severe, long periods of time may be required to restore equilibrium at the most efficient level of biological production. In many situations it is impossible to effect a return to the original condition.

Stripping land for shallow coal deposits is an inexpensive means of extracting this fossil fuel from the ground, but is it the most judicious in terms of the ecological consequences? What of destroyed food potential (crops and aquatic resources) and the general disfigurement of large areas of the landscape? Stripping by its very nature leaves the land vulnerable to erosion and acid pollution. The steep slopes of mounded rows of overburden cannot retain rainfall and soon become erosion pavements to the nearest drainage.

Fig. 7-7. Fish kill as a result of acid mine waste (sulfuric acid) released into a stream. (Courtesy Missouri Conservation Commission.)

Bituminous or soft coal usually is noted for its high sulfur content. When layers of such coal are exposed to oxidation and hydrolysis, the sulfide products are converted to sulfuric acid and iron sulfates, based on the following chemical reactions:

$$2FeS + 7O_2 + H_2O \rightarrow 2FeSO_4 + 2H_2SO_4$$

$$H_2SO_4 \text{ (sulfuric acid)} \rightarrow 2H^+ + SO_4^{-2}$$

This acid pollution is responsible for fish kills often far downstream from the actual mining operation. Fig. 7-7 provides us with an example that demonstrates again the interrelationships among ecosystems as parts of landscape function. An aerial view of strip-mine topography, shown in Fig. 7-8, emphasizes the fact that ecological changes are often extensive and, for the most part, totally disruptive, causing the initiation of new seres. Table 7-2 provides data on the extent of surface mining of soft coal in the United States, which has become a national issue in the 1970s. Approximately 1½ million acres have been denuded for bituminous or soft coal alone. Coal as a source of energy must be extracted to meet the energy needs of our technological society. What should be done? Clearly new ways must be found to utilize this resource without at the same time jeopardizing environmental quality.

Fig. 7-8. Aerial view of stripping land for coal in Boone County, Missouri. During heavy rains, acid waters in lakes overflow into stream drainages. (From Parsons, J. 1968. Arch. Hydrobiol. **65**:25-50.)

Table 7-2. Acres of land disturbed for mining of soft coal in the United States*

STATE	ACREAGE	STATE	ACREAGE
Alabama	34,900	Montana	6,820
Alaska	3,600	New Mexico	8,260
Arizona	220	North Dakota	27,200
Arkansas	3,100	Ohio	207,000
California	30	Oklahoma	13,800
Colorado	8,630	Oregon	20
Georgia	40	Pennsylvania	247,000
Idaho	10	South Dakota	310
Illinois	234,000	Tennessee	17,900
Indiana	130,000	Texas	770
Iowa	8,600	Utah	3,220
Kansas	19,700	Virginia	34,800
Kentucky	210,000	Washington	1,370
Maryland	4,610	West Virginia	196,000
Michigan	560	Wyoming	10,100
Missouri	33,500	Total	1,470,000

*From Paone, J. et al. 1974. Information Circular 8642. Bureau of Mines. U.S. Department of the Interior. Washington, D.C.

Table 7-3. Changing compositions of territorial male birds based on sightings in 100 acres of hydrosere habitat in southern Canada*†

SPECIES	STATE OF SUCCESSION		
	WET BOG	BLACK SPRUCE	BLACK SPRUCE AND HEMLOCK
Barn swallow	x		
Black duck	x		
Alder-flycatcher	x		
Bronzed grackle	x		
Red-winged blackbird	x		
Kingbird	x		
Tree swallow	x		
Song sparrow	x		
Swamp sparrow	x		
Purple finch	x	x	x
Yellow-shafted flicker	x	x	x
Cedar waxwing	x	x	x
Yellow-throat	x	x	x
Blue jay		x	
Ruffed grouse		x	
Yellow-bellied sapsucker		x	
Hermit thrush		x	
Canada jay		x	x
Brown creeper		x	x
Black and white warbler		x	x
Blackburnian warbler		x	x
Slate-colored junco		x	x
White-throated sparrow		x	x
Black-capped chickadee		x	x
Northern water thrush		x	x
Olive-backed thrush		x	x
Nashville warbler		x	x
Canada warbler		x	x
Myrtle warbler		x	x
Winter wren		x	x
Golden-crowned kinglet		x	x
Magnolia warbler		x	x
Yellow-bellied flycatcher		x	x
Arctic three-toed woodpecker			x
Scarlet tanager			x
Ruby-crowned kinglet			x
Chestnut-sided warbler			x
Red-breasted nuthatch			x
Spruce grouse			x
Veery			x
American olive-sided flycatcher			x

*Modified from Martin, N. D. 1960. Ecology **41**:126-140.

†Original data indicated actual number of males.

As the plant community undergoes species changes in primary or secondary seres, there is a concomitant change in food and habitat conditions. New niches become available, and the species composition of consumers changes accordingly. Trends in bird populations associated with a bog → spruce → hemlock succession are shown in Table 7-3. Note the familiar overlap in successional stages that is demonstrated by certain consumer species. Succession thus appears as a generally orderly process in which, as might be expected, there is a predictability of pattern and direction for both plant and animal populations. This process of change in communities comprised of producers and consumers continues until the full environmental potential is realized and no further site improvement and biotic advantages can be achieved. At this point the community is being maintained at maximum efficiency in terms of the amount of biomass supported per unit of energy produced. Successional studies of soil microorganisms have also demonstrated in principle that there are sequential patterns in the development of species populations;

Fig. 7-9. Hydrosere, or aquatic succession, in which organic deposits and erosion sediments cause a gradual filling and advancement of the sere. Emergent aquatic plants such as the tall cattail *(Typha)* shown here are important stabilizers of shorelines, as they reduce wave action and trap erosion sediments from the surrounding watershed. (Courtesy Missouri Conservation Commission.)

the duration and rate of change for any given population is tied to the heterotrophic (food) supply, which is initiated and maintained by green plants in the sere.

KINDS OF SUCCESSION

Successions that initiate from open bodies of water such as ponds, lakes, or stream margins are called *hydroseres* (Fig. 7-9). Those starting from bare rock, volcanic ash, or similarly dry sites are *xeroseres* (Fig. 7-10). The more extreme the conditions for plant growth, the longer the period of time that is required to effect successional changes. In secondary succession where soil is already present, an equilibrium will be achieved more quickly than in a primary sere where no soil is yet available. Dry substrates eventually become more moisture retentive when organic matter accumulates as a result of plant decay. It is this incorporation of organic matter with weathered rock, loess, or other geological materials that produces a mantle of soil. In addition to increased supplies of water for plant growth, nutrient storage and availability are also increased. Plants help to reduce

Fig. 7-10. Pioneer colonies of lichens on rock outcrops as an initial phase of biotic succession and soil development. A sparse grass vegetation is shown where soil is present. (Courtesy Leland Payton, Columbia, Mo.)

evaporation and stop erosion through their ability to hold the soil in place and shade the ground. The site itself becomes more stable and less subject to physical alteration.

In a hydrosere, water depth decreases, and oxygen supplies for terrestrial plants are improved as shorelines are built up by decaying plants. Through time aquatic communities evolve through wetland and marsh stages to terrestrial systems of increasing stability. This change, accompanied by the filling of lakes and ponds, is called eutrophication because of the natural enrichment in organic materials and nutrients. The cycle of youth, growth, and maturity is an intrinsic property of all ecosystems. Regardless of the substrate on which it is initiated, succession appears generally as a convergence toward a biologically stable habitat that is an intermediate stage between environmental extremes. This tendency is depicted in Fig. 7-11, which is a simplified diagram outlining several pathways the progression may take. The climax habitat thus is the most mesic possible in a given environmental potential. The range of fluctuation in physical factors is narrowed, and, as a consequence, there is more opportunity for an increasing number of species to compete in a moderated habitat.

We know that succession is a dynamic process effected by the interactions and modifying influences of living organisms that make up the community; they tend to move it toward an equilibrium state. This is the *autogenic* aspect of succession. Impinging factors that control the succession from without constitute the *allogenic* aspect of succession. When these external factors are disruptive and sustained, the regenerative or progressive

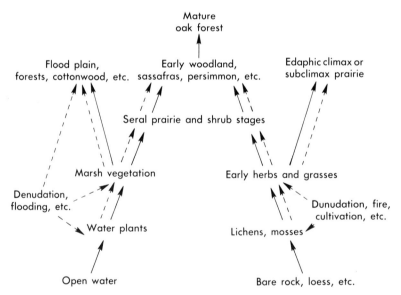

Fig. 7-11. Schematic diagram of successional pathways from dissimilar starting points toward a terminal (climax) community for central Missouri. Convergence is not always a reality, since one or more of the seres may be arrested short of the theoretical climax. Both primary (solid lines) and secondary (broken lines) seres are included.

processes of succession may be retarded indefinitely, or the allogenic effect may be made irreversible. Succession is the net change resulting from the interaction of these internal and external forces. However, under certain conditions autogenic influences may be retrogressive. The accumulation of peat on particular types of terrain in the far north often produces increasing insulation and colder soil temperatures that lead to a rising permafrost level. As a result, spruce forests retreat in these locations because of the shallow soil depth that remains above the permafrost. Only a sparse plant community can develop in its place. This interesting relationship is shown in Fig. 7-12. Allogenic factors should also be qualified, since they often cause rapid acceleration of successional processes, as illustrated by various types of pollution in lakes and streams.

EUTROPHICATION

Today we hear much about eutrophication in light of its role in the degradation of the environment. In this context eutrophication refers to the acceleration of biological processes that results from the release of nutrients and energy-rich wastes into all types of ecosystems. Successional changes are also speeded up. A biotic sequence that would normally require 10,000 years may be completed in a century. Terrestrial environments are not excluded, although eutrophication occurring as a result of pollution is conceptual-

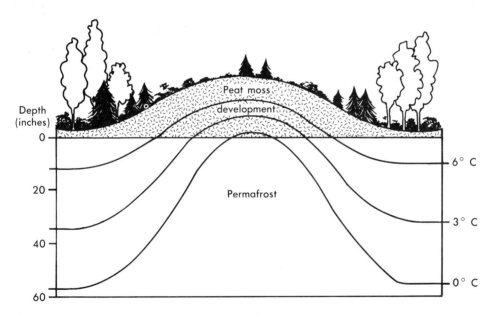

Fig. 7-12. Accumulation of peat causing bog formation in former spruce forest areas in Alaska. The insulative effect of organic deposition (sphagnum moss) causes a rise in the permafrost zone, indicated by the 0° C isotherm. The coniferous (dark) trees are black spruce, the tall open crowns are paper birch, and the understory shrubs are alder. Only soil depth and peat accumulation are to the same scale, the maximum depth of the latter being about 30 inches. (Redrawn from Heilman, P. E. 1966. Ecology **47**:825-831.)

Fig. 7-13. Massive growths of algae caused by eutrophication factors in Lake Minnetonka, Minnesota. (Courtesy Minneapolis Star, Minneapolis, Minn.)

Table 7-4. High dominance expressed by "algal blooms" measured as percentage of total biomass on a volume basis in a nitrogen-enriched farm pond*

ALGAE	JUNE 25	JULY 25	AUG. 21	SEPT. 27	OCT. 26	NOV. 14
Microcystis	99	40	1	75	—	—
Aphanizomenon	<1	—	99	11	—	—
Coelosphaerium	<1	35	<1	10	—	—
Others	—	25	—	4	100	100

*Modified from Vance, B. D. 1965. J. Phycol. 1:81-86.

ized primarily as it affects water systems. One reason, of course, is that aquatic environments are vulnerable as collecting basins in which human pollutants and waste products as well as natural sediments from surrounding watersheds accumulate.

Phosphorus and nitrogen are two principal agents in speeding up the eutrophication process. Generally these elements are not abundant in water systems, so that their addition often stimulates sharp increases in the growth rate of algae and changes in species composition. Blooms are frequently produced by one or several groups of algae. Fig. 7-13 shows massive accumulations of decaying algae along the shoreline of Lake Minnetonka in Minnesota. In the warm season of the year the blue-green algae are especially noted for their rapid growth. These microscopic organisms reach densities so great that they impart a greenish color to the water. Seasonal cycles and dominance values for several genera in an enriched pond are shown in Table 7-4. These organisms release toxic substances, often in concentrations high enough to kill fish, wildlife, and domestic animals.

Phosphorus is generally immobilized in agricultural soils; therefore the main source of phosphorus contamination is the municipal waste that is released into waterways throughout the country. Since most cities lack the facilities (tertiary treatment methods) necessary to remove phosphorus from wastes, most of it literally goes down the drain. Phosphate detergents account for nearly all of the increase in the amount of phosphorus entering sewage waste in the United States. According to the Center for the Biology of Natural Systems at Washington University, St. Louis, the use of phosphorus products rose from 11 million pounds in 1946 to 214 million pounds in 1968, an increase of 1845%. During the same period the United States population increased by 42%, suggesting a highly elevated per capita use.

Unlike phosphorus, nitrogen is not subject to chemical immobilization. It is converted by nitrifying microorganisms to highly soluble forms that are either readily carried into stream drainages or leached through the soil to underground water supplies. Agricultural fertilizers are the principal source of nitrate contamination because of their solubility and the ease with which they are transported by moving water. There has also been a tremendous increase in the use of nitrogen—more than 600% between 1946 and 1968.

Nitrogen in forms usable by plants can be synthesized as urea and ammonium compounds. The raw material is of course molecular nitrogen (N_2) from the atmosphere. Gaseous nitrogen constitutes almost 80% of the atmosphere by volume. Unlike phosphorus, there is a seemingly endless supply for synthesis. As increasing amounts of nitrogen are fixed by industrial processes, we might ask whether commensurate amounts are being released to the atmosphere from the original source, or whether we are in fact storing nitrogen in the soil and water of the biosphere. As discussed in Chapter 6, denitrification by microbial activity is a principal means of returning complex forms of nitrogen to the atmosphere. Yet we are told that nitrogen fixation by industrial methods will in 30 years exceed by five times the amount that is fixed by the nitrogen-fixing bacteria.

Sources of organic carbon such as sewage are also a significant cause of accelerated eutrophication and decreasing water quality. When enrichment occurs, usable energy is provided for aerobic bacteria, which grow rapidly. Their oxygen requirements increase as greater numbers of bacteria use greater amounts of energy. Such demands may lower the amounts of oxygen dissolved in the water to levels that are low enough to be detrimental to other forms of life. This problem is especially acute in warm water because oxygen levels decrease as water temperatures rise. The advent of thermal pollution in aquatic systems will obviously compound the eutrophication problem. Conspicuous die-offs of fish often are the result of a depleted oxygen supply. The requirement for the use of dissolved oxygen is called the biological oxygen demand (BOD). Its quantitative measure is an index of water conditions that provides a means of comparing the impact of pollution in different ecosystems. Under oxygen-free conditions, other signs of deterioration such as rising bubbles of hydrogen sulfide gas are apparent. This noxious gas is a metabolic by-product of anaerobic bacteria. The effects of organic pollution and overloaded nutrient cycles are manifold and extensive, and their impact on biological communities as well as on the condition of the physical environment may be serious.

Lake Erie is a large body of water undergoing accelerated eutrophication. Studies show that enrichment has increased certain species of plant and animal populations and decreased others. At lower depths of the lake, severe oxygen depletion has caused a deterioration in habitat for several species of fish. There have also been significant changes in species composition; the so-called rough fish are now relatively more important. This change is reflected in annual catches by commercial fisheries as follows:

Blue pike	Carp
Ciscoes	Drum
Sauger _Increasing eutrophication_→	Smelt
Walleye	Yellow perch
Whitefish	

With this shift in predator fish, species changes also occur at other levels in the food chain. One example of such a shift is shown in Table 7-5. A large species of *Daphnia* has been selectively eliminated, and the resultant increase in other, smaller species is an indication that, in this changing environment, food preferences and the competitive balance have also changed. Dominance is achieved at the expense of diversity. This example

Table 7-5. Changes in *Daphnia* species in Lake Erie resulting from shifts in predation caused by eutrophication factors*

SPECIES	SIZE (mm)	1938-1939	1948-1949
I	1.7	+	−
II	1.1	+++	++
III	0.9	++	+++

*Modified from Brooks, J. L. 1969. In Planning Committee for the International Symposium on Eutrophication. Eutrophication: causes, consequences, corrections. National Academy of Sciences, Washington, D.C.

can be extended to many other species as well. As stresses increase, growing dominance of the more tolerant survivors means that the energy flow is channeled through fewer species—a change that makes the system less predictable or stable.

Many other examples of the effects of eutrophication might be cited. Lake Zurich in Switzerland is a classic illustration, for municipal wastes from cities along its shore have lowered the oxygen level of this mountain lake so that it is becoming increasingly stagnant. San Francisco Bay is undergoing changes in biota and physical characteristics as nitrates applied to land under cultivation in the San Joaquin Valley are collected and washed into it. The Baltic Sea is losing its vitality from the burdensome demands of pollutants. Even Lake Baikal in the Siberian interior is showing the impact of eutrophication as a result of the wastes dumped into it by pulp and paper mills. It is the deepest and one of the coldest lakes in the world.

BIOMASS AND ENERGY RELATIONSHIPS

As succession advances toward a mature community state, significant changes are occurring with regard to diversity-dominance relationships, net production, structure, energy allocation, rate of organic turnover and nutrient regulation. In general, diversity tends to increase over time, and "species dominance," expressed as a fractional portion of the total community structure and process, diminishes. We can see this effect in a series of developing plant communities in an "old-field" succession, as shown in Table 7-6. In the mature community there are more than three times as many species as in the pioneer stage. Additional data on "old-field" succession in Illinois are depicted in Fig. 7-14. In the early stages of succession the degree of dominance in relation to the numbers of species indicates a geometric relationship, suggesting that the dominant species monopolizes a given percent of the resource, with successively lesser species utilizing fractional portions of the remainder. In older stages of succession there is a more pronounced flattening in the middle part of the curve, indicating an increasing number of species that are more or less equal. In other words, there are relatively fewer monopolists occupying broad niches. The successional trends exhibited here are cited in numerous examples of community development during succession. In higher latitudes, however,

Table 7-6. Species relationships with successional development of abandoned farmland from pioneer through final stages*

STAGE	NO. FOUND IN EACH STAGE	NO. INVADING EACH STAGE	NO. SURVIVING IN FINAL STAGE
I (pioneers)	27	27	8
II	42	18	10
III	56	17	10
IV	80	29	23
V	89	25	24
VI (prairie species)	94	17	17

*Modified from Smith, C. C. 1940. Ecol. Monogr. **10**:421-480.

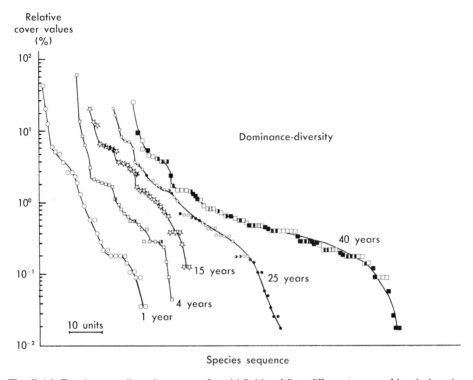

Fig. 7-14. Dominance-diversity curves for old fields of five different ages of land abandon-ment in southern Illinois. (Redrawn from Bazzaz, F. A. 1975. Ecology **56**:485-488. Copy-right © 1975 by the Ecological Society of America.)

where climatic conditions are limiting, equitability may be less in the climax, for ex-ample, a spruce stand, than in midsuccession.

Older communities generally exhibit more structural biomass. The overall effect is one of greater protection from external or allogenic fluctuations. Mature stages charac-teristically have a maximum amount of community biomass supported by energy that is community produced. An important feature of succession is the increasing ratio of total community structure to the sustaining increments of energy fed into it. The maxi-mum ratio would be found in the community that was most efficient in using available resources. In mature communities, net production is low or approaches zero. This means that total production of food and energy is balanced by the community's total respiration requirements, including both plants and animal consumers.

In early stages of succession the biomass of the community increases because productivity exceeds the rate of use. One reason is that there is relatively more photosynthetic tissue in relation to the total biomass in these developmental stages. In later stages of succession this ratio is offset by an increase in perennial biomass. At the same time the community continues to use energy in the upkeep or maintenance of this

Table 7-7. Comparison of ecosystem characteristics related to
successional development*

ECOSYSTEM CHARACTERISTICS	YOUNG STAGES	MATURE STAGES
Community structure		
Species diversity	Low	High
Total organic accumulation	Small	Large
Stratification and spatial heterogeneity	Poorly developed	Well developed
Community energetics		
Biomass support per unit of energy flow	Low	High
Net community production	High	Low
Life history		
Size of organism	Small	Large
Life cycles	Short, simple	Long, complex
Reproductive strategy	r-selected	K-selected
Nutrient cycling		
Mineral cycles	Open	Closed
Nutrient exchange, rate between organisms and environment	Rapid	Slow
Overall homeostasis		
Nutrient conservation	Poor	Good
Stability (resistance to external perturbations)	Poor	Good

*Modified from Odum, E. P. 1969. Science **164**:262-270.

structure on a continuous basis, whether photosynthesis occurs or not. As total biomass increases, the rate of net growth slows. The average rate of organic breakdown per unit of biomass decreases, causing relatively smaller releases of nutrients. In pioneer communities the nutrients are more easily dispersed because the turnover rate of the total biomass is proportionately greater for a given period of time. Stated in another way, the shorter turnover time allows a more frequent release of nutrients as well as greater chances of dispersal. Mineral cycling in these communities is more susceptible to "leaks," whereas in mature communities there is more control over the retention of nutrients within the system. A heavy-growth forest that has its nutrient capital partially stored in its biomass is a more closed and guarded unit than an annual system in which the total biomass is subject to simultaneous breakdown and decay. These community properties in relation to successional development are summarized in Table 7-7.

SUMMARY

The cycle of youth, growth, and maturity is a phenomenon of all ecosystems. An intrinsic trait or tendency of biological organization is the development of environmental equilibrium. The two forces involved in this process are autogenic, or biotic, and allogenic, the latter consisting of external influences. Where impinging factors create large stresses or fluctuations, as in cold climates, development is slow. Biotic influences become relatively more significant as external stresses are reduced. The course of succession may be seen as the net effect of these forces. Rapid or drastic environmental change disrupts succession. Communities become simplified and inherently less stable in the face of environmental fluctuations. Secondary seres are initiated. These are universal responses arising mainly from the human impact on functioning ecosystems.

However, under some conditions retardation or even retrogression may be effected by community factors. The accumulation of litter in humid grasslands and peat deposits over permafrost are examples. In the first case excessive buildup of litter from the grasses themselves is a factor in stagnation and replacement by other species. Fire is an important factor in succession under some conditions. Peat deposits resulting from slow decay in northern climates insulate the ground, preventing thawing and causing the permafrost level to rise. Soil depth becomes shallower, and the forests that cannot thrive are gradually succeeded by sparse tundra vegetation.

Generally diversity and biomass development increase with succession. Terminal phases are more efficient in conserving resources than developmental phases, for a mature community expends less energy to support a given unit of structure than a pioneer or early stage. This phenomenon occurs because, on a relative basis, metabolic requirements moderate as the nonfunctional fraction of the total biomass increases and life cycles become longer. Succession is of practical value in staging biological repair mechanisms in denuded or disrupted ecosystems. Soil is rejuvenated, nutrients come under tighter control, water is conserved, and regenerated diversity creates an increasingly more stable community.

DISCUSSION QUESTIONS

1. Define biotic succession. What is its driving force?
2. What is the value of the successional process to our human management of natural communities? Which are usually favored in this management, r-selected or K-selected species? Explain.
3. Relate biotic succession to the development and maturity of the soil profile.

REFERENCES

Auclair, A. N. and F. G. Goff. 1971. Diversity relations of upland forests in the western Great Lakes area. Am. Nat. **105**:499-528.

Bazzaz, F. A. 1975. Plant species diversity in old-field successional ecosystems in southern Illinois. Ecology **56**:485-488.

Beeton, A. M. 1971. Eutrophication of the St. Lawrence Great Lakes. In T. R. Detwyler (ed.). Man's impact on environment. McGraw-Hill Book Co. New York.

Bonck, J. and W. T. Penfound. 1945. Plant succession on abandoned farmlands in the vicinity of New Orleans, La. Am. Midl. Nat. **33**:526-529.

Booth, W. E. 1941. Algae as pioneers in plant succession and their importance in erosion control. Ecology **22**:38-47.

Bramble, W. C. and R. H. Ashley. 1955. Natural revegetation of spoil banks in central Pennsylvania. Ecology **36**:417-423.

Brooks, J. L. 1969. Eutrophication and changes in the composition of the zooplankton. In Planning Committee for the International Symposium on Eutrophication. Eutrophication: causes, consequences, corrections. National Academy of Sciences. Washington, D.C.

Brown, J. C. 1958. Soil fungi of some British sand dunes in relation to soil types and succession. J. Ecol. **46**:641-644.

Budowski, G. 1963. Forest succession in tropical lowlands. Turrialba **13**:42-44.

Carlquist, S. 1965. Island life. Natural History Press. Garden City, N.Y.

Clements, F. E. 1916. Plant succession: an analysis of the development of vegetation. Publication No. 242. Carnegie Institute of Washington. Washington, D.C.

Cooper, C. F. 1961. The ecology of fire. Sci. Am. **204**:150-160.

Cowles, H. C. 1899. The ecological relations of the vegetation on the sand dunes of Lake Michigan. Bot. Gaz. **27**:95-117.

Dansereau, P. 1957. Biogeography: an ecological perspective. The Ronald Press Co. New York.

Evans, F. C. and E. Dahl. 1955. The vegetational structure of an abandoned field in southeastern Michigan and its relation to environmental factors. Ecology **36**:685-706.

Hasler, A. D. 1970. Man-induced eutrophication of

lakes. In S. F. Singer (ed.). Global effects of environmental pollution. Springer-Verlag New York, Inc. New York.

Heilman, P. E. 1966. Change in distribution and availability of nitrogen with forest succession in north slopes in interior Alaska. Ecology 47:825-831.

Hirth, H. F. 1959. Small mammals in old field succession. Ecology 40:417-424.

Hori, S. and I. Ito. 1959. The annual succession of desmid communities in consequence of organic pollution. Japan. J. Ecol. 9:152-154.

Horn, H. S. 1971. The adaptive geometry of trees. Monograph in Population Biology No. 3. Princeton University Press. Princeton, N.J.

Humphrey, R. R. and L. A. Mehrohoff. 1958. Vegetation changes on a southern Arizona grassland range. Ecology 39:720-726.

Hutchinson, G. E. 1969. Eutrophication, past, and present. In Planning Committee for the International Symposium on Eutrophication. Eutrophication: causes, consequences, corrections. National Academy of Sciences. Washington, D.C.

Johnston, D. W. and E. P. Odum. 1956. Breeding bird populations in relation to plant succession in the Piedmont of Georgia. Ecology 37:50-62.

Kane, J. 1967. Surtsey: an island emerges. Nat. Hist. 76:22-27.

Martin, N. D. 1960. An analysis of bird populations in relation to forest succession in Algonquin Provincial Park, Ontario. Ecology 41:126-140.

Nicholson, S. A. and C. D. Monk. 1974. Plant species diversity in old field succession on the Georgia Piedmont. Ecology 55:1075-1085.

Odum, E. P. 1969. The strategy of ecosystem development. Science 164:262-270.

Olson, J. S. 1958. Rates of succession and soil changes on southern Lake Michigan sand dunes. Bot. Gaz. 119:125-170.

Paone, J. et al. 1974. Information Circular 8642. Bureau of Mines. U.S. Department of the Interior. Washington, D.C.

Parsons, J. 1968. The effects of acid stripmine effluents on the ecology of a stream. Arch. Hydrobiol. 65:25-50.

Pearson, P. G. 1959. Small mammals and old field succession on the Piedmont of New Jersey. Ecology 40:249-254.

Piemeisel, R. L. 1951. Causes affecting change and rate of change in a vegetation of annuals in Idaho. Ecology 32:53-72.

The price of strip mining. 1971. Time 97:47-48.

Shelford, V. E. 1913. Annual communities in temperate North America. The University of Chicago Press. Chicago.

Sherwood, M. and G. C. Carroll. 1974. Fungal succession on needles and young twigs of old-growth Douglas fir. Mycologia 66:499-506.

Small, W. F. 1971. Third pollution: the national problem of solid waste disposal. Praeger Publishers, Inc. New York.

Smayda, T. J. 1961. Ectocrine substances and limiting factors as determinants of succession in natural phytoplankton communities. Bact. Proc. 61:39.

Smith, C. C. 1940. Biotic and physiographic succession on abandoned farmland. Ecol. Monogr. 10:421-480.

Stockner, J. G. and W. W. Benson. 1967. The succession of diatom assemblages in the recent sediments of Lake Washington. Limnol. Oceanogr. 12:513-532.

Vance, B. D. 1965. Composition and succession of cyanophycean water blooms. J. Phycol. 1:81-86.

Viereck, L. A. 1966. Plant succession and soil development on gravel outwash of the Muldrow Glacier, Alaska. Ecol. Monogr. 36:181-199.

Watters, R. F. 1971. Shifting cultivation in Latin America. United Nations Food and Agriculture Organization. Rome.

Whittaker, R. H. 1953. A consideration of climax theory: The climax as a population and pattern. Ecol. Monogr. 23:41-78.

Wiens, J. A. 1976. Population responses to patchy environments. Ann. Rev. Ecol. Systematics 7:81-120.

Wohlrab, G. and R. W. Tuveson. 1965. Distribution of fungi in early stages of succession in Indiana dune sand. Am. J. Bot. 52:1050-1058.

ADDITIONAL READINGS

Odum, E. P. 1971. Fundamentals of ecology. W. B. Saunders Co. Philadelphia.

Rice, E. L. 1974. Allelopathy. Academic Press, Inc. New York.

PART THREE

The biomes

8 Terrestrial communities

THE EARTH'S PLANT COVER

Biomes are regional communities of native plants and animals. Certain biomes such as the arctic tundra are transcontinental in their geographical extent. The identification and classification of biomes as successional climaxes are based conveniently on the physiognomic (general form) and adaptive features of the plant cover found in different climates. Strategies for animal survival have developed according to the variable conditions of climate and vegetation. The early horse, which changed in form and eating habits from a small forest browser to a fleet-footed grass eater on the open steppe, is an example.

A second feature of individual biomes is a certain level of taxonomic continuity, which also serves as a means of distinguishing between two different biome types. The evergreen needle-leaf habit is a worldwide phenomenon; yet several biomes with these common traits can be differentiated from each other on the basis of floristic differences of the dominant tree species. The extensive Canadian forests of boreal conifers and the more temperate communities of coniferous species in the northwestern United States are but two examples, for species of conifers, and even some genera, are different from each other. Similarly there are animals of corresponding habit and niche requirements that are classified separately and that belong to different biomes. When these criteria of life form and taxa are employed in combination, the biome becomes a valid entity of regional significance.

Although biomes, or formations, reflect the temperature and moisture patterns of a given regional climate, the geology, soils, and various biotic factors, including anthropogenic fires, are also important modifying influences. Since climatic repetition occurs in a geographical pattern around the earth, a given biome type can be predicted with some assurance in specified areas. There are five well-known ''Mediterranean'' climates, for example, that have in common vegetations of generally similar structure and adaptation (coriaceous leaves, regenerative organs below ground, fire tolerance, and so on) that have evolved separately but under generally similar conditions. Usually in these cases no close phylogenetic or evolutionary ties exist. This correspondence in life form, wherever it occurs, exemplifies convergent evolution (see Chapter 1).

Where taxonomic closeness is observed, however, some kind of common lineage might be inferred and supported by the fossil record. The deciduous forests of Eurasia and eastern North America are examples of this type of ancestry, which indicates the important role of plant histories in helping to solve the anomalies and complexities of present-day plant distribution. In the following sections the principal biome types will be presented briefly. These types are based on three generalized temperature regimes: the cold (and cool temperate), temperate, and tropical regions. It is important to recognize that biome overlap does occur in so simplified a formula.

As an adaptation of Köppen's classification, the cold or cool temperate category includes those climates with an average temperature for the coolest month of $-3°$ C or less. These are designated in the literature as the D, E, and F climates of progressively greater cold. The D climate, then, represents in part transition conditions and accordingly exhibits biome overlap with the next warmer zone. The C climates have average temperatures ranging from $-3°$ to $18°$ C for the coolest month. Average temperatures

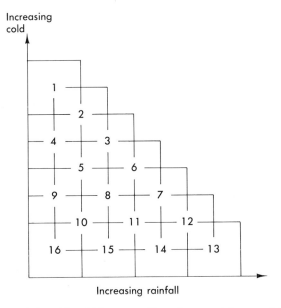

Increasing
cold

Increasing rainfall

Fig. 8-1. Schematic relationship of biome position along moisture and temperature gradients. The numbering system is as follows:

1 Tundra
2 Taiga
3 Boreal conifer forest
4 Cold desert
5 Temperate desert grassland
6 Temperate conifer forest
7 Temperate deciduous forest
8 Temperate grassland

9 Warm desert
10 Warm desert grassland
11 Open woodlands
12 Subtropical evergreen forest
13 Tropical rain forest
14 Tropical deciduous forest
15 Tropical savannah
16 Tropical scrub

greater than 18° C delineate the tropical regions, which are designated as A climates. Frost does not occur. The so-called B climates are arrayed across several temperature regimes and exhibit the increasing aridity characteristic of steppe and desert. In Fig. 8-1, 16 biome types or designates are depicted in a qualitative moisture-temperature relationship. Although this is obviously an artificial derivation, it does serve to readily orient each biome in relation to any other along selected gradients. The numbers that appear in Fig. 8-1 are used in the following discussion to designate corresponding biome types. Some are grouped where geographical and/or ecological affinity is indicated, ensuring a measure of brevity as well as convenience in presentation.

COLD-REGION BIOMES

The cold-region biomes are as follows (Fig. 8-1):

Tundra—*1*
Coniferous forest—*2, 3*

With few exceptions the general climate of high latitudes is characterized by several prominent features. These include a short season of plant growth, large annual temperature fluctuations, limited amounts of precipitation, and the drying effects of frequent winds during the winter season. Soils are also frozen for extended periods of the year, during which moisture for plants is limited or unavailable. There are extensive areas of permafrost, in which the soil and parent materials below certain levels in the profile remain frozen continuously. The thickness of permafrost in northern Alaska and Canada, for example, may be 100 to 300 meters or even more.

The depth to which surface thawing takes place during the brief summer is an important factor in determining the kind of vegetation and the degree of its structural development. In the far north, seasonal melting generally does not extend below 1 meter and in some places may be as shallow as 15 to 20 centimeters. These characteristics of climate and soils combine to produce singularly harsh environments; the most extreme of these approach the tolerance limits of plant life. It is this marginal character of the environment that is reflected, at least in part, in a generally low degree of biotic diversity and simple community structure. These two types of vegetation, both transcontinental in extent, are restricted to the Northern Hemisphere, since there are no suitable, continuous land areas at corresponding latitudes south of the equator. This land-sea relationship is shown in Fig. 8-2. The land mass of the Antarctic continent is large, but its cold summers are too severe for the development of a large-scale, organized terrestrial vegetation. Here, in essentially soilless habitats, only a few sporadic seed-producing plants, cryptogams (spore-bearing plants such as mosses and lichens), and microorganisms have been recorded.

Although they are not as extensive as boreal or northern vegetation, tundra and cold-conifer forests occur in alpine and subalpine zones of the lower latitudes, respectively. Although boreal conditions and those of high elevations are similar to varying degrees in terms of frost action and a short growing season, other meteorological and edaphic differences are apparent. The extended period of daylight so characteristic of the high-latitude summer obviously does not occur in the alpine regions of the temperate zones. Sharply demarcated microclimatic effects resulting from multiple variations in slope and aspect are also features more characteristic of mountainous terrain. In summer, nighttime temperatures are more apt to reach freezing than are those in the arctic.

Tropical alpines do not exhibit the large seasonal temperature fluctuations found in high latitudes because of the year-round constancy of the photoperiod and incoming solar energy. However, diurnal temperature variations may be impressive; particularly under clear conditions they may vary from freezing to more than 20° C between night and day. Alpine topography makes areas especially vulnerable to rapid runoff, soil wastage, and slippage. Accumulation of alluvium is therefore less marked or nonexistent when compared to that of certain riverine tracts in the high arctic. Taken as a whole, alpine tundra soils are drier than their arctic counterparts, and permanent layers of soil ice are not as common. Other differences relating to the final development of the vegetation involve the plant histories of the regions, which covers millions of years from late Tertiary times down through the Pleistocene period (a geological time scale is given in Table 4-1, p. 58).

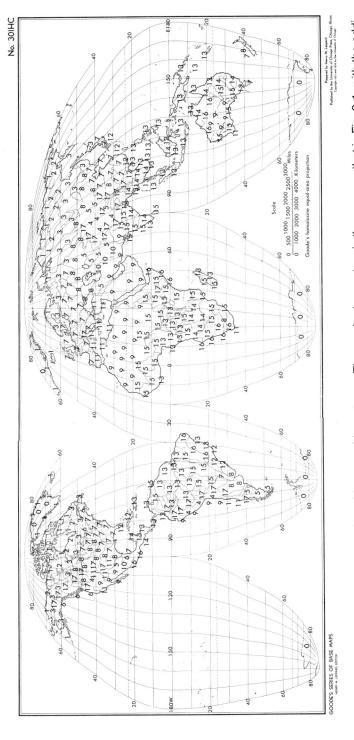

Fig. 8-2. Generalized distributions of key biomes or biome types. The numbering system is the same as that in Fig. 8-1, with the addition of *0*, ice fields, and *17*, mountainous terrain, which includes a complex of types. (Base map copyrighted by the University of Chicago, Department of Geography.)

Fig. 8-3. View of arctic tundra in Alaska. The Brooks Range is in the background. (Courtesy Atlantic Richfield Co., Anchorage, Alaska.)

Tundra

The arctic tundra is generally a treeless landform, varying in relief from a flat, alluvial expanse that is pockmarked by small pools of water to a drier, rugged upland (Fig. 8-3). It is the most northerly of terrestrial biomes, occurring extensively in North America and Eurasia. Freezing temperatures in the tundra may occur in any month. For example, at Point Barrow, the northernmost tip of the Alaskan tundra, the average period during which the minimum temperatures are at freezing levels is greater than 300 days. The season of active plant growth varies among different tundra species according to the species' relative ability to tolerate freezing temperatures. During the summer extremely long photoperiods (the hours between sunrise and sunset) occur in the arctic tundra. At 66 degrees north latitude at the full declination of the sun, "daylight" lasts slightly longer than 23 hours, so that during the principal period of growth in the high arctic, daylight is nearly continuous. Even though the season of plant growth is short, considerable solar energy is thus available for photosynthesis. Compared to alpine species in lower latitudes, many plants of the arctic tundra illustrate ''long-day'' adaptations in such responses as flowering. Precipitation in the tundra is variable, but in general it occurs in low amounts that do not exceed 25 to 50 centimeters in water equivalents per year. In temperate or tropical latitudes with longer periods of high sun and more direct sunlight, warmer average temperatures, greater heat accumulation, and increased evaporation potentials, such limited amounts of precipitation would result in environments approximating those of the steppe and the desert.

Despite its general simplicity of structure and low species diversity, communities of the northern or arctic tundra do vary in composition and growth from one locale to another. The tundra is a recurring mosaic of communities whose structure and species composition change, often within narrow spatial limits, according to the conditions of exposure, soil drainage, and other factors that are strongly affected by topographical relief. Wet, boggy conditions may be prevalent in depressions, in contrast to the drier habitats such as raised hummocks or higher ground with thin, coarse, or rocky soils.

Generalized plant assemblages range from grass and sedge-dominated communities on the deeper soils of low-lying terrain to xerophytic lichen-moss ground types on drier or more exposed sites. Low shrub, herb, and lichen mixtures are widespread over areas that are elevated and seasonally wet, including the mat-forming heaths and tussock types such as cotton grass. Near streams and in more protected situations, dwarf specimens of such tree species as alder, birch, and willow are prevalent. Along these moisture gradients, species densities and dominance shift accordingly, producing numerous intermediate combinations.

Comparable variations in the vegetation occur also in the Siberian arctic. The low, tussock- and mat-forming habit is a universal growth form in the tundra, occurring in both herbaceous and woody plants. Preponderantly high biomass, or weight ratios of underground to aerial portions of the plant are common. These traits, which emphasize low or prostrate stature and a spreading form, are ostensibly adaptations to the conditions found at the surface of the soil and immediately below it, which are relatively moderate when compared to conditions above ground, where the effects of wind and cold desiccation are more severe.

The harsh environment of the Arctic tundra is reflected in the impoverished resident fauna. Some of the characteristic mammals are the caribou, musk-ox, grizzly bear, polar bear, Arctic hare, gray wolf, Arctic fox, and lemming. These animals adjust to their environment and its cyclic food supplies in a number of ways. In winter the caribou migrate toward the south into the tree line where food is plentiful. In the tundras of Eurasia the reindeer is the ecological equivalent of the Alaskan caribou, and it also exhibits a migratory habit. The lemmings, whose populations often become exceedingly large, adapt to the winter cold by burrowing below the snow in dry hummock vegetation. Other animals are insulated by fur and fat against the cold and may burrow as well. Some forms such as the Arctic hare, Arctic fox, and some resident birds, including the ptarmigan, exhibit a white color phase in winter.

During the long Arctic summer the tundra ecosystem becomes the seasonal home of hosts of shore birds and waterfowl from the south. More than 40 species of birds migrate to these far-northern lands to breed and raise their progeny. Feeding is almost a continuous process during the long daylight hours, and the young may grow more rapidly than they would in southern latitudes. These migrant birds leave the tundra in the early fall, and only such permanent residents as the snowy owl and the ptarmigan remain during the long winter season.

Alpine tundra, herb fields, and heath vegetation occur at high elevations in both hemispheres and also extend into the tropical belt. These communities, like those in the far-northern tundras, are characterized by various morphological and physiological traits that effect a universal ecological equivalence. Generally the extensive lichen communities that characterize the arctic tundra are absent in alpine tundras, but varying degrees of taxonomic affinity with the northern tundra do exist. Numerous genera occur as dominants in both regions but generally as different species. Even in the alpine of tropical Borneo and New Guinea such familiar temperate-zone genera as sedge, rhododendron, and blueberry occur. In the cordillera of the tropical Americas, an extensive low vegetation, or *paramo,* above the montane forests occurs from Costa Rica to the Andes. This type includes both herbaceous and woody vegetation. In these are found certain ''temperate'' genera, reflecting also a degree of floral interchange. Generally, however, the alpine types of the Southern Hemisphere have a different origin from those of the Northern Hemisphere, as shown by the numerous genera of plants aligned with Australia, New Zealand, and the outlying islands of the South Pacific.

As is true of plants, there are few species of mammals that are common to the arctic and alpine tundras. The caribou is common in the alpine areas of the Alaskan mountains but not in mountains that are more southern. Other native animals are the mountain goat, Dall's sheep, collared pika, and hoary marmot. The mountain sheep, common pika, and yellow-bellied marmot are found in the southern alpine. Many animals of the montane forests move up into the alpine tundra during the summer.

Much discussion has been devoted to the factors that differentiate the tundra from impinging boreal forests. The presence of a permafrost at a shallow depth during the summer is inimical to the development of root systems and therefore to tree growth. Marginal conditions for the growth of boreal forest proper exist in those regions or localized situations where the soil is frozen at a level of less than *1 meter* below the surface during summer thawing. One cause of soil erosion in the tundra is melting of the surface zone of soils. Any factor that promotes greater melting thus creates an increased erosion hazard. Removal of plant cover and surface litter or mulch results in a loss of insulation and in greater heat absorption. The present construction of the Alaskan pipeline poses serious threats to soil stability. Not only is plant cover being stripped from the soil, but oil heated to approximately 65° C would cause the permafrost to thaw to an unnatural degree.

The tundra biome is fragile and always vulnerable to biotic changes, some possibly irreversible. Large-scale incursions that drastically upset the vegetation and soil equilibrium and consequently upset the dependent food chains initiate long-lasting, deleterious effects on the total environmental complex from which recovery may be impossible.

Coniferous forest

The coniferous forests of the Northern Hemisphere extend from central Alaska across Canada and through northern Europe and Siberia. This biome type is one of the most extensive of the world's forest communities (Fig. 8-2). Bordered on the north by tundra, its southern limits merge with temperate grasslands and deciduous forests, ranging from about 60 degrees north latitude on the western sides of the continents to about 45 degrees north latitude on the east. This dipping effect from west to east is mainly the result of the effects of maritime influences on continental climates. At these latitudes, warm ocean currents push warmer temperature belts and therefore plant zones northward on the west sides of continents, while cold, pole-fed currents depress zonations southward on the eastern sides. Average annual temperatures that mark the boundaries of boreal forest vary from one locale to another. Numerous attempts have been made to classify climatic zones in relation to vegetation patterns. In general, isotherms aid in depicting the temperature characteristics of the boreal forest. For the warmest month, temperatures of about 10° C as a minimum define the northern limit; temperatures of between 0° and −5° C as a minimum for the coldest month delineate the southern boundary.

The term "taiga" has been interpreted in various ways. Some authors view the taiga as restricted to poleward positions of the coniferous belt, where trees become stunted and more scattered and are interspersed with tundra plants. Others define the term more broadly, so that taiga embraces the boreal forest as a whole. The taiga also has been classified as a separate vegetational unit cognate with the tundra and boreal forest biomes that it separates. Whatever interpretation is employed, the tundra-boreal forest transition, or lichen-woodland, suggests conditions that are marginal for the development of the dense, high forest types found farther south. The desiccative effects of the cold, dry winds coupled with thin or frozen soils are significant factors in its distribution, for, as a consequence, root systems are shallow and the trees themselves are low and sprawling. This forest-tundra transition occurs over wide areas in Canada and Alaska and especially in eastern Asia. Its counterpart in lower latitudes is a timberline type found at high elevations.

The fact that trees generally grow more slowly in cold climates can be determined by counting the number of annual rings in the cross section of a stump. Such ring counts have indicated that small specimens with boles less than 5 centimeters in diameter may be 100 years old! The ecological conditions implicit in these low growth rates are reflected in certain adaptive morphological and physiological features. The needle-shaped leaves of conifers provide a reduced surface area, resulting in a decreased transpiration potential. This is significant especially when warming of the atmosphere creates vapor pressure deficits during periods when the soils remain cold and water movement into root systems is sluggish, if not curtailed. Even so, needles frequently turn brown from excessive water losses, causing "red belts" on mountain sides. Evergreens are also capable of initiating photosynthesis as quickly as rising temperatures permit. This is an advantage over deciduous trees, which must first produce a new set of leaves before resuming photosynthesis on a full scale. Yet in the most extreme environments, where cold temperatures and extensive and prolonged soil freezing combine to produce severe stresses, the deciduous habit, as manifested in the needle-shedding larch species, is an important adaptation. Among the conifers, the Siberian larch, found in eastern Siberia, has the most northerly of the earth's tree distributions.

In the boreal forests of North America the conifers with transcontinental distributions are black spruce, white spruce, balsam fir, eastern larch (or tamarack), and jack pine. Black spruce commonly occurs in low, wet, poorly drained ground, sometimes in pure stands but also mixed with balsam fir and tamarack. In drier ground these species appear as scattered trees with other conifers and several cold-hardy deciduous trees, including birch, aspen, and willow. White spruce is one of the *most widely distributed* of the boreal conifers, occurring typically on well-drained soils. White spruce, black spruce, and tamarack form the northernmost outliers of forests in transition with tundra vegetation. Jack pine is the only pine species that has a transcontinental range in the New World. Farther south, in the lake state region of Minnesota, Wisconsin, and Michigan, it is an important seral species found on dry, sandy soils or in the aftermath of forest fires.

Subalpine forests in the middle latitudes are ecological equivalents of the boreal conifers. They occur in transition with alpine tundra, where growth conditions become marginal for trees. Here specimens are stunted and scattered, often forming a *krummholz,* or zone of wind-deformed trees. These communities extend downward and become more

Fig. 8-4. Subalpine forest and alpine tundra in the Rocky Mountains, Boulder County, Colorado. The elevation at camera level is 11,200 feet. Engelmann spruce and alpine fir are the principal conifer species. (Courtesy United States Forest Service.)

luxuriant, closed-canopy forests as conditions ameliorate. The principal genera, spruce, fir, and pine, are represented by different species than those in the boreal zones. In the subalpine zone of the Rocky Mountains, for example, the principal conifers are Engelmann spruce and alpine fir (Fig. 8-4). Other conifers in this forest region include bristlecone pine, limber pine, whitebark pine, and lodgepole pine. The latter is an important seral species, often occurring in pure stands following fire. Zones of overlap between the subalpine and boreal forests do occur, however, as in the case of the alpine fir, which extends as far north as Alaska on a diminishing elevational gradient.

In North America, animal ecologists recognize two animal biosociations, or major climax communities, in the coniferous forests. These are the boreal forest, or cold-conifer region community, and the montaine, or more temperate community type of lower latitudes. Fauna in the two divisions tend to overlap, as shown by certain species such as the snowshoe rabbit, porcupine, red squirrel, gray wolf, and black bear. There is an increasing number of mammals, birds, and invertebrate species in the more moderated conditions of these forests, as compared to tundra environments.

Fig. 8-5. The moose, among the largest of American herbivores, in typical habitat of willow thicket with spruce forest in the background. (Courtesy Robert Breitenbach, University of Missouri, Columbia.)

Animals characteristically found in the boreal forest are the moose, an inhabitant of aquatic seral stages (Fig. 8-5); the woodland caribou; and a number of mustelids such as the wolverine, ermine, least weasel, fisher, and American martin. Other species of the boreal forest ecosystem include the lynx, several shrews, and a number of rodents.

Animals characteristic of the American montane, or temperate, biosociation include those forms common to both communities as well as the elk (or wapiti), mule deer, mountain lion, bobcat, grizzly bear, a few mustelids, and a larger number of rodents.

The boreal forests of Eurasia harbor similar forms; some of the larger animals are conspecific with those of the North American forests.

TEMPERATE-REGION BIOMES

The temperate-region biomes are as follows (Fig. 8-1):

> Broad-leaved forest—*7, 12*
> Grasslands—*5, 8, 10*
> Desert—*4, 9*
> Brushlands and woodlands—*11*
> Coniferous forest—*6*

The principal biomes or biome types of temperate latitudes are represented by several distinctive vegetational life forms. These are often found in repetitive patterns in similar climates around the world. These large-scale communities range from 45 to 60 degrees on their poleward margins toward the perennially warm sectors of the earth, where frost does not occur except in mountainous terrain. Mean annual temperatures across these latitudes range from about 5° to 25° C. Such conditions constitute a significant moderation in temperature effects and frost action compared to the tundra and cold-forest environments discussed in the preceding section. Temperate climates are nonetheless characterized by conspicuous annual variations in temperature, frost of varying duration, and generally well-marked precipitation patterns. The seasonal periodicity of meteorological events is significant in the adaptive phenologies of temperate plant communities such as the synchronous dropping of forest leaves in autumn or the breaking of bud dormancy in spring. Warmer temperatures, together with growing seasons that are longer than those in the high latitudes, provide extended periods of photosynthetic activity that result in a larger energy base through more biomass accumulation and greater elaboration of community structure. This trend extends into the frost-free zones where maximum complexity is achieved in mixed rain-forest communities.

The effects of temperature and frost action are not the sole dominant features of temperate climates as they are in tundra and boreal forest. In the middle latitudes and in the tropics a general division of biome types is seen where moisture lines become visibly more influential on a regional basis. In the continental United States the annual rainfall ranges from approximately 150 centimeters (60 inches) in the eastern mountains to 10 centimeters or less in the desert southwest. Decreasing moisture gradients show broadly generated forest → grassland → desert sequences at similar latitudes. This shift in emphasis from the latitudinal banding of arctic tundra and boreal forest to a longitudinal orientation as a function of moisture in the continental United States is illustrated in Fig.

8-2. Similar parallels for the middle latitudes are seen in west-central Europe and in eastern Asia. The pattern of vegetation is determined by continental moisture patterns that, in turn, are modified by the effects of higher elevations. In the mountains, moisture stresses are ameliorated, at least at midelevations, by the presence of cooler temperatures, increasing amounts of precipitation, and reduced evaporation. In arid and semiarid zones, particularly, the change from desert vegetation to woodland communities is especially noticeable.

Broad-leaved forest

The deciduous forests of eastern North America extend from the Atlantic seaboard to the mid-Mississippi Valley and from Canada to central Florida. Prior to the westward migration and settlement, this vast region was a mosaic of stable communities characterized by the unifying feature of a single seasonal leaf drop. The region has ecological equivalents in west-central Europe and in western Asia, primarily China and Japan. These forests are closely related through several common genera such as oaks and maples. Millions of years ago, when moist, mild conditions existed in what are now the boreal and tundra climates, they were once part of an extensive, contiguous distribution across the Northern Hemisphere. In the Southern Hemisphere, temperate deciduous forests are not nearly so extensive; they share no close taxonomic relationships to the northern forests. The closest analogs are the southern beech communities of South America. However, most of the species are evergreen, and it is speculated that the few deciduous species that are present developed in response to cooler and/or drier habitats.

The typical deciduous forest climate, unlike continental grasslands, characteristically has a generally uniform distribution of precipitation throughout the year. Amounts of winter precipitation tend to remain high and in some locales may actually exceed the amounts received in summer. In addition, subsoil moisture is recharged to capacity under normal rainfall conditions during the winter when transpiration losses are minimal. Toward the northern and western peripheries in the United States, however, the amount of rainfall diminishes conspicuously during the colder months.

On its northern margin, the deciduous forest of eastern North America is in transition with boreal forest communities. In this extensive mixing of coniferous and deciduous genera there is a gradual shift in species importance along north-south climatic gradients. Thus there is a generally continuous change in forest composition as one passes from one recognizable biome to another. Within this transition, however, there are several coniferous species, including northern hemlock and several pines. The latter are jack pine, red pine, and eastern white pine, all of which are important successional species. They may often be followed by terminal or climax forests of deciduous genera. The validity of the concept of a forest continuum as opposed to the association or community-type concept has been the subject of debate for many years.

On the south, the deciduous forest biome merges with a complex of forest types; some are seral, or successional, while others are more or less stable climax communities. The successional types include pine scrub and flatwoods and are generally fire maintained. An important southern pine species is the longleaf pine. Cypress swamps are limited by flooding and are succeeded by mixed deciduous forest with improved drainage. Climax forests of the lower south include those of broad-leaved evergreen, in which magnolia and swamp bay are dominant trees. The evergreen condition may be an adaptational advantage in regions of rapid decay rates. Compared to the massive, one-pulse leaf drop of a deciduous forest community, the release of nutrients by limited, year-round leaf fall is gradual, and under these conditions, nutrients would tend to be conserved.

In the central United States the deciduous forest is in transition with the tallgrass prairie, which occurs as a broad, wedge-shaped region that tapers toward the east. The general outline of this peninsular region, which includes a large portion of the corn belt, corresponds generally to the area of increasing incidence of drought and high evaporation potentials that lies in the path of the dry prevailing westerlies. These broad regional relationships between vegetation and climate are shown in Fig. 8-6. Within this region itself, however, stable forest and grassland communities have remained side by side for thousands of years, delineated by variable soil and geological factors. These factors often create sharp lines of demarcation between areas environmentally suited to either forest or

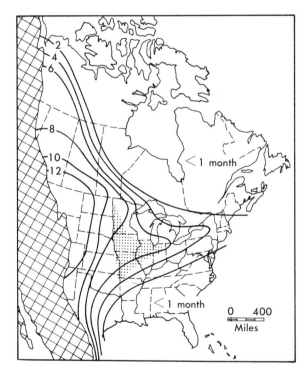

Fig. 8-6. Diagram of wind patterns in the Middle West in relation to the "prairie peninsula" region (shading). The numbers at the upper left indicate average number of months per year with mean air flow from the eastern base of the Rocky Mountains (cross-hatching). In eastern Canada and in the southeastern United States the yearly average is less than 1 month. (Modified from Borchert, J. R. 1950. Ann. Assoc. Am. Geog. **40**:1-39.)

grassland. The location of a "tree line" in the prairie border is difficult to place, since forest species are able to penetrate deep into the grassland climate as a result of geological dissection and stream topography, which provide a water supply ample for growth. With the decreasing precipitation to the west, there is a gradual dropping out of forest genera until only several species of oak, hickory, and ash occur at the extreme distribution limits of eastern forest trees. In addition, the most persistent species are reduced to shrub or small-tree status. The climatic criterion for cessation of closed forest communities in exposed upland situations is a diminishing rainfall gradient that approaches the 60- to 70-centimeter (25- to 30-inch) isohyet.

The forest biome in the United States occurs across a climatic gradient that is broad in terms of both temperature and rainfall. This gradient is modified by paleoecological events, geological materials, and soils. Heterogeneity of vegetational development in terms of composition, density, and size of mature individuals would be expected. Several regional divisions or associations of the biome are commonly employed. Population gradients would suggest difficulties in delineating various associations because of the continuous variation in composition. However, central tendencies of forest tree assemblages may be identified on a broad geographical basis; between these assemblages, intermingling and shifting dominance of species occurs. Three important designates that will be discussed are the mixed mesophytic, beech-maple, and oak-hickory forests.

Mixed mesophytic forests. Mixed mesophytic forest communities are best represented in the Cumberland Mountains of eastern Kentucky and Tennessee. The wide selection of sites available in this dissected topography results in a richly diverse assemblage of forest genera. Two typical forest trees that often attain large sizes in sheltered coves are the tulip tree and yellow buckeye. The climate of this Appalachian forest has remained unchanged for tens of thousands of years, as the region lies south of the glacial boundary. As a consequence, the area has had a long, uninterrupted history of plant occupation and protection.

Beech-maple forests. With climate amelioration following the recession of the last glacier, eastern forest species invaded the till plain. American beech and sugar maple became important dominants in this northern portion of the biome. Since the presence of these forests on the till plain is postglacial, dating from 10,000 to 15,000 years ago, the beech-maple communities are more youthful than those south of the glacial boundary. Toward the northeast, into Wisconsin and Minnesota, beech diminishes in abundance, and in its place basswood becomes an important associate of sugar maple. Farther south, these sugar maple–basswood forests occur only as postclimax types in coves, on north-facing slopes, and in similar situations where temperature and moisture relationships are more favorable for these species. In central Missouri, for example, large sugar maples of great age and with straight, massive boles still occur with ecological associates such as the basswood, red elm, and northern red oak in select topographic situations. These forests are protected from drying winds and are amply supplied with moisture stored in deep soil profiles.

Oak-hickory forests. The principal forest genera toward the west and extending into the drier portions of the biome are various species of oak and hickory. In the Ozarks

Fig. 8-7. View of deciduous forest in the Ozark region of Missouri. White oak is a principal species. (Courtesy United States Forest Service.)

region of Missouri and Arkansas and northward and westward through the forest-prairie transition, oak-hickory communities are the predominant types. A representative view of a mature white oak community is shown in Fig. 8-7. West and southwest into the drier grasslands, the outliers of the oak-hickory association in Oklahoma and Texas consist of only a few species, including post oak and blackjack. Diminishing rainfall and greater evaporation potentials are selective factors in the westward depletion of the eastern forest. Farther north and as far west as Wyoming, bur oak is the only species of eastern oak, occurring as a small tree or shrub. In western regions, lack of moisture and winter desiccation caused by cold winds are factors that restrict forest development.

Some of the mammals characteristic of the deciduous forest are the opossum, raccoon, gray squirrel, and gray fox as well as the mountain lion, bobcat, and black bear, which also occur in adjacent biomes. Bird species are abundant, as are reptiles and amphibians. The moist forest communities support salamanders and tree frogs. Insect life is varied and abundant. On the forest edge, animals, particularly birds, can be separated into eastern and western forms. The cottontail rabbit, woodchuck, and fox squirrel occur typically in the east. In the gulf coast states, certain animal populations appear to have their centers of concentration in the region and are represented by mammalian forms such as the cotton and rice rats and the swamp rabbit.

Grasslands

Temperate grasslands are widespread throughout the world, particularly in the Northern Hemisphere, in climates with strong continental influences. In addition to the prairies and plains of North America, other natural grasslands of the temperate zone include the steppes of Eurasia, the South African veld, and the pampas of South America. The principal features of a grassland climate are large annual temperature variations, high-evaporation potentials, and diminishing rainfall, particularly in comparison to deciduous forest climates. Whereas patterns of precipitation in the latter are generally predictable, those of the continental grasslands are less reliable, and drought is more common. Rainfall varies from 25 to 30 centimeters (10 to 12 inches) in drier portions of the grass-land biome to 100 centimeters or even more in regions of forest transition such as the easternmost parts of the "prairie peninsula" region. With only a few exceptions, grass-land climates generally experience the main precipitation period in the spring and early summer months, when the community produces most of its annual growth. The world's grasslands are among its most important resources in terms of tillable soils and native forages. They are also probably the most exploited in terms of intensive cultivation and overgrazing.

Several features of plant structure and ecological adaptation are characteristic of the grassland community. The dominant growth form is the grass plant. Within this grass-dominated system, numerous but less abundant herbaceous plants and various semishrubs also occur. Several species of trees are found sporadically, as they are restricted to areas that have a favorable soil-moisture balance. Prairie grasses and herbs develop and complete their cycle of growth and reproduction through staggered periods of the season. There are morphological variations as well, ranging from the deep-rooted habit of certain grasses and legumes to the shallow rhizomes, tubers, or bulbs of many other species. Temporal spacing of physiological activity and varying structural aspects enable numerous species to occupy common ground with minimal competition.

During dry periods or extended drought,* many grasses respond to moisture deficiences by cessation of growth; their aerial parts become dried and essentially nonfunctioning. Even during periods of temporary stress such as that occurring on a hot, windy day, the leaves will roll inward from the edges. This phenomenon is caused by the collapse of special cells (bulliform cells) and creates a less exposed leaf surface, thus

---

*We can describe drought as a phenomenon whose duration is a departure from the established norm, thus setting it apart from rainless periods that can be expected on the basis of long-time averages.

Fig. 8-8. Parkland communities of grass vegetation and pine forest in South Dakota. The bison, once numerous in North America, are limited to a few small herds.

reducing moisture loss. Most species in the native grassland are perennials; there are relatively few annual plants in undisturbed grassland communities. Based on the life-form spectrum, a majority of species are classified as hemicryptophytes, or half-hidden plants, that is, plants with renewal buds on the surface or in the soil (see Chapter 1). In addition to the grass family, several others, including the pea and sunflower families, have significant representation.

In North America the grassland is one of the most extensive biomes in the interior of the continent, ranging from the oak-hickory forest on the east to the base of the Rocky Mountains and the Black Hills region, where it merges in open, parklike situations with lower montane forests that consist mainly of ponderosa pine (Fig. 8-8). On the north it is in a parklike transition with the Canadian boreal forest. To the south, the grasslands extend to the Mexican deserts. Isolated grasslands in mountain valleys occur westward to the Palouse region of native wheat grasses in the Pacific Northwest.

Two major subbiome types or regional associations are readily discernible in the midcontinent. These are the tall grass prairies of the eastern sections, of which the "prairie peninsula" is an important transition zone with oak woodland, and the short grass plains in the western parts. Important grasses of the prairies are the big and little bluestems. The former is a coarse, deep-rooted species occupying lower slopes and swales. Its flower stalks reach 2 meters (6 to 7 feet) or more in height. The latter occurs

Fig. 8-9. Tall grass prairie in Missouri. A principal grass is big blue stem, shown in flowering condition. (Courtesy Missouri Conservation Commission.)

more prominently on upper slopes and shallower soils. Both are important forage species. There are other grasses, including Indian grass, wild rye, and switch grass. Slough grass and bluejoint may occur in wetter areas. In these tall-grass communities fire is an important ecological factor. It maintains a vigorous growth of grasses and arrests tree invasion from adjacent drainages. Where fire and other forms of utilization such as grazing by large herbivores are excluded for extended periods, grass species are gradually replaced by woody plants and various broad-leaved herbs. One factor in the deterioration of the grass community is the accumulation of litter, which over a period of years causes stagnation in growth of the principal grasses, declining competition, and an increasing dominance of other plant species. Most prairies have disappeared as they gradually are broken for cultivation, but some relics have been preserved. These provide opportunities for scientific studies and educational programs as well as for public enjoyment (Fig. 8-9).

Under the influence of diminishing rainfall within the grassland biome, plant composition and growth form change, so that short grasses are the principal species in the drier plains section. Here litter accumulation is less than in the tall-grass prairies, and trees are less prevalent; fire is not as crucial to grassland stability as it is in the more moist sections of the biome. In this steppelike climate of seasonal temperature extremes, low rainfall, and prevalent wind, buffalo grass and blue grama are well known. Before these areas were settled, these and other grasses were the mainstay of the large bison herds of the plains. Within this rainfall gradient, where it is modified by local soil and topographical conditions, there is a transition belt or mixed prairie, in which species from both east and west intermingle. A shift in geographical position from this zone marks the inception of the increasing dominance of one group and the "dropping out" of species from the other. Yet this middle zone has a certain identity of its own, with certain grass dominants such as the needlegrass and wheatgrass species.

Within the broad geographical extent of the grassland biome in North America, we would expect to find faunal differences such as those that have been observed in plant composition. This is illustrated by the occurrence of the white-tailed jack rabbit in the north and the black-tailed species in the southern region. Similarly, the desert cottontail rabbit on the western borders of the biome is a different type than the eastern cottontail.

The bison, pronghorn, badger, and coyote typically inhabit the North American grasslands. Before the massive westward migrations of settlers, the bison population was numbered in the millions. One estimate is as great as 60 million animals! Today the bison have been reduced to fragmentary herds. At one time the species was in actual danger of extinction because of the exploitive practices of the white man.

Animals that live in similar habitats throughout the world often function in similar ways, although taxonomically they may be quite unrelated. This tendency is best illustrated in the grasslands by several species of predators, including the coyote and cougar in North America; the red wolf and pampas cat in South America; the Cape hunting dog, lion, jackal, and cheetah in Africa; and in Australia the Tasmanian wolf, which is the only marsupial in this group of predators.

Deserts

Extensive regions of the earth are classified as desert. For the midlatitudes, most of northern Africa, trans-Arabia, parts of the Asian interior, Australia, southern Africa, the west coast of South America, and the United States and Mexican deserts make up an impressive aggregate. Together with tropical scrub and thorn vegetation, about one fifth of the earth's land surface is classified as arid. By definition, such regions receive no more than 25 centimeters (10 inches) of precipitation each year and frequently the average annual rainfall is less than that for long periods of time.

The reasons for deserts are varied. Mountain barriers, proximity to cold ocean currents, and remoteness from moisture are factors that are not limited to any latitudinal zone. Another significant influence is the subtropical convergence effect as a result of warm, dry, descending air masses. It is in these affected areas, on the poleward side of the tropical belts, that the great deserts are concentrated (see Chapter 2).

Plant and animal adaptation in the desert is perhaps more striking than that observed in any other environment. In the desert, where scant, irregular rainfall, typically high summer temperatures, drying winds, and rapid water losses are commonplace, a most critical factor in biotic survival is adaptation to a minimal water economy. Accommodations are revealed in a number of interesting ways.

In desert animals, estivation, nocturnal activity, and burrowing are manifestly well suited to reducing vapor losses, escaping the heat of daylight hours, and enjoying the higher humidities of the below-ground environment. The kangaroo rat, a familiar rodent of our desert Southwest, is able to exist without a visible supply of water at its disposal. It is capable of utilizing the water derived through the metabolism of carbohydrates and fats in the seeds that are its principal source of food. Another factor in the ability of many animals, including birds, reptiles, and herbivorous mammals, to survive in the desert is the physiological adaptation that permits removal and reabsorption of water from wastes in the digestive and urinary tracts. Rodents, well adapted to desert conditions, are numerous; examples are the deer mouse, desert wood rat, and kangaroo rat. Other desert inhabitants include the black-tailed jack rabbit, coyote, and kit fox. The collared peccary and the mountain sheep are characteristic species, as are the mule deer and the mountain lion. Lizards and snakes are frequently observed; the horned toad (a lizard) and the sidewinder rattlesnake are common and represent typical adaptations to desert living.

Adaptations exhibited by desert xerophytes are similarly varied. These may take the form of microphylly (small leaves) or succulence, which permits water storage and retention. Leaf shedding during drought, the development of waxy cuticles and sunken stomata, and protoplasmic changes resulting from dehydration and increased osmotic values of the cells are still other adaptive characteristics. Some species are deep rooted, relying on moisture from low levels, whereas others have shallow root systems that capitalize on the surface moisture of light showers before it can evaporate. Still other plant species might be called drought evaders. Examples are the annual species for which the deserts are well known (see Chapter 1 for distribution of life-form types in different climates). During periods of rain there is a flush of growth, and life cycles and seed production are rapidly completed. The parent plant dies, but the seeds are left in a state of dormancy and will germinate when conditions again become favorable for another brief period of growth.

In general there are two basic types of annual plants: those germinating in early summer and completing their life cycle the same year and those germinating in the fall but flowering and setting seed the following spring. The first are called summer annuals, the latter winter annuals. An interesting report by Mulroy and Rundel presents different photosynthetic strategies for each. The summer annuals are characterized by the so-called C_4 pathways of carbon fixation, and the winter annuals tend to be C_3 plants. For summer rainfall regions such as the Chihuahuan Desert, most of the annuals are of the former type. Because of a greater photosynthetic efficiency, these plants complete their life cycle for growth and seed production in a relatively brief period. The C_3 plants are mainly winter annuals in regions where conditions are not as acute during the critical period of development (cooler temperature and conservation of moisture). In the Mojave Desert, with mostly winter rain, all annuals reported were of the C_3 type.

In certain parts of the Northern Hemisphere, particularly in the Asian interior, there are extensive areas of dry mountain ranges, plateaus, and salt basins. Because of their comparatively high elevation and continental position, these regions experience extremes in seasonal temperatures. In summer, daily temperatures may average 15° C but reach highs of 30° C or even more, reflecting strong diurnal variations. These regions are isolated from moisture by great land distances, or they are walled in by even higher mountain ranges, so that the rainfall is meager, often less than 10 centimeters (4 inches) annually.

The vegetation is sparse, and the composition of these plant communities varies widely and depends on features of local soils, geology, and drainage patterns. Salt deserts are a common feature in basins lacking outlets, and chenopodiaceous species of varying salt tolerance such as saltbush, glasswort, and sea-blite are characteristic. Some salt plants, or halophytes as they are called, exhibit a succulent habit and maintain high osmotic pressures (effected by high salt content of the cells) as a means of resisting water losses to the environment.

Other important inhabitants of the desert-steppe complex include grasses and composites. One of these ecological entities is the Gobi Desert in central Asia, a vast landform ranging in elevation from 1000 to approximately 2000 meters. The nearest equivalent of the Gobi Desert (cold desert type) in the United States is the Great Basin desert in Nevada and parts of Idaho and Oregon, in which the sagebrush is a dominant species over wide areas.

Farther south in the United States in areas of lower elevations, average temperatures are much higher, and frost on a sustained basis is less common. Differences in soil texture, topographical position, and even the timing of seasonal precipitation are factors in the development of a variable vegetation pattern. Creosote bush is one of the most common shrubs, occupying desert flats throughout the warm desert region. The familiar sahuaro is our largest cactus. At maturity it may reach a height of 10 to 15 meters and weigh several tons. It is found mainly in Arizona. Unfortunately the numbers of this spectacular desert plant are decreasing. One factor is predator control, which has resulted in increased rodent consumption of the supply of seed available for natural regeneration. Disease, range deterioration as a result of domestic grazing, and excessive use of the plant in landscaping desert suburbia are still other factors in this species' decline.

The sahuaro occurs prominently at more elevated positions than creosote bush and on better-drained soils. These two species, then, become focal points within a topographical gradient, each with a spectrum of ecological associates (Fig. 8-10). Yet, as in most adjacent communities, considerable species overlap does occur. In the more western parts of the American desert, in southern California (Mojave) and Baja, most of the annual precipitation occurs during the winter. An especially notable desert plant of the Mojave region, and one that is largely restricted to this area, is the Joshua tree. In these areas the succulent character becomes less prevalent; the giant cacti are not native here. Eastward, in the direction of the Texas and Chihuahuan deserts where there is an increasing incidence of summer rains, there is a great variety of succulent types. An important group consists of the prickly pear and cholla cacti. Sisal, century plants, and lecheguilla are also conspicuous members of this easternmost section of the continental deserts. However, the

Fig. 8-10. Sahuaro cactus (top) and creosote bush communities, the latter characteristically on the lower ground and heavier soils. (Courtesy R. R. Humphrey, University of Arizona, Tucson.)

massive size observed in the Arizona and Sonoran tree cacti, including several other species besides the sahuaro such as the organ-pipe cactus, is absent here. For the reader interested in further study of the desert environment and its remarkable adaptations, several references are listed at the end of this chapter.

Brushlands and woodlands

On the poleward sides of steppes and deserts or at higher elevations, vegetation assumes a more woodland character. Increasing rainfall is a factor. Woody species are generally small, even scrubby, and in many situations shrub species may be the principal representatives. The criterion employed here to distinguish between woodland and forest is a greater spacing in woodlands, which produces noncontiguous canopies and a conspicuous herbaceous ground cover that can thrive because of resultant increased light levels at the ground. Examples are the scrubby taiga or lichen-woodlands of the boreal regions and the dry forest and tree savannahs of tropical environments, in which a grassy sublayer is common.

In the upper strata of these semiarid communities, either needle-bearing or broad-

Fig. 8-11. Foothill communities in the Great Basin desert of Nevada, consisting principally of piñon pine and juniper and an understory of grasses. (Courtesy United States Forest Service.)

leaved species may be represented as dominants. Examples of adaptive traits are sclerophylly (hard, thick leaves or needles), cutinized surfaces, and sunken stomata. Piñon pines and several juniper species form extensive foothill communities in our desert southwest, as shown in Fig. 8-11. The dry wood and brushland equivalent may be found in many regions. In Australia, gum (eucalyptus) trees form extensive communites subject to periodic fires and the uncertainties of rainfall.

A special vegetation within this general classification is the so-called Mediterranean type. These woody communities occur in warm, temperate regions with dry summers and winter rainfall. Five such climates are found only on the western sides of continental land masses in latitudes between approximately 30 and 40 degrees in both the Northern and Southern Hemispheres. These are the Mediterranean region of southern Europe and north Africa, the southwestern tip of Africa, southwest Australia, western South America, and the coastal region of southern California. In southern California extensive areas are occupied by sclerophyllous vegetation. Various oaks are important species of open woodland communities in the coastal ranges. Manzanita, snowberry, and chamiso are widespread in the more shrublike communities, or chaparral. Because of the dry summers, such vegetation is subject to periodic burning. Fire may be a selective factor in the development of chaparral and its ecological equivalents in the other Mediterranean climates where it is also a recurring factor. Many species of these areas exhibit fire resistance through the survival of their underground systems, even though the aerial portions of the plant are frequently killed. Top growth is resumed, however, by prolific sprouting from these subterranean structures. Various other species, not necessarily sprouters, maintain themselves as abundant seed producers. In a California study Wells showed that the principal seed-producing plants were nonsprouters and that the genera represented had differentiated with the largest number of species. This suggests that the sprouting characteristic is a conservative trait when compared to sexual reproduction (seed production) and the subsequent increase in possibilities for the evolution of more species. It is thought that the traditional use of fire by humans in some regions may be an important factor in the development and maintenance of the "Mediterranean" type.

Coniferous forests

The temperate coniferous communities are ecological intermediates between the semidry woodlands and the cold-climate forests. Conditions of increased rainfall and moderated temperatures provide suitable environments for the development of extensive forests that are of great commercial importance. As a consequence, these forests are subject to harvest and silvicultural management the world over.

In the western United States and throughout the montane, these forests present economic opportunities that are unexcelled anywhere. The spectrum of species diversity and community structure includes the well-publicized coast redwoods, Sierrra redwood, Douglas fir, western red cedar, western white pine, and Sitka spruce of the Pacific Northwest; the ponderosa pine is the most widely distributed coniferous species of the western montane (Fig. 8-12). Animals characteristic of this region have been discussed in the previous section on coniferous forests (pp. 193-194).

Fig. 8-12. Old-growth stand of ponderosa pine, Long Valley Experimental Forest, Flagstaff, Ariz. (Courtesy United States Forest Service.)

Eastern tree species include the white and red pine. Throughout the south, several species that form extensive forests range in ecological adaptability from the loblolly pine of wetter areas to the drought-tolerant, fire-resistant longleaf pine on sandy flats. Another relative is the shortleaf pine that is found in a wide variety of sites but generally in the dry upland. Unlike the western conifers, the eastern species are likely to constitute seral communities that are followed in the normal course of succession by deciduous forest. Their degree of permanence in many situations depends on various factors, including periodic fire and/or the presence of droughty soils. In the west the temperate conifers are often the terminal community, as there is no universal broad-leaved type as a potential successor.

The pines are the most widely distributed and most numerous coniferous species. Like the oaks, they are largely restricted to the midlatitudes of the Northern Hemisphere. In Europe and Asia Scotch pine is an important forest constituent (also in the boreal forest) and is the most widely distributed of any pine species. In eastern Asia a group of

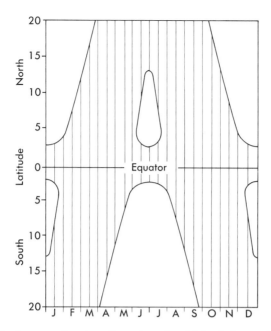

Fig. 8-13. Schematic diagram showing alternation of wet (shaded areas) and dry (unshaded areas) seasons in tropical zones. Note occurrence of two dry seasons, one characteristically longer than the other.

temperate conifers not native to America includes such forest trees as cryptomeria and dawn redwood.

These coniferous genera are essentially unknown below the equator, but there are ecological equivalents. The so-called southern conifers form communities generally comparable to the conifers of the Northern Hemisphere in their latitudinal range and ecological requirements. Such forests and savannah types occur in Eastern Australia, Asia, and South America. Two coniferous genera are fairly well known. These are the plum pines and the auracarians. The plum pines have the more northerly distribution.

TROPICAL-REGION BIOMES

The tropical-region biomes are as follows (Fig. 8-1):

> Forests—*13, 14*
> Savannah and desert scrub—*15, 16*

It is well known that freezing temperatures do not occur at lowland elevations in the equatorial regions. Variations in seasonal temperatures are relatively small in comparison to fluctuations recorded in temperate regions. Yet in extensive areas of the tropics there is indeed a pronounced seasonality manifested in plant response and animal activities. This is a result of the wide variation in the amount and monthly distribution of rainfall.

For the tropical latitudes as a whole there is a general pattern of wet and dry seasons.

Fig. 8-13 illustrates these relationships and also the relative duration of each as a function of increasing latitude. Note that the dry period expands as distance from the equator increases, so that aridity becomes even greater.

Forests

There are three well-defined regions of equatorial rain forest in Africa, Asia, and South America, indicated by Fig. 8-2, *13*. Here vegetation has evolved in its most complex and diversified form. Many species are broad-leaved, and although there is perpetual dropping of leaves, the community retains its evergreen appearance. Rainfall generally occurs in every month, and annual accumulations may be as great as 500 centimeters (200 inches) or more. Everywhere these magnificent forests are being rapidly depleted or even destroyed.

Traditionally forests were cleared and burned for a "shifting" type of cultivation. Family-size units of land are periodically cleared for food production. Owing to the intensive leaching and rapid decay of organic matter under open conditions, soils become depleted of nutrients in a few years. Such areas are abandoned for a time, and successional development of new vegetation takes place as adjacent forests are cleared for crops and the process is repeated. Exposure of such soils often leads to intensive weathering and the formation of laterites (see Chapter 3). Land clearing is a destructive process not only in terms of soil and minerals but also from the standpoint of plant production. A 1971 report by the Food and Agriculture Organization of the United Nations indicates that the results of such practices rarely compensate for the value of the timber that is lost.

What are the characteristics of a typical rain forest? What are its outstanding features? Are there differences from one continental biome to another? Unlike temperate-zone forests, rain forests usually contain a greater array of forest species, and species dominance is less pronounced. There are also many more epiphytes and vines. A view of the forest edge is shown in Fig. 8-14. These highly diverse forest systems occur over wide areas on what might be described as normal soil development. Where strong local influences such as river-bank flooding or poor drainage occur, there is a trend toward fewer species and the emergence of a select few as dominants. A consideration of the principal taxa and their sociological relationship within the forest community is beyond the scope of this book. Each rain-forest region (in America, Africa, and Asia) is unique in its own right in terms of biotic representation and historical and geographical variables, but where undisturbed communities still exist, all share a complexity of structure and a great diversity of species. For more detailed information, several references are cited at the end of this chapter; the work by Richards is most instructive.

In those tropical areas of diminishing rainfall the evergreen character of the true rain forest gives way to a more deciduous habit and a decreasing diversity of species. In these "monsoon" types the tropical dry season induces leaf shedding. This is exemplified in such communities as the teak forests of Burma. With increasing dryness, forests become more sparse and eventually assume the character of open woodlands. Leguminous trees become relatively more important as the populations of other families more characteristic of rain forests decline. Large areas of deciduous woodland, which receive as little as 50 to

Fig. 8-14. Plant diversity in a rain forest in southwestern Costa Rica. Annual rainfall in this region is approximately 450 centimeters (about 180 inches).

100 centimeters (20 to 40 inches) of rainfall annually, occur in eastern Brazil, south-central Africa, and Asia.

Speciation of animals in the tropical forests has paralleled that of the plants; there are a strikingly large number of forms living in a diversity of habitats. Animals are distributed vertically in a rain forest much as they are in a lake. As many as six strata from above the forest canopy to the ground level have been identified. Rain forests of the wet tropics are characterized by arboreal animals that seek food and shelter at different levels in the canopy. Monkeys, the orangutan, gibbons, tree shrews, fruit-eating bats, colorful birds, various reptiles, amphibians, and a host of insects are combined in a diverse array of niches and functional relationships. Monkeys, amphibians, and reptiles have adapted to the tree habitat through structures such as prehensile tails and suction pads on fingers and toes. Indeed certain frogs are so well adjusted that they lay their eggs in trees!

Savannah and desert scrub

Savannah vegetation becomes more prevalent along a continuing gradient of diminishing rainfall from forest environments. This community type is essentially a grassland, in

Fig. 8-15. Acacia savannah in Serengeti National Park, Tanzania. The bottom photograph is of the burned grasses in the dry season. (Courtesy Dennis Herlocker, Serengeti Research Institute, Arusha, Tanzania.)

Fig. 8-16. Herd of zebra, Serengeti National Park, Tanzania. During the dry season, animals move from the short grass plains to the taller grass areas and more wooded sections where forage and water are available.

which there are varying densities of tree species. Acacias are an important tree component in many savannah regions, lending a characteristic aspect to many landscapes around the world (Fig. 8-15). Where adequate fuel is provided through grass growth, fires are an almost universal phenomenon. If fires are excluded as a factor, trees often become more numerous, suggesting that fire is of importance as a selective agent in the maintenance of the grass vegetation. As rainfall diminishes even further, tree-savannah communities yield to shorter grasses and a plainlike aspect that is essentially devoid of trees. These grassy areas of the tropics are important sources of forage for both native herbivores and livestock.

On the East African savannah there is a diversity of grazing ungulates that is unrivaled anywhere in the world. In the Serengeti region, there are large concentrations of gazelles, zebra, and wildebeests. Many other species are found in fewer numbers, including the giraffe, eland, and hartebeest. Generally these animals migrate annually in response to the seasonal availability of forage and water (Fig. 8-16). Associated with these herds is a host of predators, including the lion, cheetah, leopard, hunting dog, hyena, and jackal (Fig. 8-17). In the more wooded savannahs and along the water courses, elephant, rhinoceros, buffalo, impala, and dik-dik are to be found.

Fig. 8-17. Zebra kill, Serengeti National Park, Tanzania. Large carnivores and scavengers (left) are important consumers in this grassland food chain.

This notably rich fauna faces increasing competition from agriculture and domestic animals for water and grass. The idea of harvesting wild animals on a sustained-yield basis as food for people has been given considerable attention, and this may well be the solution to saving the once-great herds of herbivores from extermination.

Tropical scrub exists under conditions of minimal rainfall, as little as 10 to 15 centimeters annually. Adaptations to a stringent water economy are well developed. Many species are spinescent, or thorny, and succulence is common. The grass cover is poorly developed under these conditions of unreliable as well as marginal amounts of rainfall. Some plants adapted to these harsh conditions of high temperature and low moisture remain dormant for extended periods of time. The tropical scrub *caatinga,* found in northeast Brazil, is typical of this community type. Its analogs occur within every tropical belt.

SUMMARY

A remarkable pattern of life-form units, or biomes, has evolved across a wide spectrum of seasonal temperatures and rainfall values. Each has developed its own unique biota of plants and animals and its own environmental equilibrium. Adaptations are expressed in numerous ways. People potentially have a use for each region. The application of sound management procedures in the use as well as in the perpetuation of these biomes is based on these inherent characters. A satisfactory method for one locale may be detrimental in another. We can cite numerous examples of misuse, from the large-scale land clearance in the wet tropics to road building on thin tundra soils. The utilization of biome resources over time will be more successful if the ecological dictates by which the evolution of each biome has been shaped are taken into account.

DISCUSSION QUESTIONS

1. The terrestrial biomes range in usable resources from sparse tundra lands to intricately complex rain forests. Discuss the dictates of their ecological value and judicious land-use practices.
2. Discuss the ecological effects of fire in grasslands. What other biome type exhibits a fire-selection history?
3. The Alaskan pipeline has been completed. What are some adverse effects on the tundra ecosystem visualized by environmentalists?

REFERENCES

Alexrod, D. I. 1966. Origin of deciduous and evergreen habits in temperate forests. Ecology **20**:1-15.

Billings, W. D. and H. A. Mooney. 1968. The ecology of Arctic and alpine plants. Biol. Rev. **43**:481-529.

Black, R. F. 1950. Permafrost—a review. Geol. Soc. Am. Bull. **62**:839-856.

Bliss, L. C. 1956. A comparison of plant development in microenvironments of Arctic and alpine tundras. Ecol. Monogr. **26**:303-337.

Borchert, J. R. 1950. The climate of the central North American grassland. Ann. Assoc. Am. Geog. **40**:1-39.

Britton, M. E. 1967. Vegetation of the Arctic tundra. In H. P. Hanson (ed.). Arctic biology. Oregon State University Press, Corvallis, Ore.

Burtt, B. D. 1942. Some East African vegetation communities. J. Ecol. **30**:65-283.

Campbell, D. H. 1944. Relations of the temperate floras of North and South America. Calif. Acad. Sci. Proc. **25**:139-146.

Churchill, E. D. and H. C. Hanson. 1958. The concept of climax in Arctic and alpine vegetation. Bot. Rev. **24**:127-192.

Costin, A. B. 1967. Alpine ecosystems of the Australasian region. In H. E. Wright and W. H. Osburn (eds.). Arctic and alpine environments. Indiana University Press. Bloomington, Ind.

Daubenmire, R. 1953. Alpine timberlines in the Americas and their interpretation. Butler Univ. Bot. Studies **11**:119-136.

Golley, F. B. and E. Medina (eds.). 1975. Tropical forest ecosystems. Springer-Verlag New York, Inc. New York.

Hansen, H. C. 1953. Vegetation types in northwestern Alaska and comparisons with communities in other arctic regions. Ecology **34**:11-140.

Hedberg, O. 1964. Features of afroalpine plant ecology. Acta Phytogeog. Suecica **49**:1-144.

Holdgate, M. W. 1967. The Antarctic ecosystem. Royal Soc. Lond. Philosoph. Trans. S. B. **252**:363-383.

Hustich, I. 1953. The boreal limit of conifers. Arctic **6**:149-162.

Igoshina, H. N. 1969. Flora of the mountain and plain tundras and open forests of the Urals. In B. A. Tikhomirov (ed.). Vegetation of far north of the U.S.S.R. and its utilization. Israel Program for Scientific Translations. Jerusalem.

Köppen, W. 1936. Cited in C. P. Wilsie (ed.). 1962. Crop adaptation and distribution. W. H. Freeman & Co., Publishers. San Francisco.

Llano, G. A. 1965. The flora of Antarctica. In T. Hatherton (ed.). Antarctica. Praeger Publishers, Inc. New York.

Llano, G. A. (ed.). 1972. Antarctic terrestrial biology. Antarctic Research Series Vol. 20. American Geophysical Union. National Academy of Sciences, Washington, D.C.

Marr, J. W. 1948. Ecology of the forest-tundra ecotone on the east coast of Hudson Bay. Ecol. Monogr. **18**:117-144.

Maycock, P. F. and J. T. Curtis. 1960. The phytosociology of boreal conifer-hardwood forests of the Great Lakes region. Ecol. Monogr. **30**:1-35.

Meggers, B. J. et al. 1973. Tropical forest ecosystems in Africa and South America: a comparative review. Smithsonian Institution Press. Washington, D.C.

Michelmore, A. P. G. 1939. Observations on tropical African grasslands. J. Ecol. **27**:282-312.

Monk, C. D. 1966. An ecological significance of evergreenness. Ecology **47**:504-505.

Monk, C. D. 1968. Successional and environmental relationships of the forest vegetation of north central Florida. Am. Midl. Nat. **79**:441-457.

Mooney, H. A. and E. L. Dunn. 1970. Convergent evolution of Mediterranean-climate evergreen sclerophyll shrubs. Evolution **24**:292-303.

Mulroy, T. W. and P. W. Rundel. 1977. Annual plants: adaptations to desert environments. Bioscience **27**:109-114.

Nichols, G. E. 1935. The hemlock-white pine-northern hardwood region of eastern North America. Ecology **16**:403-420.

Pewe, T. L. 1966. Permafrost and its effects on life in

the north. In H. P. Hanson (ed.). Arctic biology. Oregon State University Press, Corvallis, Ore.

Quaterman, E. and C. Keever. 1962. Southern mixed hardwood forest: climax in the southeastern coastal plain, U.S.A. Ecol. Monogr. 32:167-185.

Rawitscher, F. 1948. The water economy of the vegetation of the "Campos Cerrados" in Southern Brazil. Ecology 36:237-268.

Remmert, H. and K. Wunderling. 1970. Temperature differences between arctic and alpine meadows and their ecological significance. Oceologia 4:208-210.

Sakai, A. 1970. Mechanism of desiccation damage of conifers wintering in soil-frozen areas. Ecology 51:655-664.

Shantz, H. L. 1954. The place of grasslands in the earth's cover of vegetation. Ecology 35:143-145.

Stoddart, L. A. 1941. The Palouse grassland associations in northern Utah. Ecology 22:158-163.

Stonehouse, B. 1971. Animals of the Arctic. The ecology of the Far North. Holt, Rinehart and Winston, Inc. New York.

Tedrow, J. C. F. and H. Harries. 1960. Tundra soil in relation to vegetation, permafrost, and glaciation. Oikos 11:237-249.

Tikhomirov, B. A. 1969. The interrelationships of the animal life and vegetation cover of the tundra. Israel Program for Scientific Translations. Jerusalem.

Transeau, E. N. 1935. The prairie peninsula. Ecology 16:423-437.

Vesey-Fitzgerald, D. F. 1963. Central African grasslands. J. Ecol. 51:243-273.

Vipper, P. B. 1968. Interrelations of the forest and the steppe under the conditions of the mountainous country in south-western Transhaikalia. Bot. Zh. 53:491-504.

Weaver, J. E. and W. E. Bruner. 1954. Nature and place of transition from true prairie to mixed prairie. Ecology 35:117-126.

Welch, J. R. 1960. Observations of deciduous woodland in the eastern province of Tanganyika. J. Ecol. 48:557-573.

Wells, P. V. 1968. The relation between mode of reproduction and extent of speciation in woody

genera of the California chaparral. Evolution 23:264-267.

Zaborski, B. 1955. U.S.S.R., Central Siberia. In G. H. T. Kimble and D. Good (eds.). Geography of the northlands. John Wiley & Sons, Inc. New York.

ADDITIONAL READINGS

Allee, W. D. et al. 1949. Principles of animal ecology. W. B. Saunders Co. Philadelphia.

Barnard, C. 1964. Grasses and grasslands. Macmillan Inc. New York.

Braun, E. L. 1950. Deciduous forests of eastern North America. The Blakiston Co. Philadelphia.

Brown, G. W., Jr. 1968. Desert biology. Academic Press, Inc. New York.

Chapman, V. J. 1860. Salt marshes and salt deserts of the world. John Wiley & Sons, Inc. New York.

Dansereau, P. 1957. Biogeography, an ecological perspective. The Ronald Press Co. New York.

Jaeger, E. C. 1967. The North American deserts. Stanford University Press, Stanford, Calif.

Kendeigh, S. C. 1961. Animal ecology. Prentice-Hall, Inc., Englewood Cliffs, N.J.

Knight, C. B. 1965. Basic concepts of ecology. Macmillan, Inc. New York.

Larson, P. 1970. Deserts of America. Prentice-Hall, Inc. Englewood Cliffs, N.J.

Life Nature Library, a guide to the natural world. Vols. 1 to 25, 1961-1965. Editors of Life. Time, Inc. New York.

MacArthur, R. H. 1972. Geographic ecology. Harper & Row, Publishers. New York.

McGinnis, W. G. 1968. Deserts of the world. University of Arizona Press. Tucson, Ariz.

Richards, P. W. 1964. The tropical rain forest. Cambridge University Press. New York.

Walton, K. 1969. The arid zones. Aldine-Atherton Publishing Co., Inc. Chicago.

Weaver, J. E. 1954. North American prairie. Johnsen Publishing Co. Lincoln, Neb.

Weaver, J. E. and F. W. Albertson. 1956. Grasslands of the Great Plains. Johnsen Publishing Co. Lincoln, Neb.

9 The world ocean

THE LAST DOMAIN

The oceans cover about 130 million square miles, or 71% of the earth's surface. Ranging from a comparatively thin layer over continental shelves to deep trenches whose depth is measured in miles, the world ocean comprises an estimated 330 million cubic miles of water. Such a quantity is difficult to comprehend. If these blocks of water, 1 mile on a side, were laid end to end, they would span three times the distance from the earth to the sun. Despite its vastness, the ocean is a finite entity vulnerable to the debilitating effects of pollution and to the overexploitation of its marine life.

The traditional human habitat is the land. As our numbers grow, the pressures we exert on terrestrial ecosystems also increase. Thus the battle for living space, minerals, and food will be joined in an ever more urgent confrontation with the sea. Visualized as a rich domain, the ocean looms pervasively as a safety valve for the population explosion and a safeguard against future starvation. The seas will be farmed, it is said. Modern methods will be employed. There will be purpose and planning, much as in the approach to problems of terrestrial agriculture. Selection for high yields, fast growth, and short life cycles, chemical control of pest and predator, and fertilization of crops will be practiced. There will be sea pastures and feeding enclosures for preferred fishes and other edible species.

Such optimism demonstrates confidence in our technology, but are our expectations fully justified? Are there not ecological values and functional properties of the ocean that impose adamant restraints on the extent to which it may be manipulated? Are there indeed limits to the oft-cited "bounty of the sea"? A recent report by Holt on the state of ocean fisheries points up the finite character of their productive potential. Many stocks that were considered to be underfished as recently as 1949 are overexploited and in danger of being depleted today. The anchoveta fishery off the coast of Peru is one example, and the once abundant tuna is another. A list of these stocks is given in Table 9-1, indicating that no part of the world ocean has escaped the pressures of contemporary harvests. Large baleen whales of the southern ocean, a source of many products, are being hunted to the point of extinction. The blue whale, largest mammal on earth, is a dramatic case in point. Species in the oceans of the Northern Hemisphere have been in a state of serious decline for many years. However limitless we would like to imagine the ocean as being, our use of it indicates that it is not immune to overuse.

Today there are more hungry people than ever before in our history. It is incumbent on us to correct this problem, if at all possible. Under this pressure, we tend to rationalize the exploitive practices seen all around the world. We must remember, however, that although domestication and technological development increase the food supply to a point, these practices also reduce diversity. As a result, nature's checks and balances, manifest in predator-prey and host-parasite relationships, are also eliminated. In their place we devise, as poor substitutes, chemical controls for prolific pests and disease organisms. By using them, however, we also poison our food and the environment around us. It is conceivable that if people were removed from the biosphere, together with those domesticated plants and animals that have no survival value in the untended state, the rest of life would still prevail. Could we say the same if the reverse were true?

Table 9-1. Ocean fisheries that have become depleted by excessive harvests since 1949*

STOCK	LOCATION
Anchoveta	Peruvian coast
Cod	North Atlantic
Flounder	North Atlantic
Herring	North Atlantic
Ocean perch	North Atlantic
Hake	Atlantic midlatitudes†
Pilchard	South Atlantic, African coast
Tuna	All oceans

*Modified from Holt. S. J. The food resources of the ocean. Copyright © 1969 by Scientific American, Inc. All rights reserved.
†Depletion of cod and herring began much earlier in the North Sea. Plaice and haddock, not included above, were overfished on both sides of the North Atlantic before 1949, as was hake in the waters around the British Isles.

An acceptance of the real world and its finite character is a first rule of ecology if there is to be a successful man-environment relationship. The growing dominance of any one species, in this case ourselves, subverts the stability of the entire system. The ocean is no exception. We know that its physical domain is impressive, that it is indeed rich in certain minerals and other resources, and that its food potential can no doubt be expanded. But do we know also that the ocean, *whose interplay with terrestrial functions is a necessity for the preservation of the biosphere,* can be destroyed as a living system? Without the ocean, ours would be a sterile planet.

STRUCTURE AND PROCESS

Let us briefly consider some of the descriptive and functional aspects of the ocean biome so that we may develop a better appreciation of its limitations as well as its potentials. The shallower waters are classified as the *neritic zone.* This is the inshore region overlying the continental shelf, shown in Fig. 9-1, a composite diagram. Generally its depth is less than 200 meters (approximately 600 feet). This figure serves as a kind of dividing line between the so-called continental seas and the deep, relatively steep-sided oceanic basins. These basins constitute the region of open ocean, or *oceanic zone,* in which there are even deeper trenches, some with a depth in excess of 35,000 feet. There is a vertical subdivision of the ocean as well: the *pelagic,* or open water, *zone* and also the bottom, or floor, of the ocean, called the *benthic zone.* The deep sections of the ocean constitute 90% of the total surface area; about 10% may be classified as coastal water. A very small area, less than 1%, is designated as zones of upwelling.

Photosynthesis in the sea is restricted to a relatively shallow surface layer. Light diminishes in intensity as it penetrates the water, finally reaching a depth at which attenuation is complete. The depth at which the intensity of light is reduced to 1% of its surface intensity is defined as the *euphotic zone.* This depth will, of course, vary from place to place in the ocean, since it depends on the transparency of the water and the angle

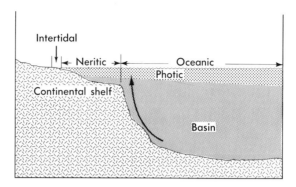

Fig. 9-1. Diagrammatic representation of the ocean profile. The neritic zone, which is the most productive part of the sea, constitutes less than 10% of the total area. The zone of up-welling, indicated by the ascending arrow, is also fertile but very limited in extent.

of incidence of the impinging radiation. The attenuation of light is shown in the following expression:

$$\frac{I_D}{I_0} = e^{-kD}$$

where

I_D = Intensity at depth D in some selected unit of measurement such as meters or feet
I_0 = Intensity at zero depth
e = Base of natural log = 2.71
k = Extinction factor, or percent reduction in intensity per unit D

From this we see that the intensity of light remaining is inversely related to depth and to the extinction value. The greater the screening effect (as a result of suspended particles or pollutants), the higher the extinction value and thus the shallower the depth at which a given light reduction occurs. As visible light passes through water, its spectral composition, or wavelength, also changes. The red, or long, wavelengths are screened out first, followed progressively by the shorter wavelengths. Blue light penetrates to the greatest depth, giving the transparent ocean its characteristic color.

In highly transparent waters the euphotic zone extends to a depth of 100 meters or more. If we use 100 meters as a basis of calculation, only about 3% of the total ocean volume would be included in the euphotic zone. This small fraction points up how critical this surface region is as a source of potential photosynthesis by marine plankton—the diatoms, dinoflagellates, and other microscopic algae. We might assume that the figure of 3% is really a high estimate, since not all of the ocean's surface waters are of equal clarity. Coastal waters are more turbid and are subject to more of the pollution effects from land. Yet as we will see in the next section, the inshore waters are also commonly the most productive. The clear blue waters that have a high degree of transparency are lacking in the nutrients that, in addition to light, must also be present.

Thus there is a delicately balanced interplay in the logistics of nutrient supply and

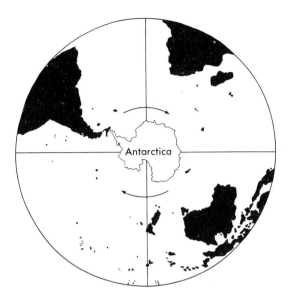

Fig. 9-2. Antarctic Current showing global circulation from west to east. This southernmost of oceans is the last refuge of the great blue whale, the largest living creature. (Modified from von Arx, W. S. 1957. In L. H. Ahrens et al. (eds.). Physics and chemistry of the earth. Pergamon Press, Ltd. Oxford, England.)

adequate light penetration for the development of high-yielding food chains. When either nutrients or light is inadequate, the system has a low potential. As the oceans become more polluted from a wide variety of substances, including oil, the integrity of the photosynthetic zone is expected to diminish. Without this life-giving layer of green plants, the ocean's remaining biota would cease to exist, as we know it.

Not all life, of course, is limited to these zones of active plant metabolism. There are numerous and varied forms of pelagic animal life that are found at considerable depths in the ocean. Others live on the bottom (benthic species) and feed on dead organic matter of varying particulate size that drifts down from above. Still others such as the bizarre lantern fishes and various crustaceans live in deep water by day for protection from predators and come to the surface to feed under cover of darkness. These massed organisms constitute a deep-scattering layer (or layers) that is so described because sound waves emitted from scientific instruments on surface vessels are returned as a soft, or diffuse, echo. It has been discovered that these pulses are reflected by aggregated layers of marine life rather than by the ocean bottom itself. There is yet much to be learned about life in the deep.

Ocean waters have a pattern of circulation that is maintained by global rotation and the prevailing winds. These vast directional flows would be similar to those of atmospheric currents if there were no land masses, a fact shown by the Antarctic Current, which flows from west to east around the globe where no extensions of continents block its path (Fig. 9-2). The temperature of ocean currents influences the moisture-retaining capacity of accompanying air masses, for as air is cooled, its ability to retain water decreases. On warming, it expands; and its moisture-retentive capacity increases. Air with a temperature of 50° C at saturation could hold approximately 10 times as much water as it could at 10° C. These relationships for a barometric pressure of 76 centimeters of mercury are as follows:

TEMPERATURE (°C)	SATURATION VAPOR PRESSURE (SVP) (mm Hg)
0	4.75
10	9.14
20	17.37
30	31.52
40	54.86
50	91.95

Impinging winds in turn affect, in some cases quite strongly, the precipitation pattern of adjacent land masses over which the air passes. Cold waters might be expected to bring about dry climates. As these cold currents underrun air of warmer temperature, the convective process leading to cloud formation at upper levels is curtailed. Foggy conditions near the surface, however, are common, but little condensation or precipitation results. A good example is the Peruvian, or Humboldt, Current that flows northward along the west coast of South America, causing a pronounced aridity in this region. The temperature of the ocean surrounding the Galapagos, situated on the equator, is lowered 10° to 15° by this flow of water from the southern ocean. The so-called Gulf Current, which actually originates off the northern coast of South America, circulates warm waters and carries a moderating influence into the high latitudes of western Europe. In the Pacific the Japan Current has similar effects on the coastal region of southern Alaska, where temperatures are considerably warmer than the latitude might lead one to expect. In both regions there is a considerable annual precipitation.

The phenomenon of coastal *upwelling* occurs on the western sides of continents. This is the upward movement of cold water from lower depths to replace the warmer surface waters that are being driven away from the land by the prevailing winds. This upwelling process carries mineral nutrients into the euphotic zone (Fig. 9-1), where they are available to marine plankton. It is this combination of nutrient-rich waters at the surface where light is adequate for photosynthesis that provides the elements necessary for a highly productive food chain. When prolonged lulls interfere with the replacement of warm water by nutrient-loaded cold water that can hold more oxygen than warm water, there is a spectacular die-back in populations of plankton as well as fish. The deleterious effect extends to the fish-eating sea birds. Off the coast of South America such a phenomenon of wind failure allowing warmer, nutrient-poor waters to spread southward from the equator is called "El Nino." In such years the deposition of bird guano is measurably reduced, and fish-eating birds such as cormorants perish by the thousands. These results indicate how the disruption of the food chain at one point can adversely affect it at another.

A more significant factor in reduced guano production is overfishing. According to Orr and Marshall, more than 20 tons of fish taken by sea birds is required to produce 1 ton of guano that is then deposited on the land. This estimate takes into account the excrement that is lost at sea. The same quantity of fish, however, will yield 3 tons of fishmeal. With this differential it is not difficult to see why fishing will prevail over the guano fertilizer business. The international balance of payments, if the past record is any witness, is not based on the stability of food chains or on other ecological values. Here is an example of man substituting himself for still another consumer—in this case the millions of sea birds of this coastal ecosystem.

The ocean is a storehouse of chemical wealth. Nearly all the earth's elements have been identified in it. This is not an unexpected finding, since all the materials in the exposed part of the earth's crust are subject to erosion and ultimate deposition in the sea. Some elements occur in large concentrations per unit volume of water and others as traces. In addition, the sediments on the ocean floor also contain a part of the total.

The saline character of the ocean is well known. The concentration of dissolved salts, occurring in various combinations as chlorides, sulfates, bromides, and bicarbonates, is about 3.5% by weight of water. Scientists conventionally express salinity as units of salt per thousand units of water. This, of course, provides a figure that is 10 times the previous percent and is written as 35 ‰ S. Common table salt, or sodium chloride (NaCl), comprises nearly four fifths of the total dissolved material. In solution it dissociates as positive and negative ions as per the following equation:

$$NaCl \rightarrow Na^+ + Cl^-$$

Quantities per unit volume for the five most abundant elements are shown in Table 9-2. For NaCl alone, the yield of salt is about 140 million tons per cubic mile of seawater. This large amount of dissolved chemicals in the ocean, equivalent to about 1.3 pounds in each 5 gallons, is an obvious deterrent to the use of seawater to make agriculture possible in arid lands. In Chapter 3 it was noted that a salt load exceeding 300 ppm for river water used in desert irrigation is damaging to sensitive crop plants. This concentration is less than $1/100$ the amount found in seawater. Thus the successful use of seawater for the irrigation of arid lands will require some method of desalinization. One such method employs freezing. At present, however, the use of seawater for either municipal or agricultural purposes is limited because the cost of extracting the salt is prohibitive and more economical methods are required for large-scale operations.

We have all observed that plants wilt and even die when subjected to high salt concentrations in water or soil. Under such conditions, osmotic potentials are high in the plant's

Table 9-2. Five most plentiful elements in seawater*

ELEMENT	AMOUNT (tons/mile³)
Chlorine (Cl⁻)	91,136,000
Sodium (Na⁺)	49,920,000
Magnesium (Mg⁺⁺)	6,400,000
Sulfur (SO₄⁻²)	4,200,000
Calcium (Ca⁺⁺)	1,900,000

*Modified from Goldberg, E. D. 1963, In M. N. Hill (ed.),. The sea. John Wiley & Sons, Inc. New York.

external environment. Water is effectively "sucked out" of nonadapted plant cells by simple diffusion. The concentration of water molecules per unit volume is higher inside the cell than outside, so the water molecules tend to move toward the lower concentration that is caused by the presence of dissolved salts or ions in the external medium, either water or soil.

The osmotic pressure of a 1 molar solution of undissociated molecules such as sugar is equivalent to approximately 1700 centimeters of mercury in an evacuated column, or slightly more than 22 atmospheres of "suction." Since this value is proportional to the degree of ionization, a completely dissociated 1 molar solution (one in which all the molecules are separated into positive and negative ions) would have twice as much osmotic pressure, the equivalent of 44 atmospheres. Since the average concentration of dissolved, or highly dissociated, salts in seawater is about a 0.5 molar solution, we can see that its osmotic value is equivalent to approximately 22 atmospheres of force.

Marine organisms are adapted to their salty environment, which in a sense is "physiologically dry" because of the water stress factor; they are able to maintain a high internal osmotic pressure to offset water losses. A simple example illustrates the comparative differences in osmotic pressure between the organism and the environment it occupies. Pure water freezes at $0°$ C. The more impurity (salt) it contains, the lower its actual freezing point. Since a 1 molar solution freezes at $-1.84°$ C (Δ °C), we have a convenient basis for comparing pressures and determining an osmotic potential. Let us say the osmotic pressures of a fish and the seawater in which it lives are Δ 1.0 and Δ 1.8, respectively. The osmotic potential (OP) is calculated as follows:

$$OP = \frac{1.8 - 1.0}{1.84} \times 22 \text{ atm} = 9.5 \text{ atm}$$

Marine organisms are under a constant water balance stress, which many freshwater species with much lower osmotic pressures could not tolerate.

The oceans are the base for the circulation of water throughout the biosphere. The hydrological cycle (see Chapter 2) is a repetitive process of loss and renewal that is carried out on a vast scale between land and sea. The atmosphere is an intermediary in this system of gas-liquid interchange that converts a saline water source to meet the freshwater needs of land-bound ecosystems. The cycle is closed by gravity, as surface runoff and groundwater seepage are returned to their original source. We can calculate a turnover time for ocean-borne water condensing over the continents and ultimately returning to its origin via runoff from the land. The analogy is similar to that presented in Chapter 6 in the discussion of organic turnover in balanced ecosystems. The return of water from the continents via streams has been estimated by Weyl to be about 1 billion kilograms per second. This is the equivalent of approximately 7500 cubic miles per year. The total volume of the oceans, as noted earlier, is about 330 million cubic miles. Thus the replacement time can be calculated from this relationship between input and total volume as follows:

$$\text{Turnover time} = \frac{330 \times 10^6 \text{ mile}^3}{7.5 \times 10^3 \text{ mile}^3} = 44{,}000 \text{ years}$$

A steady state is assumed in this interpretation, that is, the amount of water leaving the land is replaced through precipitation by a similar quantity from the ocean. What effect would surface pollution such as oil have on the hydrological cycle and the turnover time of water in the oceans? A speculative question, perhaps, but one that nonetheless could complicate the distribution of water in the biosphere.

The oceans contain 50 to 60 times as much carbon dioxide (CO_2) as the atmosphere. As more CO_2 is added to the atmosphere (from the combustion of fossil fuels, for example), some is absorbed by the oceans. A recent report showed that not all the CO_2 from fossil fuels can be accounted for in the atmospheric reservoir. A fraction apparently is absorbed by the ocean ecosystem, which thus serves as a kind of buffer or safety valve by taking up part of the excess. By contrast, oxygen (O_2) is much less soluble in water. Thus the ocean contains only a small amount ($^1/_{100}$) compared to its atmospheric counterpart (see Chapter 2). The self-cleansing ability of the ocean to rid itself of pollutants through the action of decomposers depends on the supply of O_2 regenerated by the photosynthetic activity of marine phytoplankton.

Biological decomposition and degradation of wastes is slowed under anaerobic conditions or those of O_2 depletion. The problem is further complicated by the slow exchange of O_2 between the atmosphere, where it is abundant, and the ocean, where it is not. Poisons such as hydrogen sulfide (H_2S) build up and further complicate the picture of growing deterioration until a point is reached at which biological restoration becomes difficult, if not impossible. Pollution may disrupt vital processes such as photosynthesis. The sudden impairment of the photosynthetic process would soon leave the ocean in a biologically depleted condition.

POLLUTION

It must be admitted that we still treat the ocean as though it were the city refuse heap. In a recent magazine article it was noted that four fifths of the cities of Italy have no sewage-treatment programs. As a result, raw wastes and garbage pollute the waterways running to the sea. It is estimated that 85% of Italy's shoreline is polluted. The rivers of North America annually carry over 600 million tons of wastes to the oceans. Despite the vastness of the oceans, we must wonder whether these bodies of water can sustain such inputs of foreign material indefinitely. New York City has its own ''dead sea.'' The daily depositions of tons upon tons of garbage over the past decades have been too much for the self-cleansing and detoxifying processes of an area only a few miles from shore, and now there is no life in the bottom waters of this dumping ground.

Oil spills from offshore drilling and tanker accidents occur with increasing frequency. Bird life on both coasts of the United States is affected by oil spills (Fig. 9-3). Even without these incidents, oil pollution of the ocean is ever present. Consider that an ''empty'' tanker of the super class still retains 3000 gallons of crude oil on the walls of its hold. Hundreds of tankers are routinely flushed at sea because of the danger of explosion from gases developing inside the oil storage tanks of the ships. In addition, the number of such ships is increasing. They are also being made larger, and some are now twice as big as the *Torrey Canyon,* from which oil spillage caused widespread damage to the southern

Fig. 9-3. Oil-stricken birds unable to fly as a result of oil spills, California (left) and Cape Cod (right). Many birds die as a result of ingesting oil while trying to clean their feathers. (Courtesy Wide World of Photos.)

British coast in 1967. The damage is thus magnified when a spill does occur. In 1955 there were 2500 tankers running the seas; in 1966 there were 3600. By 1983 there will be 4300 ships carrying oil around the world.

The Baltic Sea has basin characteristics, including deep water in some parts, that contribute to poor circulation. It is vulnerable to stagnation. With 20 million people living in the highly industrialized nations on its shores, it has become highly polluted. At lower depths, hydrogen sulfide is beginning to appear in enormous quantities with ever-increasing frequency and duration. Such stagnation periods, prolonged by pollution, threaten the deep-water areas and may turn them into a kind of biological desert that is devoid of oxygen and of life with the exception of anaerobic bacteria. These bacteria can use certain compounds from which oxygen can be derived for their own metabolism. By reducing sulfates, for example, hydrogen sulfide is produced, and when its bubbles rise to the surface, it is a sign of oxygen depletion at lower levels. The Black Sea characteristically has hydrogen sulfide in its basin because the deeper waters do not mix with those of the surface, but industry also has an impact on hydrogen sulfide

Table 9-3. Quantitative relationships of oxygen and hydrogen sulfide in the Black Sea*

DEPTH (m)	OXYGEN (% by volume)	HYDROGEN SULFIDE (% by volume)
10	0.51	–
25	0.74	–
50	0.67	–
75	0.55	–
100	0.23	–
150	0.02	–
200	–	0.09
300	–	0.23
400	–	0.42
600	–	0.51
800	–	0.61
1000	–	0.60

*Modified from Raymont, J. E. G. 1963. Plankton and productivity in the ocean. Pergamon Press, Ltd. Oxford.

levels. Table 9-3 shows the distinct relationship between the presence of this poisonous gas and oxygen deficiencies. Organic pollution, however, greatly increases the chances of this phenomenon in any aquatic system.

Heavy metals such as lead and mercury are being concentrated in the sea. The *Second Report of the Council on Environmental Quality* indicates that the upper layer of the oceans is being contaminated with industrial lead. Goldberg reports that we are putting as much lead into the oceans as comes from natural processes. It is estimated that 250,000 metric tons are ''washed out'' of the atmosphere over the oceans each year. Its source is the automobile that uses antiknock gasoline. Mercury has been reported in swordfish at levels that necessitated its removal from the market. It is also found in canned tuna. Industrial discharges of mercury into Minamata Bay, Japan, contaminated seafood and caused the so-called Minamata disease. Deaths were attributed to the consumption of fish contaminated with mercury. Other victims suffered loss of vision, muscular weakness, and paralysis. A characteristic feature of the Minamata disease is damage to brain cells. In addition, it has been found that mercury poisoning causes birth defects. Released into the environment as mercuric chloride, a factory waste product, it is converted by bacteria to water-soluble methyl mercury, which spreads throughout the food chain and ultimately reaches the human consumer.

In another locale, there have been sharp declines in the sardine harvests in the coastal waters off central California. Once measured in the hundreds of thousands of tons, this fish resource has virtually disappeared. Many believe that the declines are the result of DDT from the San Joaquin Valley draining into San Francisco Bay. And so it goes, one report after another of how we have implicated ourselves in the pollution of the seas, the depletion of their oxygen supply, and in the arrest of their various food-producing capacities. Fig. 9-4 shows that even small concentrations of DDT can depress the rate of photosynthesis in marine phytoplankton under laboratory conditions. Even though the

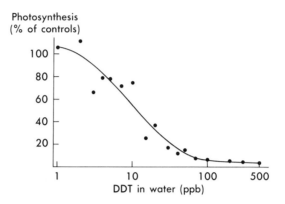

Fig. 9-4. Effect of DDT on photosynthesis in marine algae. (Redrawn from Wurster, C. F., Jr. 1968. Science **159:**1474-1475.)

Table 9-4. Selected lists of marine animals in which DDT, its metabolites, or related chlorinated hydrocarbons and polychlorinated biphenyls (BCPs) have been detected*

ORGANISM	LOCATION
Invertebrates	
Bat starfish, common starfish, kelp crab, purple snail, purple urchin, sea cucumber, squid	California coast
Soft clam	New York
Oyster	Maine, Gulf of Mexico
Vertebrates	
Pinfish, mullet, dolphin	Gulf of Mexico
Bermuda petrel	Atlantic
Brandt's cormorant, brown pelican, Caspian tern, common murre, night heron, western gull, black petrel, eared grebe, elegant tern, fish bat, Hermann's gull, osprey, peregrine falcon, shearwaters	California (U.S.A.), Baja California
Brown booby, frigate bird	Panama
Peregrine falcon	Cornish coast (Britain), Alaskan tundra
Northern fur seal	Pribilof Islands
Adelie penguin, Emperor penguin, skua, crabeater seal, Weddel seal	Antarctica

*Data from Anas, R. E. and A. J. Wilson, Jr. 1970. Pesticides Monit. J. **3:**198-200; Cade, T. J. et al. 1971. Science **172:**955-957; George, J. L. and D. E. H. Frear. 1966. J. Appl. Ecol. **3:**155-167; Hickey, J. J. and D. W. Anderson. 1969. In J. J. Hickey (ed.). Peregrine falcon populations, their biology and decline. University of Wisconsin Press. Madison, Wis.; Risebrough, R. W. et al. 1967. Nature **216:**589-591; Risebrough, R. W. et al. 1968. Nature **220:** 1098-1102. Sladen, W. J. L. et al. 1966. Nature **210:**670-673; and Wurster, C. F., Jr. and D. B. Wingate. 1968. Science **159:**979-981.

~~data are~~ based on studies with only a few species, we should not minimize the implications. A more recent study by still other investigators verifies that DDT slows photosynthetic rates of certain marine algae. Concentrations of this pesticide in seawater of about 7 ppm caused a 50% inhibition. The presence of DDT and other chlorinated hydrocarbons in marine life is universal, as shown in Table 9-4. Perhaps no single factor attests more vividly to the universality of oceanic pollution than our use of DDT and other chlorinated hydrocarbons. However, we continue to tolerate their global use at the expense of biotic diversity and ecosystem function. The tendency toward the development of irreversible conditions becomes more evident as time goes on, giving rise to the expression that the "seas are dying." These are serious problems, long recognized, against which positive action to protect the oceans is finally being taken. The oceans represent vital functions of the global ecosystem. Any disruption of energy (food) production and nutrient cycling is a threat to ecological stability on land as well as in the sea. The hydrological cycle, for example, becomes an active mechanism for the transfer of particulates and vaporized substances to points on the globe far removed from their source. Airborne substances, including oxides of nitrogen, mercury, lead, radioactive materials, and chlorinated hydrocardons, are involved in the "rainout effect."

PRIMARY PRODUCTION

What percent of the world's photosynthesis takes place in the seas? Estimates vary, sometimes rather widely, but, in general, primary production of the ocean equals or may exceed that attributed to terrestrial ecosystems. Estimates as high as 70% have been cited. Even if we adopt the more conservative estimates shown in Table 5-5 (p. 99), the ocean is involved in a significant amount of organic energy and oxygen replenishment.

Not all parts of the ocean are equally productive. Fig. 9-1 depicted the neritic and oceanic portions as the ocean's main physical divisions. According to Ryther, the neritic, or coastal, zone is twice as productive per unit area as the open sea, yielding approximately 200 to 250 grams of photosynthate per square meter, or about 1 ton of production per acre. Values for the open sea would approximate 1000 pounds per acre of ocean, an amount roughly equivalent to the annual production by desert vegetation.

In areas of upwelling such as that off the coast of Peru, photosynthesis rates are the highest, being about six times the rate of the broad ocean. Here the production by microscopic plants, the diatoms, and other marine algae would average about 3 tons per acre, a yield comparable to that of tall grassland in a humid climate. However, since these areas are somewhat localized and constitute such a small fraction of the total ocean, the contribution to the overall production of the ocean is quite minor, only 1 part in 160. The great bulk of the ocean, about 90%, has the lowest photosynthetic yields. As we fly over the Atlantic or over almost any other ocean, we indeed view an area that is a biological desert in terms of primary productivity. What accounts for these characteristics in the broad sense? The phytoplankton of the sea have nutrient requirements similar to those of grasslands, forests, or cornfields. When certain essentials are in minimal supply, photosynthesis and growth are curtailed. In the open sea there is limited availability of nitrates and phosphates. In the neritic zone and in those areas of upwelling, nutrients are more abundant as a result of runoff from the land or transport from the ocean basin. The combined area of these nutrient-rich zones is only about 10% of the whole ocean. Let us examine some of the food chain characteristics of the ocean that limit its productivity in terms of food for our own species.

FOOD CHAINS

The relative complexity, or number of trophic levels, of the food chain reflects the variations in primary production that occur in the seas. Attributes of the producers and consumers themselves may also be distinctive.

Some very large carnivorous consumers are supported by short food chains. Here we may cite the example of the large baleen or filter-feeding whales of the high-latitude oceans. In Antarctic waters there is an abundance of nutrients that result from a mixing of the bottom with the photic zone. Through rapid photosynthesis and growth, the marine phytoplankton, primarily populations of diatoms, become a rich food source for a small crustacean called the krill. Known by its scientific name as *Euphausia superba,* the krill constitutes the principal diet of baleen whales. Among these is the blue whale, the largest animal on earth. This food-chain relationship can be expressed as follows:

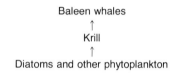

Tremendous quantities of krill are consumed. According to Nemoto, more than 1 ton is sometimes found in whale stomachs. Total consumption of krill by the Antarctic whales before their stocks began to decline has been estimated at amounts as great as 77 million tons per year.

As the number of great whales is reduced through hunting, what will happen to the large quantities of unused krill, to other food chains, and to the stability of the Antarctic ecosystem in general? People will undoubtedly become harvesters. The Soviet Union, one of the principal fishing nations, is developing methods for taking krill from the sea. Depletion of the krill population would have serious effects on other species such as seals, fish, and penguins, which also consume this species. As in all simple ecosystems with generally low diversity, the numbers of food options and substitutions are limited. The trophic position of krill may be likened to the central importance of lemmings in the high Arctic, where a host of consumers trace their dependency to this little herbivore. As we increase our predation in the Antarctic, we will cause the evolution of even more simplified ecosystems through shortened and simplified energy pathways. In the process, we move closer to the phytoplankton base of the food pyramid.

The relatively short food chain common to the southern ocean bordering Antarctica is not unique to that region. According to Ryther, food chains tend to be shorter in coastal waters as compared to the open sea. The underlying factors are nutrient availability and the larger size of plankton cells. Greater cell size (colonial forms also occur) provides a

Table 9-5. Comparison of phytoplankton production and fish harvest in different zones of the ocean*

QUANTITY MEASURED	OCEAN ZONE		
	OPEN OCEAN	COASTAL AREA	UPWELLING
Percentage of total ocean	90.0	9.9	0.1
Average photosynthate (tons/acre/yr)	0.5	1.0	3.0
Average number of trophic levels from plant production to fish production $(T_1 - T_x)$	5.0	3.0	1.5
Percentage of total photosynthesis	81.5	17.9	0.6
Percentage of total fish production	0.8	55.1	44.1

*Original data from Ryther, J. H. 1969. Science **166**:72-76; modified from Crisp, D. J. 1975. In D. E. Reichle et al. (eds.). Productivity of world ecosystems. Proceedings of the Symposium. Seattle, Wash. 1972. National Academy of Science. Washington, D.C.

suitable food for larger herbivores and carnivores such as fishes, which in turn are usable for human consumption.

Coastal waters, zones of convergence, and areas of upwelling are the mainstays of the fishing industry. In these environments, important species such as the herring can utilize phytoplankton as part of their diet. Such an abbreviated energy structure provides the consumer with a greater supply of food and the potential for developing greater population densities. By contrast, plankton size is very small in the nutrient-poor open ocean. As a consequence, the consumer size tends to be reduced, and a greater number of trophic levels or links in the food chain would be required to effect a certain body size. An energy loss occurs at each transfer level,decreasing the overall efficiency of that system. These interrelationships of primary production (photosynthesis), trophic efficiencies, and the development of important fisheries are shown in Table 9-5. We see that more than 80% of the photosynthesis of the ocean supports less than 1% of the fish harvest. Therefore food chains based on less than one fifth of the phytoplankton production support more than 99% of the commercial fisheries. This fact may initially seem incongruous, but it serves to bring home some basic ecological facts concerning the ocean environment. The ocean is a significant source of oxygen, not because of a broadly based high rate of photosynthesis but because of its sheer size as part of the total land-sea relationship. Only limited areas are productive of human food, depending on the nutrient supply and the timing of its availability. "A place for everything and everything in its place" must be a guide to future management.

In the high latitudes such as Antarctica, there is a single pulse of high phytoplankton photosynthesis. This occurs in the relatively short spring-summer period. As we move into more temperate zones and longer seasons, two periods begin to emerge. These are the spring and autumn pulses of plankton growth. Continuing into the tropical latitudes, these periods coalesce into a less distinctive pattern spread throughout the year but centered on an intermediate "winter" period. The spring break typical of the temperate zone is initiated earlier, and the fall activity begins later under these conditions. Periodicity and fertility both set obvious limits on the food potential of the ocean. Ricker estimated the yearly food harvest from the sea at about 60 million tons at a time when the world population was between 3 and 4 billion. Since that time, harvest estimates have been revised upward; the population, however, has increased also (to over 4 billion) so that our annual energy requirement is still many times what the ocean alone could supply under present conditions. In addition, this ocean harvest is not all usable. There are losses in processing, and part of the catch is diverted to other uses such as pet foods and fertilizers. Ultimately only about one fourth becomes consumable protein.

THE INTERTIDAL ZONE

At the fringes of the ocean there is a relatively narrow band of brackish water. This is a mixing zone of salt water and fresh water such as those found at the mouths of rivers. It is an in-between world of ocean and land—partly both but distinctly neither. Here are the estuaries, bays, coastal marshes, and mangroves that are subject to the ebb and flow of the tide (Fig. 9-5).

Much concern has been expressed over the increasing depredation of littoral environments. Estuaries are nutrient-rich areas and are among the most productive of biotic communities. Eugene Odom states that 20 times as much organic matter is produced per unit area in these systems as in the open ocean beyond the continental shelf. On the

Fig. 9-5. An expanse of mud flat at low tide, with red mangrove forest *(Rhizophora mangle)* in the background at Golfito, southwestern Costa Rica.

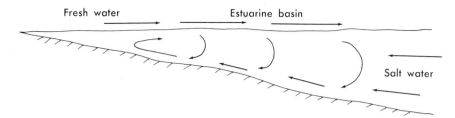

Fig. 9-6. Diagrammatic representation of mixing of salt water with incoming fresh water in an estuarine environment. The salt water is heavier and promotes a circulation with the less dense fresh water that enters the estuary as an upper layer. Salinity gradients occur from top to bottom as well as horizontally with the ebb and flow of the tide.

average, their productivity exceeds that of agricultural land. In a broad sense we might visualize these intertidal communities as dynamic sieving systems that catch, hold, slow down, and utilize nutrients and organic debris accumulated by the river systems of the continents. The effect of trapping nutrients and organic matter within the estuary zone is diagrammed in Fig 9-6. A circulation is caused by differences in density between fresh water flowing into the estuary basin and salt water of the incoming tide. With low levels of fresh water, the dilution effect is lessened, and salinity effects within the estuary are increased.

Various benthic species, or bottom feeders, reside in the intertidal zone. Oysters, clams, and other shellfish filter and ingest plankton and detritus from large quantities of moving water. They are food concentrators, converting such energy sources as dead organic matter into sources fit for human consumption. By the same mechanism, they also accumulate DDT, other pesticides, and toxic materials in amounts that may be thousands of times higher than those in the water itself. The concentration of such substances limits their commercial value. Along the same lines, the factors that contribute to the nutrient enrichment of estuarine waters also make them particularly vulnerable to contamination from agricultural runoff that contains sediments and chemicals and from industrial and municipal wastes. Shrimp, crabs, and lobsters, too, are consumers of commercial importance in this highly complex food web.

Estuaries and salt marshes are the spawning grounds for various ocean fishes. For these organisms such as the sea bass a critical phase of the life cycle must be spent in the estuarine environment. Interdiction of the life cycle through habitat destruction seriously threatens many species today. Yet real estate developers and industrialists recklessly ignore the ecology of these interlocking biotic communities as they continue to fill and build, dredge and drain.

In addition to their value as spawning grounds and nursery beds, the littoral zones are suitable environments for wildlife and in some instances are unique sanctuaries for particular species. In California alone the destruction of estuaries runs into the thousands of acres. It is estimated that 67% of those areas suitable for wildlife habitats have been lost

Table 9-6. States leading in destruction of estuaries classified as wildlife
and nutrient areas*

STATE	AREA LOST (thousands of acres)	TOTAL AREA (%)
California	256	67.0
New York	20	15.0
New Jersey	54	13.1
Connecticut	2	10.3
New Hampshire	1	10.0
Texas	68	8.2
Florida	60	7.5

*Modified from Commission on Marine Science, Engineering and Resources. 1969. Panel report. United States Government Printing Office. Washington, D.C.

through dredging and filling. According to a government report on marine science and engineering, 83% of the estuarine marshlands in San Francisco Bay have been destroyed. During the last 20 years, more than 7% of this country's important estuarine acreage has been destroyed, primarily to provide space for the construction of housing developments. Table 9-6 shows the extent of the losses in those states that are leading in the destruction of coastal habitats. These acreage reductions infringe not only on native wildlife and waterfowl but also on the supply of seafoods, which are important sources of protein. Some species will be phased completely out of commercial existence if the exploitive practices and extensive pollution by both the private and public sectors continue!

SUMMARY

Despite its size and great wealth of minerals, food, and energy, the ocean is vulnerable to depletion. Although this tremendous volume of water is in a constant state of motion and mixing, only about 3% at any one time supports the growth and production of phytoplankton. This is the upper layer of about 100 meters, or the euphotic zone, in which light is adequate for photosynthesis. At lower depths, and even on the ocean floor itself, numerous organisms are supported in a complex web of grazing and detritus food chains.

The food potential of the ocean varies widely. About 10% of the total area is relatively productive. The other 90% can be considered a kind of biological desert. The main reason is that the open ocean is quite low in certain nutrients for plant growth, phosphates and nitrates being in especially short supply. Only on the coastal shelf and in zones of upwelling is photosynthesis adequate to support rich food chains and large fish harvests. About 99% of all commercial fish are derived from such areas.

The ocean is being subjected to severe stress from pollutants such as oil, pesticides, and industrial poisons. The prospect of radioactive contamination from atomic testing should not be excluded. Noted oceanographers state that in the next quarter of a century some areas will be "dead seas" as a result of this gigantic influx of waste from various sources. No place in the ocean is free of certain pollutants. DDT is a prime example. It is found in Atlantic sea birds and Antarctic penguins alike. Its widespread occurrence indicates the interrelated nature of the biosphere, in which vast global cycles are implicated. How we safeguard the integrity of photosynthesis and the life of the sea has profound effects on terrestrial ecosystems as well.

DISCUSSION QUESTIONS

1. What are the principal factors limiting productivity of the open seas? Explain.
2. Discuss the estuarine environment and its value to man and other species. What are serious threats to the integrity of estuaries in a technologically oriented society?
3. Discuss the causal factors of upwelling. Why are they important in food production? Explain.

REFERENCES

Anas, R. E. and A. J. Wilson, Jr. 1970. Residues in fish, wildlife and estuaries. Pesticides Monit. J. **3:**198-200.

Arx, W. S. von. 1957. An experimental approach to problems in physical oceanography. In L. H. Ahrens et al. (eds.). Physics and chemistry of the earth. Pergamon Press, Ltd. Oxford, England.

Arx, W. S. von. 1962. An introduction to physical oceanography. Addison-Wesley Publishing Co. Reading, Mass.

Bowen, V. T. et al. 1969. Strontium-90 concentrations in surface waters of the Atlantic Ocean. Science **164:**825.

Bowes, G. W. 1972. Uptake and metabolism of 2,2-bis-(p-chlorophenyl)-1,1,1,-trichloroethane (DDT) by marine 9-phytoplankton and effect on growth and chloroplast electron transport. Plant Physiol. **49:**172-176.

Cade, T. J. et al. 1971. DDE residues and eggshell changes in Alaskan falcons and hawks, Science **172:**955-957.

Cloud, P. 1969. Mineral resources from the sea. In Committee on Resources and Man. Resources and man. W. H. Freeman & Co., Publishers. San Francisco.

Commission on Marine Science, Engineering and Resources. 1969. Panel report. United States Government Printing Office. Washington, D.C.

Council on Environmental Quality. 1971. The second annual report to the President. United States Government Printing Office. Washington, D.C.

Crisp, D. J. 1975. Secondary productivity in the sea. In D. E. Reichle et al. (eds.). Productivity of world ecosystems. Proceedings of the Symposium. Seattle, Wash. 1972. National Academy of Sciences. Washington, D.C.

Cronin, L. E. and A. J. Manseuti. 1971. The biology of the estuary. In P. A. Douglas and R. H. Stroud (eds.). A symposium on biological significance of estuaries. Houston, Texas. 1970. Sport Fishing Institute. Washington, D.C.

Geological Survey. 1970. Mercury in the environment. Geol. Surv. Prof. Paper No. 713. United States Government Printing Office. Washington, D.C.

George, J. L. and D. E. H. Frear. 1966. Pesticides in the Antarctic. J. Appl. Ecol. **3:**155-167.

Goldberg, E. D. 1963. The oceans as a chemical system. In M. N. Hill (ed.). The sea. John Wiley & Sons, Inc. New York.

Goldberg, E. D. 1970. The chemical invasion of the oceans by man. In S. F. Singer (ed.). Global effects of environmental pollution. Springer-Verlag New York, Inc. New York.

Greze, U. N. 1970. The biomass and production of different trophic levels in the pelagic communities of south seas. In J. H. Steele (ed.). Marine food chains. University of California Press. Berkeley, Calif.

Gullard, J. A. 1970. Food chains studies and some problems in world fisheries. In J. H. Steele (ed.). Marine food chains. University of California Press, Berkeley, Calif.

Hickey, J. J. and D. W. Anderson. 1969. The peregrine falcon: life history and population literature. In J. J. Hickey (ed.). Peregrine falcon populations, their biology and decline. University of Wisconsin Press, Madison, Wis.

Holt, S. J. 1969. The food resources of the ocean. Sci. Am. **221:**178-182, 187-194.

Isaacs, J. D. 1969. The nature of oceanic life. In D. Flanagan (ed.). The ocean. Scientific American Publications. W. H. Freeman & Co., Publishers. San Francisco.

McLuksy, D. S. 1971. Ecology of estuaries. Heinemann Educational Books Ltd. London.

McVay, S. 1966. The last of the great whales. Sci. Am. **215:**13-21.

Munk, W. 1955. The circulation of the oceans. Sci. Am. **193:**96-102.

Nemoto, T. 1970. Feeding pattern of baleen whales in the ocean. In J. H. Steele (ed.). Marine food chains. University of California Press. Berkeley, Calif.

Odum, E. P. 1971. Fundamentals of ecology. W. B. Saunders Co., Philadelphia.

Orr, A. P. and S. M. Marshall. 1957. The fertile sea. Fishing News (Books), Ltd. London.

Raymont, J. E. G. 1963. Plankton and productivity in the ocean. Pergamon Press, Ltd. Oxford, England.

Ricker, W. E. 1969. Food from the sea. In Committee on Resources and Man. Resources and Man. W. H. Freeman & Co., Publishers. San Francisco.

Risebrough, R. W. et al. 1967. DDT residues in Pacific sea birds: a persistent insecticide in marine food chains. Nature 216:589-591.

Risebrough, R. W. et al. 1968. Polychlorinated biphenyls in the global ecosystem. Nature 220: 1098-1102.

Ryther, J. H. 1969. Photosynthesis and fish production in the sea. Science 166:72-76.

Sladen, W. J. L., C. M. Menzie, and W. L. Reichel. 1966. DDT residues in Adelie penguins and a crabeater seal from Antarctica. Nature 210:670-673.

Steeman, N. E. 1960. Productivity of the oceans. Ann. Rev. Plant Physiol. 11:341-361.

Weyl, P. K. 1970. Oceanography. An introduction to the marine environment. John Wiley & Sons, Inc. New York.

Wolne, P. R. 1971. The estuarine environment. In R. S. K. Barnes and J. G. Green (eds.). The estuarine environment. Applied Science Publishers, Ltd. Barking, Essex.

Wurster, C. F., Jr. 1968. DDT reduces photosynthesis by marine phytoplankton. Science 159: 1474-1475.

Wurster, C. F., Jr. and D. B. Wingate. 1968. DDT residues and declining reproduction in the Bermuda petrel. Science 159:979-981.

ADDITIONAL READINGS

Clarke, G. L. 1954. Elements of ecology. John Wiley & Sons, Inc. New York.

Cousteau, J.-Y. 1965. World without sun. Harper & Row, Publishers, New York.

Hedgepeth, J. W. 1970. The oceans: world sump. Environment 12:40-47.

Horsfield, B. and P. B. Stone. 1972. The great ocean business. The New American Library, Inc. New York.

Marx, W. 1967. The frail ocean. Ballantine Books, Inc. New York.

Risebrough, R. W. 1971. The death of the oceans. In T. R. Harney and R. Disch (eds.). The dying generations. Dell Publishing Co., Inc. New York.

Steele, J. H. 1974. The structure of marine ecosystems. Harvard University Press. Cambridge, Mass.

Wenk, E., Jr. 1969. The physical resources of the ocean. Sci. Am. 221:166-176.

PART FOUR

Populations

10 Population ecology

THE POPULATION

The abundance or rarity of bird species in an oak-hickory forest, of annual plants after a desert rainstorm, or of any species anywhere is determined by the number of progeny produced and the mortality occurring in the population. A population can be defined as all the individuals of a single species occurring in a given area. This is a broad definition because a population can include any number of individuals in areas ranging in size from the Pacific Ocean in the case of the bluefin tuna to the small volume of water trapped inside a pitcher plant leaf for a mosquito larvae population. In an effort to put some functional significance into the definition, most field ecologists choose to delineate a population as a group of individuals that have a higher probability of mating with each other than with individuals of another group.

Why do we study populations?

Population studies are important not only because they promote our understanding of a particular species but also because each species is an independent component of the community or ecosystem, and its role in energy flow or nutrient cycling is largely determined by its abundance. Populations also deserve consideration as the genetic unit on which evolution acts, resulting in the adaptations that all species have accumulated to allow them to survive in their natural habitat. Some examples of population studies may be useful at this point.

In East African national parks the elephant has become a major problem species. Its natural habitat has been so reduced by human activities that exceedingly high densities of the species are forced into wildlife refuges. High elephant concentrations can cause massive destruction of vegetation, including large trees (Fig. 10-1). This habitat alteration must be stopped either by moving or by killing excess elephants. Studies on elephant population ecology have revealed the best way to stop the destruction. Since elephant social organization is centered around female herds (a matriarch, her daughters, and granddaughters [Fig. 10-2]), less damage is done to the social organization of the elephants by removing whole herds rather than by taking a few (perhaps critical) individuals from each. This practice has the advantage of yielding important data on herd size and composition, the age of individuals, and the number of past births for each individual as determined by the number of placental scars.

Blazing star, a perennial forb of the family Compositae, is found in the tall grass prairies of the Midwest. The ages of individual plants can be determined by counting the annual rings in the underground stems. The oldest plants of this species can live 34 years if they escape fire, desiccation, and insect damage. In undisturbed sites there is a high proportion of older individuals with very few young plants. A lack of new recruits to the population means that it will eventually die out in these habitats. Young plants occur primarily in disturbed areas. The species therefore is a fugitive, constantly moving to newly disturbed areas where it will live and reproduce until the last individual dies in the then mature vegetation.

Fig. 10-1. Elephant damage in Tsavo National Park, Kenya. During the dry season when lush vegetation is scarce, elephants will push over trees with their tusks and eat the wet pulpy wood as well as the twigs and branches. **A,** Baobab tree. **B,** A species of the genus *Commifera*.

Fig. 10-2. A small herd of African elephants in Tsavo National Park. Most herds consist primarily of females—a matriarch and her daughters and granddaughters—and a few immature males.

POPULATION STRUCTURE

We are frequently struck by the great changes in density (number of individuals per unit area) of some pest species such as starlings, Norway rats, and houseflies and some endangered species—all of which have shown large population changes in the last two centuries. Most populations, however, tend to remain stable in size over long periods, neither continually growing nor declining to zero. Four processes alone determine the size of populations: births, deaths, immigration, and emigration. The simplest way to study a population would be to determine the number of individuals affected by each process. But this approach treats all individuals as equals and ignores the fact that certain types of individuals such as those of different age or social position are more likely to die or to produce offspring than others. By determining how many individuals are in each category, a population structure can be derived. Different types of population structure are possible, depending on the nature of the catgeories, and each will reveal something different about the population and how it functions.

Age structure

In all populations from the smallest bacteria to the largest whales, individuals age with time, and their liklihood of surviving changes with time. Categorizing individuals by age is called *life table analysis* and is probably the best-understood type of population structure. Life table methods were originally developed by insurance actuaries to enable them

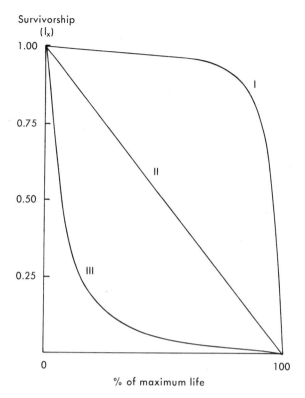

Survivorship
(l_x)

Fig. 10-3. Survivorship curves showing the proportion of a cohort that live to a given age. Curve I is typical of human populations; curve II is found rarely but is observed in some bird and plant populations; and curve III, the most commonly observed in nature, is illustrated by oyster population data.

to predict how long an average life insurance buyer would live and to calculate the cost of his/her policy accordingly.

The easiest way to study age structure is to follow a group of newborn individuals, *a cohort,* through their lives, recording when each produced offspring and when it died. In many cases only females are studied to avoid the complicating factor of uneven sex ratios.

Mortality or the inverse, survivorship, is usually the first data gathered when constructing a life table. Since it is not the individuals that have died, but those that have survived that are important in maintaining the population, we will be concerned here with *age-specific survivorship* (l_x). This term refers to the proportion of the cohort that lives to be age x. The survivorship pattern of different species can be compared by plotting l_x against age (Fig. 10-3). Curve 1 represents a human population that shows high survivorship early in life but high mortality late in life. This pattern is fairly rare in nature and is exhibited by species that carefully nurture and protect the young, thereby increasing their chance for survival. Curve II shows a situation in which a constant percent of the original cohort dies at each age. The data for this curve came from a wildflower population, but

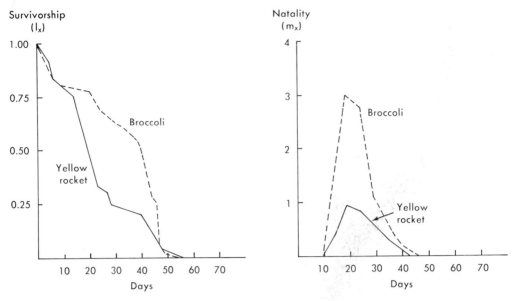

Fig. 10-4. Age-specific survivorship and natality for the cabbage aphid feeding on a cultivated plant, broccoli, and a wild plant, yellow rocket. Both plants are members of the plant family Cruciferae. (Redrawn from Root, R. B. and A. M. Olson. 1969. Can. Entomol. **101:**768-773.)

this pattern of equal probability of death for both young and adults is also found in some tree and bird populations. Curve III is by far the most commonly observed survivorship pattern, with high mortality of young but relative safety after a certain age and size are attained. The third curve was constructed from data on an oyster population and is typical of most plants and animals that produce large numbers of offspring.

Survivorship curves are also used to compare the effect of different environmental conditions on a single species. Root and Olson found that cabbage aphids grown on cultivated plants of the family Cruciferae such as broccoli had a much higher survival rate than when grown on a wild crucifer such as yellow rocket (Fig. 10-4). On both plants, however, the maximum life span was about the same.

Natality, or birthrate (m_x), is a second important life-history parameter. The age-specific birthrate may be defined as the number of daughters born to the average female of age x. In most species the production of offspring varies with age; there is usually a prereproductive period followed by a rise in the number of offspring produced until a peak or plateau is reached. Man is one of the few species that has a postreproductive period.

The environment, or host plant in the case of the cabbage aphid, also exerts an effect on m_x values (Fig. 10-4). The onset of reproduction in the cabbage aphid occurred at the same time, whether it fed on broccoli or yellow rocket, but at peak reproductive age the cabbage aphid produced three times more young on broccoli than on yellow rocket.

With information on both survivorship and natality, it is possible to calculate the rate at which a population will change in size. The number of daughters that will replace a mother in a population is the *net replacement rate* (R_0); it will be a function of how many daughters a female will have at each age and of the likelihood that she will survive to that age. If a female is replaced by exactly one daughter ($R_0 = 1$), the population will be stable in size, but if R_0 is greater than 1 ($R_0 > 1$), the population will grow and if R_0 is less than 1 ($R_0 < 1$), it will decline. The amount by which R_0 exceeds 1 gives a rough indication of how fast the population is growing.

Returning to the cabbage aphid experiment, R_0 of aphids on yellow rocket is 5.58, whereas on broccoli it is 34.20. Thus each female leaves over six times more progeny while feeding on the cultivated plant than on its wild relative. The reason for this great disparity lies with the plant, not with the aphids' taste preferences. All members of the family Cruciferae contain mustard oils that act as a chemical deterrent to feeding herbivores. These chemicals taste bitter to humans and presumably are unpleasant to other herbivores as well. Gradually we have selected the mustard oils out of the cultivated broccoli plants by saving seeds of plants with the best taste. In effect we have removed the broccoli's chemical defense system, making it a desirable food plant for humans and aphids as well.

Obviously populations with different survivorship and natality grow at unequal rates; however, two populations of the same size with identical l_x and m_x can also grow at different rates for a limited period of time. This could occur if the two populations had different age distributions. Suppose there are two towns of identical size, one composed of young married couples, the other largely a retirement village. Even if all members of both towns had the same likelihood at birth of surviving and reproducing at age 20 years, the town composed of young married couples would grow more rapidly at first. However, with sufficient time both towns would eventually reach and maintain the same age distribution, as long as l_x and m_x did not change. This is called a *stable age distribution;* by definition it is a specific proportional distribution of ages characteristic of a particular l_x and m_x schedule. The low survivorship and high natality of human populations in many developing countries result in a pyramidal stable age structure (Fig. 10-5) in which most of the population is young. In more developed countries high survivorship and low natality produce populations with an almost rectangular stable age structure in which all ages are equally abundant (Fig. 10-5). A stable age distribution is usually assumed when dealing with life-history and population growth rate equations. Since there is no stable age distribution in the United States at this time, we are now approaching a net replacement rate of 1, and yet the population is still growing. There is a bulge in the age distribution over ages just entering the prime childbearing years, and this results in more births than predicted from a population of 220 million (Fig. 10-5).

Most organisms have morphological and physiological adaptations that match the environmental stresses placed on them. The heavy winter plumages of black-capped chickadees and their ability to lower their body temperatures in order to conserve energy are adaptations to survive the cold temperatures and low food availability of northern latitude winters. Natural selection also affects life-history parameters so that the number and spacing of births suits the environment. The individual producing the most viable

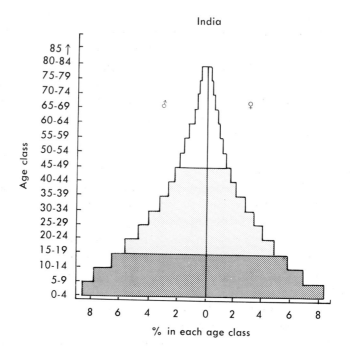

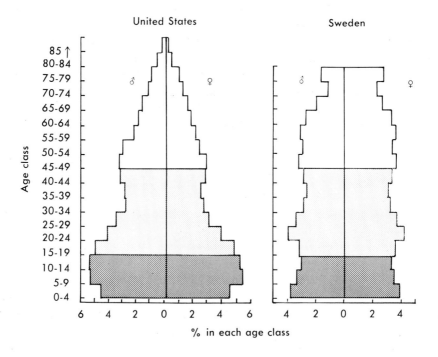

Fig. 10-5. For legend see opposite page.

offspring will be the one whose traits will be passed most frequently to the next generation. The net replacement rate of a population can be increased in three ways: females can reproduce for a longer period in their lives, they can reproduce at more frequent intervals, or they can produce more offspring during each reproductive effort. For each of these mechanisms the optimal level arrived at by selection is usually less than the theoretical maximum.

Increasing the reproductive life span by producing offspring earlier can have a dramatic effect on population growth rates. A woman who had nine children, one a year beginning at age 30 years, would contribute less to population growth than a woman who had only five children but started at age 16 years. This operates on the same principle as compound interest: the sooner children are produced, the sooner they, in turn, produce offspring. It is not surprising, then, that people interested in slowing human growth rates strongly advocate late marriages and late childbearing. For example, the average age at which marriage occurs in the slowest growing countries such as Ireland is over 30 years. Natural selection should favor early first age of reproduction to maximize R_0, but natural populations do not continually decrease this parameter. Instead a lower age limit is reached below which the female is either too young or too small to be successful at bearing young and caring for them.

Another means of increasing R_0 is to maximize the number of young produced per reproductive bout described as the clutch size in birds and invertebrates or the litter size in mammals. Natural selection should favor a large clutch size; however, there is an upper limit beyond which a particular young's survival is jeopardized by parental care split too many ways. C. M. Perrins demonstrated this with swifts. He manipulated the normal clutch size (two or three eggs) by removing eggs from some nests and adding eggs to others. Perrins' results clearly demonstrated that a three-egg clutch was optimal and that one larger resulted in a decrease in the number of chicks fledged as a result of inadequate feeding and growth. His findings can be tabulated as follows:

Clutch size	1	2	3	4
No. of young fledged/nest	0.93	1.91	2.62	1.83

Energy devoted to reproduction may be utilized in different ways. Energy can be allocated either to offspring biomass, producing either many small progeny or fewer large or

---

Fig. 10-5. Age structure distribution for three different countries in 1970. Pyramidal shape (India) is typical of developing countries; rectangular shape (Sweden) is characteristic of developed countries with a near-stable population. The dark stippling refers to the prereproductive period, the light stippling to major reproductive ages, and unstippled areas to the postreproductive period. (Redrawn from Miller, G. T., Jr. 1975. Living in the environment: concepts, problems, and alternatives. Wadsworth Publishing Co., Inc. Belmont, Calif.; data for India from International Demographic Statistics Center, U.S. Bureau of the Census; data for United States from U.S. Bureau of the Census, Current Population Reports, series P-25; data for Sweden from U.N. Population Division, working paper no. 11.)

well-developed progeny, or to parental care and other reproductive behaviors. The optimal reproductive strategy is one that allows survival of the greatest number of offspring in that environment. For example, in habitats in which the environment is variable, most mortality will be caused by physical factors, as conditions favorable to survival are "patchy" in either time or space—either they occur for only a short period before becoming unsuitable or they occur as small scattered areas. In either case an organism has small chance of finding a favorable area but, having found it, will face only minimal intraspecific competition. The optimal reproductive strategy for a population living in these conditions is to divide the available reproductive energy into as many offspring as possible, gambling that a few will find a suitable area. Such species are referred to as "r" selected; they tend to have rapid development of young, early reproduction, small body size, and single reproduction during the short lifetime. Most successional plant species utilize this strategy, since areas of disturbance in the climax vegetation are only present for a short time. Among animals, marine intertidal invertebrates often produce hundreds of thousands of eggs so that a few may find a temporally bare area to settle and become established, and many insects that feed on short-lived resources such as annual plants or rotting carcasses also employ this reproductive strategy. At the opposite end of the gradient is the reproductive strategy of organisms that live in stable environments. Favorable conditions are easy to find, but they are usually crowded. In such situations fewer offspring are produced, but more energy goes into each one to give it a head start in a highly competitive situation. This strategy is termed "K" selection and is usually characterized by slower development of young, delayed reproduction, large body size, and repeated reproduction during the relatively long lifetime. It is important to note that r selection and K selection are relative terms. A species is only r selected or K selected in terms of some other species along the gradient, since there are no absolute criteria for labeling a species.

Social structure and systems

We have seen that age is an important factor in determining survivorship and natality; similarly the individual's position in the social system can also influence these population parameters. In most ant and bee nests a single female (queen) does all the egg laying and survives much longer than the rest of the females (workers). In populations of sage grouse, a large bird of the sagebrush cold desert, the dominant 10% of the males accounted for more than 75% of the copulations each year. These examples illustrate how important an individual's position in a social system is in determining survivorship and reproductive success.

There are many different types of social systems, each shaped by natural selection and the interaction of organisms and their environment. The definition of society as a cooperating group of individuals of the same species is broad enough to cover a large diversity of interactions, ranging from relatively unorganized groups such as loose feeding aggregations to highly organized territorial and dominance hierarchy systems. The latter have the greatest impact on survival and reproduction and deserve closer examination. In territorial systems the objective is to secure and defend an area for exclusive use. This can be an area for feeding, nesting, mating, and so on, depending on the species. In dominance systems the objective is to achieve status in the group and gain preferential treatment in feeding, mating, and other activities.

Comparative sociobiology studies the relationship of social systems and environmental conditions by examining two closely related species that live in different environments. The African lion and the American mountain lion are the dominant cats in their respective ranges. Both have about the same size home range of 40 square miles, but they live in entirely different types of habitat. African lions are found in open savannahs, frequently near water holes that attract prey during the dry season. The mountain lion inhabits forested country.

The social organization of the African lion is centered around a pride of 4 to 37 lions, primarily composed of adult females and their young and one or two adult males. All the hunting is done by the females, who may stalk their prey as a unit in a cooperative effort. Once a kill has been made, the adult males feed first, followed by the adult females and finally by the young if any food remains. A pride is advantageous for hunting in this habitat because of the following: (1) In open country where the prey has an unobstructed field of vision, hunting as a group is twice as successful as hunting alone because the prey can be surrounded and escape routes cut off. Bigger and more dangerous prey such as cape buffalo can be killed by predators hunting in groups. (2) In an area with many other predators and scavengers, young lions would be in danger if left alone while their mothers hunted. By acting cooperatively, all the young can be left together with a few adults during the hunt. (3) A group of lions, unlike a single individual, can successfully protect a kill from scavengers such as hyenas and vultures. (4) If a lion becomes temporarily sick or lame and unable to hunt, it can still feed at the kills of the group. (5) When a female with cubs is killed, the cubs can be adopted and cared for by other females in the pride. The pride as a social system can only be sustained by the high concentration of prey that occurs on the African plains.

The mountain lion, on the other hand, has none of this elaborate group behavior because the habitat makes such behavior impossible or unnecessary. The heavy vegetation and rough topography of typical mountain lion habitat limit the visual field of the prey, making an ambush tactic a productive way to hunt for a single cat. The prey (usually deer) may be scattered in distribution and are relatively small in size. Not enough energy can be gained from such prey to sustain more than a family group in an area. The lack of numerous other predators and scavengers means that a single mountain lion can guard a kill and that it is relatively safe to leave a litter unguarded in a den while the mother hunts.

An even more striking example of the influence of environment on social systems is illustrated by comparing populations of the American turkey from two different habitats. In the eastern United States, the turkey is a woodland bird that employs a harem type of social system. A single male will attract several females and will remain with them until they are bred. Adult males are antagonistic to each other during the breeding season. In the grass- and bushlands of southeastern Texas, an unusual variation of this social system occurs. All the surviving males from a single clutch form a sibling group when they are 6 to 7 months old. This group is inseparable for life and has an internal dominance hierarchy that, once established, does not change. When the breeding season begins in February, the males assemble at a display ground where they court interested females. The courting involves a display of synchronized strutting by all the sibling group members. The females are attracted to the dominant group, usually the one with the most members, and mate with the dominant male of this group. Thus most of the mating is done by a single male. Why do Texas turkey follow such an unusual pattern? The answer probably lies with the infrequent rainfall of the area, which is immediately followed by vegetation growth and an insect bloom. The group courtship approach greatly stimulates the females, preparing them rapidly for breeding. They can thus better match their short-lived food resource with the reproductive period. The rapid group courtship also reduces the danger from predators by decreasing the exposure time in the open and by providing more watchful eyes when the birds are occupied with courtship behavior.

Both the lion and the turkey social systems clearly illustrate not only the influence of environment on social systems but also how an organism's social position influences, if not determines, the survivorship and reproductive activity of individuals. To fully understand the dynamics of a population, it is necessary to determine the different types of social position and the number of individuals filling each role.

Spatial structure

Organisms that live in favorable habitats survive longer and produce more offspring than those from suboptimal areas. Two components determine environmental favorability: factors *extrinsic* to the population such as physical conditions (temperature, moisture, pH, and so on) or other species (predators, parasites, competitors, and so on) and factors *intrinsic* to the population such as the number and spacing of individuals of that species within an area. In this section we will be concerned only with intrinsic factors. If there are too many individuals dividing the available resources, reproduction will decrease, even in physically favorable areas. On the other hand, if individuals are so widely spaced that pollination or mating encounters occur infrequently, reproduction will also decline. An individual's spatial relationship with other conspecific individuals, called a *dispersion pattern,* is thus an important factor in determining reproduction and, in some cases, survival. Spatial patterns of plants and sessile animals are relatively easy to study because the organisms are stationary and the patterns are constant. Mobile animals pose a much greater problem. Each animal has its own home range (the area in which it normally lives), but since home ranges overlap in some species, the dispersion pattern can change.

There are three basic types of dispersion patterns: random, clumped or aggregated, and regular (Fig. 10-6). You could study the dispersion of dandelions in your lawn by throwing a barrel hoop onto the lawn several times and counting the dandelions inside the ring for each toss. *Random* dispersion patterns occur where there is no spatial relationship between individuals. Thus random dispersion of dandelions means that each sample unit (one barrel-hoop toss) has an equal probability of having a dandelion, and the presence of a dandelion in a sampling unit does not change the probability of finding another in the same sample. This type of pattern can be described statistically by a Poisson distribution,

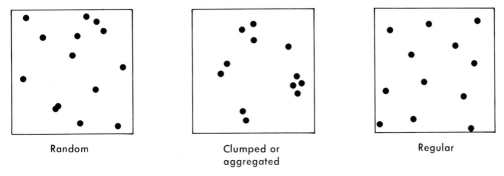

Random   Clumped or   Regular
                          aggregated

Fig. 10-6. Dispersion patterns illustrating the spatial distribution of organisms in random, clumped or aggregated, and regular patterns.

which tells how many samples should have a certain number of individuals (0, 1, 2, and so on). A Poisson distribution has an interesting property: the mean number of individuals per sample unit equals the variance of the individuals found in each of the samples. A ratio of the variance to the mean will therefore equal 1. This ratio is a simple means of assessing the type of dispersion of organisms whose spatial distribution is unknown.

The dandelion may not be randomly dispersed. In fact, there are likely to be more individuals in some areas of the lawn than others, that is, the presence of an individual increases the likelihood of finding another in the same sample unit. Such a distribution is called *aggregated, clumped,* or *contagious.* Because a few sample units will have many individuals but most will have none, the variance of number of individuals per sample unit will be greater than in a random distribution. As a result the variance to mean ratio is greater than 1. Aggregated distributions arise for several reasons: (1) Some areas may be more favorable than others, attracting organisms to settle there. Red-winged blackbirds aggregate in marshes and avoid nearby upland areas. (2) Settling may be random, but differential mortality may result in occupation of only certain areas. (3) If offspring have low mobility, they may tend to aggregate around the parent. Many milkweed bug nymphs are frequently found on the same plant because of limited movement from the oviposition site. (4) Organisms that are passively moved by wind or water currents may accumulate in certain areas. Coconuts floating in the ocean may be carried to some beaches more frequently than others. (5) Social behavior may cause clumping of individuals who are attracted to each other for mating or group protection. Individual sardines have a much lower chance of avoiding predators than those living in large schools.

A *regular* dispersion pattern occurs when individuals are evenly spaced throughout an area. Returning to the lawn of dandelions, regular dispersion means that the likelihood of finding an individual in the sample unit decreases if one individual is already present. Since most sample units will contain the same number of individuals, the variance of numbers per sample unit will be small. Thus the variance to mean ratio will be less than 1. Two basic mechanisms are responsible for regular distributions: (1) A random settling pattern can become regular if individuals too close to others die. This may explain the regular distribution of creosote bush in some Arizona deserts because it allows each plant the maximum area from which to collect the sparse rainfall. (2) Social behavior is a more common cause of regular dispersion. For example, the territories maintained by most temperate-forest birds space individual pairs evenly, ensuring an adequate area for feeding and reproduction. Individuals that are unable to secure and maintain a territory suffer high mortality and fail to reproduce. The muskrat population living in Iowa marshes demonstrates that a regular spatial distribution, which arises from individuals holding territories, affords the best opportunity for survival and reproduction. In a typical marsh territories are maintained only in the central, physically most favorable areas. A limited number of territories can be established in this restricted space. When the number of young muskrats exceeds the space in the central part of the marsh, they leave and take up a nomadic existence on the marsh margins and surrounding country. Errington found that territorial muskrats were resistant to mink predation and disease, although a very high percentage of the nonterritorial individuals died from predation, exposure, and disease. A superficial examination of muskrat populations would have concluded that predation controlled muskrat numbers; in fact, the mink is only the unwitting executioner of the muskrat territorial system.

DISPERSAL

The number of individuals found in an area is influenced by movements as well as births and deaths. Movement of an organism away from an area is called *dispersal*. Day-to-day foraging movements such as cabbage white butterflies flying from flower to

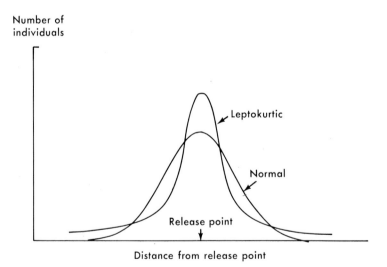

Fig. 10-7. Dispersal curves showing the distribution of an organism away from a release or starting point. Species that move randomly have a normal distribution; directed movements of dispersing individuals result in a leptokurtic distribution.

flower can be called dispersal only when the animal becomes lost or is blown or washed away from its normal foraging area. Such random movements show a *normal distribution* from the starting point (Fig. 10-7) and are indicative of accidental or unpredictable movements with no specific destination. Random movements can be very important in establishing new populations in isolated areas.

Most species also make intentional movements from an area, both to colonize new habitats and to relieve overcrowded conditions. These two purposes are interrelated; there would be no reason to disperse if something were not to be gained. In most cases individuals leave overcrowded or harsh conditions to find a better environment. If leaving a population meant certain mortality, dispersal ability would be rapidly selected out of a population, as evidenced by island insect species that have lost their wings.

Intentional dispersal means that some individuals travel much farther than would be expected by random movements; however, most individuals are relatively sedentary. This behavior results in a *leptokurtic dispersal curve* (Fig. 10-7). In many populations dispersal is a genetic trait passed from generation to generation. Since not all individuals disperse, there are *dispersal genotypes* and *sedentary genotypes* in a population, each with different inclinations for movement. The proportion of each will largely be determined by the nature of the habitat in which the species lives. Unstable environments such as successional vegetation, ephemeral ponds, or seasonally arid habitats will favor a high proportion of dispersal genotypes to allow escape. In more stable environments there will be a smaller proportion of dispersers. These individuals will tend to leave the population even when the population is at low density. It is not until the population reaches higher densities that the normally sedentary individuals will be induced to move away from an area. Obviously population growth will depend on the proportion of each genotype present and the reproductive capacities involved.

GROWTH OF POPULATIONS

If we assume that the number of individuals entering a population (immigration) equals the number leaving (emigration), population growth is the result of the increase of births over deaths. Population size at some time in the future (N_t), equals the number in the population at the beginning of the time period (N_0) plus the difference between births (B) and deaths (D) during the time interval. This relationship can be expressed as follows:

$$N_t = N_0 + (B - D)$$

From this we can calculate the rate of increase per unit of time of the population (r). This is an important factor because it enables us to make predictions about future population size that are based on the assumption that survivorship (l_x) and natality (m_x) remain constant. The rate of growth of any population over a time period is calculated as follows:

$$r(\%) = \frac{N_t - N_0}{N_0} \times 100 = \frac{B - D}{N_0} \times 100$$

As long as r is positive and constant, the population will grow in ever-increasing increments.

To illustrate this point, let us assume a starting population of 50 individuals. During the interval under observation (t), there are 10 additions and 5 losses, giving a net increase of 5 and an N_t of 55. Applying the previous formula, the rate of increase is 10%, calculated in the following manner:

$$r(\%) = \frac{55 - 50}{50} \times 100$$

Carrying the example through nine additional periods of equal duration, the result is a final population of 130 individuals. We note that even though the rate of increase is kept the same, the individual increments themselves become greater with each successive period. The reason, of course, is that at the beginning of each new period there is a larger initial population than during the preceding period; it is this larger population to which the rate is being applied. Remember that this takes into account the losses taking place during every period. From a starting point of 50 we observe the following changes in the population through the 10 periods:

TIME PERIOD	NET INCREASE	POPULATION SIZE (N_t)
1	5.0	55.0
2	5.5	60.5
3	6.1	66.6
4	6.7	73.3
5	7.3	80.6
6	8.1	88.7
7	8.9	97.6
8	9.7	107.2
9	10.7	118.0
10	11.8	129.8

Thus these changes are not arithmetical and would not provide a straight line when plotted against time. Rather the increases are exponential, as depicted in Fig. 10-8. This means that even the slowest breeding animals such as elephants could produce enough offspring to cover the earth in less than 100 generations as long as l_x and m_x remained the same.

How can we calculate population size from a working formula instead of laboriously working out each increment? Knowing N_0 and r, we can predict the final population size for any number of periods. The change in the population size through a moment in time $\frac{dN}{dt}$ is the product of r and N where r is now the instantaneous rate of increase. The formula is as follows:

$$\frac{dN}{dt} = r \times N$$

Note that this differential equation is similar to the preceding expression in which we determined the rate of growth. After integration the following equation is used to make estimates of future growth for selected periods of time:

$$N_t = N_0 e^{rt}$$

where

$$e = \text{Base of natural logarithm} = 2.71$$
$$\log_e 2.71 = 1.0$$

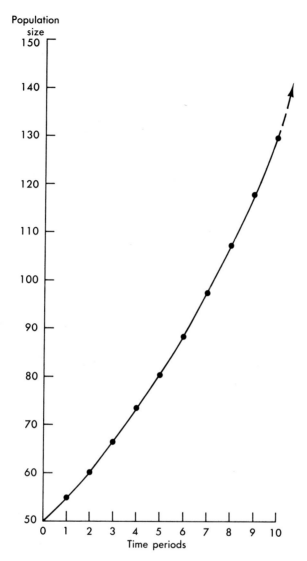

Fig. 10-8. Exponential growth curve. Each succeeding increment is larger than the preceding one, resulting in logarithmic increase. The plotted points are based on the text example for 10 growth periods, p. 254.

Using the data from our previous example in this equation, we can easily calculate the population size after 10 periods as follows:

$$\log_e N_t = \log_e 50 + 1.0(0.10 \times 10) = 4.91$$

$$N_t = 135.9$$

The difference between this exact answer and our earlier estimate results from the continuous integration of N_t in the calculus-based formula. Applying similar procedures to our own population, we can determine how many people will be living in the United States in the year 2000 by using an intrinsic growth rate of 0.87% per year and a current population figure of 220 million. At that time the estimated population of the United States will be 275,450,000 persons.

Populations do not continue to grow exponentially for long. Various limiting factors become more severe as population density increases. A mortality factor that influences a greater percentage of the population as it increases can regulate numbers by increasing deaths in proportion to births. At high densities the rate of increase will decrease and become negative until the population recedes to some lower number. Populations can be regulated by other species, since predation, parasitism, disease, and competition tend to become more severe at high densities. Dense populations are more frequently limited by exhaustion of their food resources. Assuming there is some maximum number of individuals that an environment can support, we can add resource limitation to our population growth equation by inserting a density-dependent factor $\frac{K - N}{K}$ as follows:

$$\frac{dN}{dt} = \left(\frac{K - N}{K}\right) r \times N$$

where

K = the carrying capacity of the environment, or the number of individuals of that species the environment can permanently support

This equation is called the *logistic equation* and graphs as a sigmoid curve (Fig. 10-9). As the population grows, N approaches K, (K − N) approaches zero, and population growth stops. When N = K, the rate of increase is zero (zero population growth), and no further change in population size occurs.

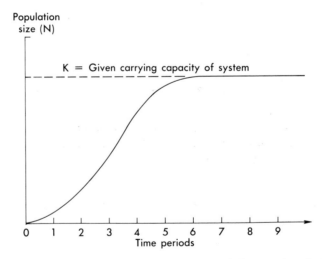

Population size (N)

K = Given carrying capacity of system

0   1   2   3   4   5   6   7   8   9
Time periods

Fig. 10-9. A sigmoidal growth curve described by the logistic equation. As the size of the population approaches the environmental carrying capacity (K), the rate of growth decreases until the size of the population becomes stable at K.

The logistic equation is too simplistic to accurately describe most real populations because (1) K is not constant, (2) real populations do not respond to changes in density instantaneously, and (3) not all members of the population have the same impact on the resource base. The logistic equation is also based on the assumption that the resource is constantly being renewed. In some situations densities exceeding K can damage the resource base of an animal (vegetation or prey population) and make resource renewal impossible. When this occurs, the population does not gradually decrease to K as the logistic equation predicts, but "crashes" to low numbers or extinction.

INTERACTING POPULATIONS

Although the preceding equations show us growth characteristics for populations either under ideal conditions or under conditions in which only members of the same population inhibit growth, we know that in the real world these conditions rarely exist. The various resources that one population finds valuable for growth are often also suitable for other, coexisting populations. When one or more of these resources comes in limited supply, two populations may compete for the resource. The population that most efficiently uses the resource will be best able to grow and maintain itself, to the demise of the other population. A population that grows very large may in effect be making it easy for its predators to focus on it, learn to catch it, multiply rapidly, and eventually reduce its numbers. Thus the interactions between predators and prey affect growth rates. In response to competition and predation, sometimes two different populations have formed ecological partnerships in which each benefits by the presence of the other.

Population interactions such as these can be put into three major categories. Competition refers to a situation in which two populations are attempting to use the same resources in such a way that each is hurt by the presence of the other. Predation involves one population feeding on the other. Obviously the population doing the feeding (the predator) benefits from the relationship, whereas the prey suffers. Mutualism refers to the situation in which two different populations have evolved interactions for the benefit of each.

As with single populations, growth rates for interacting populations can be expressed by equations. Although a veritable science of its own has grown up around the mathematics of competition, predation, and mutualism, we will use just one set of equations to examine all these phenomena. Thus for two interacting species we have the following:

$$\frac{dN_1}{dt} = r_1 N_1 \left(\frac{K_1 - N_1 \pm aN_2}{K_1} \right)$$

$$\frac{dN_2}{dt} = r_2 N_2 \left(\frac{K_2 - N_2 \pm bN_1}{K_2} \right)$$

where

a = effect of species 2 on species 1
b = effect of species 1 on species 2

Under this simplified format, if species 1 and 2 are competing for a resource, they have negative effects on one another, and we obtain the factors $-aN_2$ and $-bN_1$. If species 2 is a predator, it would be benefiting from the densities of species 1 ($+bN_1$), whereas species

1 would suffer from the effects of increased numbers of the predator $(-aN_2)$. Two species in a mutualistic interaction would both gain in the union, resulting in factors of $+aN_2$ and $+bN_1$.

Obviously the preceding equations are quite simplistic and show only the rudiments of the interactions. Nevertheless they show us the basic ways in which one population can affect another and will serve as a foundation for our discussions of competition, predation, and mutualism. In addition to these three categories, other types of interactions between populations occur. These include parasitism, where one population exploits a host population but does not kill individuals, and commensalism, where a population benefits from interactions with a second, but the second population is unaffected. These two interactions will not be discussed further here.

Competition

Competition occurs any time two individuals are attempting to use the same resource and this resource is in limited supply. It may take two basic forms. If the two individuals actually meet and battle over the resource, it is called a contest, or interference competition. More commonly, the first individual to find the resource utilizes it, making it unavailable for the second individual. This is termed a scramble, or exploitation competition. If the two individuals are of the same species, we call the interaction intraspecific competition. This form of competition may be one of the important density-limiting factors observed in the logistic equation for single-species growth rates.

If the competing individuals are of two species, we have interspecific competition. We have already seen the effects of competition on growth rates of two populations. Because competition has negative effects on both species, natural selection tends to favor those individuals that best minimize the effects of competition. Thus we rarely see actual competition in action. Instead we see what are apparently the results of thousands of years of competition by coexisting species. Although we can look at a species in two areas with differing amounts of apparent competition and observe differences in its density, food habits, foraging behavior, and so on, most field evidence for competition is of an indirect nature.

For this reason laboratory experiments have played an important part in the development of competition theory. The first experiments were done by Gause with carefully controlled cultures of the protozoan *Paramecium*. After growing two species separately under ideal conditions, he put them together. One species always outcompeted the other, and eventually only it would occur (Fig. 10-10). Similar studies on the flour beetle *(Tribolium)* yielded similar results. From these studies, Gause developed what is now called the *competitive exclusion principle*. It states that two species with identical or nearly identical ecologies cannot live together in the same place at the same time. If two species attempt to do so, one should eventually (as in the laboratory experiments) cause the extinction of the other. From this we assume that coexisting species differ in enough ecological criteria to ensure their continued coexistence.

We can ask, then, what are the ways in which coexisting species differ to achieve this ecological separation, or ecological isolation, and just how different must coexisting

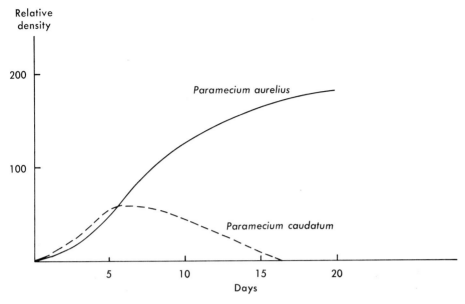

Fig. 10-10. Competition effects between two species of paramecium. After both populations reach a certain size, one is driven to extinction under experimental conditions. (Redrawn from Gause, G. F. 1934. The struggle for existence. The Williams & Wilkins Co. Baltimore.)

species be? To the first question there are probably as many specific answers as there are resources that competing species might find limiting. Speaking more generally, species may be ecologically isolated in three basic ways: range, habitat, or foods.

By possessing different geographical *ranges,* two species can have identical ecologies and continue to exist. Such ecological equivalents can readily be seen by examining the distribution patterns of related species, many of whom have east-west or north-south replacement pairs.

Species living in the same general area may differ by *habitat.* These differences may vary from very simple to exceedingly subtle. Two species with similar ecologies might easily be successful if one lived in forest and the other lived in fields, for they would never really be in the same place at the same time. Other cases are not as simple. One of the most famous of all ecology papers concerns five warbler species that occurred together in northern spruce forests. At the time when the competitive exclusion principle was first being proposed, many scientists pointed to these warblers and suggested that, in fact, these species were doing the same ecological things in the same place at the same time. Only after the late Robert MacArthur looked at these species closely could we see that each species differed in the portion of the spruce tree in which it concentrated most of its foraging activity (Fig. 10-11). Thus in certain cases even very subtle habitat differences are sufficient to ensure ecological isolation.

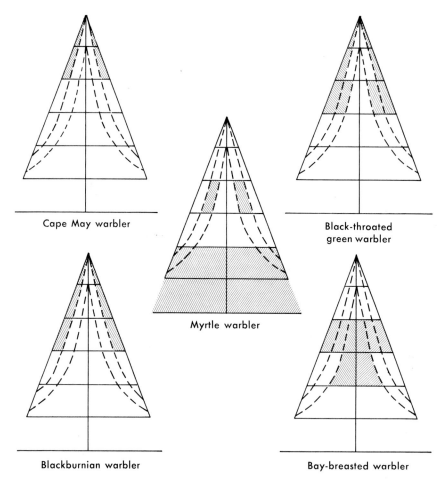

Fig. 10-11. Relative feeding positions of five coexisting warblers in a spruce forest. Each tree is divided into zones for observational purposes. The shaded areas indicate those zones that account for over 50% of the feeding time for each warbler. (Redrawn from MacArthur, R. H. 1958. Ecology **39**:599-619.)

Similar species living in the same habitat may differ in some aspect of the *foods* they use. This ecological isolating mechanism also varies from quite obvious to very subtle. We can easily see how two similar species could coexist if one ate insects and the other seeds. For species sharing the same type of resource, however, some ways of dividing it must be developed. The means vary from different techniques of finding and capturing the food, for example, catching insects in the air versus picking them off leaves, to specializing on different sizes of food. This last category leads to cases of similar, coexisting species differing by size, where the size of each species is related to the average size of its food. The allowable degree of similarity by size varies with habitats, but we assume it reflects the required degree of ecological isolation for continued coexistence.

Although examples can be found in which each of these mechanisms is working alone to allow the ecological isolation of competing species, in many cases competitors may be separating their environment in several ways at the same time. Thus although small communities of birds on islands seem to be organized around size differences, diverse tropical bird communities seem to have exceedingly complex means of ecological isolation.

Now that we have seen how competition limits the similarity of coexisting species, we might ask just how important competition really is? Although early work focused on the ways similar species survived competition, some recent studies have expressed the idea of competition between species that are very different in many ways, or *diffuse competition*. Researchers studied bird distributions on several Andean mountains and discovered that over 70% of the birds' altitudinal ranges were limited by the action of competition. By comparing large and small island populations, other workers found species that were as much as 10 times more abundant on small islands with few competitors. Some apparent cases of competition between insectivorous birds and lizards have been observed, which suggest that we may be just beginning to understand the role of competition in the natural world.

Predation

Predation has some similarities with competition, but in this case the predator benefits from the interaction (in fact, it survives by it), whereas the prey suffers from it and could even become extinct because of it. Our simplified set of equations showed this general relationship; more complex predator-prey equations are characterized by their cyclic nature. Mathematical ecologists have had a difficult time finding predator growth rate equations that can hold up with prey growth rate equations in a stable manner. They have found comfort in the fact that certain simple predator-prey systems such as the lynx-hare system are cyclic in nature. Recent work has suggested that prey cycles are controlled by various environmental factors other than predation, with the predator populations simply responding to prey densities with cycles of their own.

As with competition, much of our early understanding of predator-prey relationships came from laboratory experiments conducted by Gause. In one of the first of these, he put *Paramecium* in an aquarium with predatory *Didinium*. In time the *Didinium* ate all the *Paramecium* and starved to death. Realizing that this experiment was similar to putting lions and zebra in the same cage at a zoo, Gause added some sediment to his aquarium so the prey could have a place to hide. The *Didinium* ate all the available *Paramecium* and starved to death; those *Paramecium* hiding in the sediment soon emerged and spread throughout the aquarium. Only by adding a few immigrants to the system at regular intervals could Gause achieve the cyclic populations the equations had predicted. Huffaker attempted to improve on Gause's experiments by using mites who fed on oranges and who were, in turn, fed on by a predatory mite. Only by constructing an incredible maze of several hundred oranges and rubber balls with many Vaseline barriers could he achieve a system in which the prey kept one step ahead of the predator.

These laboratory experiments might suggest that prey extinction through predation is a common phenomenon, although we know this is not the case in nature. Yet we can intuitively see how predators affect which prey breed (those that avoid the predators), whereas in times of low prey density the prey can have strong selective effects on the predators by affecting which of the predators are successful. Thus these laboratory experiments may at best emphasize the evolutionary nature of predator-prey interactions. For example, when predators develop better systems of prey capture, they may have a large effect on prey densities. Although prey densities may decline, those prey left to breed have some characteristic that has allowed them to escape the predators' new system. If prey offspring have these same characteristics, predator and prey populations will soon return to stable densities. A similar pattern could occur if the prey evolved some escape mechanisms first; eventually the predator would evolve ways of capturing the prey. The overall result is a continual cycle of adaptation of both predator and prey to one another. The amount of change allowed in prey is limited by environmental factors and the various kinds of predators they face. Predatory species must balance the advantages of specializing on a prey with the advantages of being able to harvest several kinds of prey. Because most predators eat many kinds of prey and most prey face numerous predators, most predator and prey populations seem to be relatively stable. Nonetheless the effects of predation on the characteristics of both prey and predator species are pronounced.

To understand just how strong these predator pressures must be, we can examine some of the many techniques that possible prey have adapted to avoid becoming actual prey. A relative measurement of the energy and resources put into these predator avoidance devices gives some idea of the predator pressures involved. One of the most common techniques used is simple escape, which may vary from jumping like a grasshopper to looking like the bark that is a part of the prey's background. Exotic forms of escape include flash coloration of the wings, including a moth showing orange underwings that "disappear" when it lands and warning colorations such as the large "eyes" of other insects.

Other prey populations have adapted an outbreak strategy of predator avoidance. They are absent for many years, thus keeping their predators at relatively low densities. Periodically they all hatch out and become extremely abundant. During this outbreak, predator populations increase but not fast enough to have much effect on the prey. Good examples of this strategy are the pine sawfly and the 17-year cicada.

Instead of attempting to avoid predation in an active way, certain prey have tried to become undesirable to predators through physical or chemical means. Physical means include thorns on trees, hair on leaves, and so on; chemical defenses may include being distasteful or even poisonous. These organisms often possess bright orange or red warning coloration that in a sense advertises their distastefulness and provides a distinct image to those predators that make the mistake of feeding on them. Because these "learning mistakes" do occur, distasteful organisms often converge in color to reduce these losses and leave a distinct set of colors or patterns that a predator learns not to touch. This convergence is called Müllerian mimicry. As might be expected, certain tasteful organisms that resemble distasteful ones might converge on the Müllerian mimics to avoid predation. These are called Batesian mimics. They cannot become too common without destroying the protection they have gained by the mimicry. In the tropics, mimicry complexes of both tasteful and distasteful butterflies are incredibly intricate and serve as further evidence of the importance of predator-prey interactions in shaping the characteristics of many organisms.

Mutualism

The last interaction that we will discuss is mutualism, which occurs when two populations both benefit by some type of relationship. Although both competition and predation are rather orderly phenomena, mutualism involves a wide variety of biological interactions that may include such diverse organisms as bacteria and vertebrates. Although once considered almost a biological curiosity, mutualism is now known to be a common interaction in nature, particularly in tropical regions.

The forms of mutualism range from cases in which two species may benefit from the interaction but can survive without it to cases of obligate mutualism, where neither species can live without the interaction. A good example of the latter is the relationship between termites and a flagellated protozoan that lives in the termite gut. The termite utilizes wood as an energy source, but only the protozoan can digest it. Together they thrive, but neither could survive alone. Other examples of mutualism include the combination of algae and fungi to form lichens, the bacteria that digest cellulose in the rumen of many large vertebrates, and the nitrogen-fixing bacteria that are found in association with many plants.

Although the previous examples are of two species that actually coexist, other forms of mutualism may involve species that are only briefly in contact. Chief among these are many interactions between flowering plants and their potential pollinators or seed dispersers. In these complexes, a plant may be able to attract a specialized type of pollinator or predator and ensure high fertility or good seed dispersal. To accomplish this, the plant may develop highly concentrated nectar, specially designed flowers, or exceptionally nutritious fruit. The animal responding to these offerings may benefit by having a better and a more readily available food source. With the coevolution of two species over time, highly specialized relationships can develop.

In fact, mutualistic systems can develop that involve many species. Earlier in the book, you saw how 18 species of the genus *Miconia* space their fruiting period throughout the year in Trinidad. A small family of birds called manakins have evolved an intricate mutualistic interaction with *Miconia*. The manakins serve as seed dispersal agents for the *Miconia,* which grows in disturbed habitats and thus needs to be dispersed quickly to compete with other species for space. To ensure this dispersal, the *Miconia* has developed fruit providing a completely balanced diet for the manakins. The spacing of *Miconia* species throughout the year has several advantages. It ensures that there will be a year-round food supply for the manakins so that there is a stable population of dispersers. It also helps to prevent overloading the dispersal ability of the manakin population. Because of this close relationship with their food source, manakins are relatively free from competition with other birds and spend little time looking for food. This in turn has led to an unusual breeding system in which the males participate in elaborate group displays.

•   •   •

We are really just beginning to understand the extent and variety of interactions among populations on the earth. Although the mutualism between manakins and *Miconia* seems intricate, it involves members of only two families of organisms. Recent observations in the tropics suggest that many species of flowering trees that compete in some ways may also be spacing their reproductive seasons much as the *Miconia* did. Presumably each species benefits by the maintenance of a large number of pollinators and seed dispersers. The result, however, is a highly complex system of species exhibiting competition, predation, and mutualism. Although each species has its own population growth characteristics, we can see that the evolution of these characteristics and their expression in nature are the result of a wide variety of intricate interactions.

SUMMARY

Populations are the building blocks of ecological communities and ecosystems. Each population is an interbreeding group of individuals. A species is the sum of its populations, and it is at the population level that interactions occur that determine the survival of a species and the ecological characteristics it possesses.

Each population has various structural characteristics that reflect the way it has adapted to its environment. The size of a population is determined by the number of births versus the number of deaths. Age structure is important because age influences survivorship and natality. The age-specific birthrate and mortality schedules are modified to fit the environment and to ensure that the maximum number of viable progeny are produced. Social position is a second determinant of survival and reproductive success. Social systems are influenced in part by habitat and physical conditions. An individual's spatial relation to other conspecific individuals is a third important variable affecting mortality and reproduction. A random, clumped or aggregated, or regular distribution is the result of an organism's relationship to resources and to other species. In addition to births and deaths, emigration and immigration account for changes in population size. Dispersal is movement out of a population and serves to alleviate density stresses and to colonize new habitats.

In an unlimited environment a population will grow exponentially. When resources are limited, population growth should decrease as the size of the population approaches the carrying capacity. This results in a sigmoidal curve, the logistic growth curve. These growth curves are models and may be unrealistic but they do help our understanding of how natural populations function.

Although one can look at the characteristics of an individual population, many of these traits are the results of interactions that the population has had with coexisting populations of other species. When two different populations are using the same resources, a competitive situation exists that has adverse effects on both. If competing species are too similar in the use they make of these resources, one may become extinct.

Some populations decline by being the prey of predator populations. Distinct predator-prey interactions have led to the adoption of numerous predator-avoidance devices by the prey; the evolutionary response is often an increased prey-catching ability in the predator. The end result is a never-ending cycle of adaptation and counteradaptation.

Other populations have engaged in ecological alliances, or mutualism, where both benefit by interacting. These relationships *are most common in stable environments* and involve such interactions as those between plants and their pollinators, plants and nitrogen-fixing bacteria, and cellulose-digesting protozoans and termites. All of these interactions result in the coevolution of two or more populations, so that each develops the characteristics that best aid its survival. By examining populations and the communities they form, we can acquire an idea of the complex interactions that have led to the diversity of the natural world.

DISCUSSION QUESTIONS

1. Discuss the several age class structures in human populations (Fig. 10-5). Which age class structure will take the longest period to reach zero population growth? Explain.
2. Develop the equation for the logistic growth curve. What is its relationship to a geometrically expanding population?
3. What differences might you expect between tropical and temperate regions in the degree of mutualistic interactions present? Explain.

REFERENCES

Cody, M. L. 1974. Competition and the structure of bird communities. Princeton University Press. Princeton, N.J.

Deevey, E. S. 1947. Life tables for natural populations of animals. Q. Rev. Biol. **22:**282-314.

Errington, P. L. 1963. Muskrat populations. Iowa State University Press. Ames, Iowa.

Gause, G. F. 1934. The struggle for existence. The Williams & Wilkins Co. Baltimore.

Gilbert, L. E. 1975. Ecological consequences of coevolved mutualism between butterflies and plants. In L. E. Gilbert and P. H. Raven (eds.). Coevolution of animals and plants. University of Texas Press, Austin, Tex.

Hespenheide, H. 1971. Food preference and the extent of overlap in some insectivorous birds with special reference to the Tyrannidae. Ibis **113:**59-72.

Holling, C. S. 1959. The components of predation as revealed by a study of small-mammal predation of the European pine sawfly. Can. Entomol. **91:**293-320.

Howell, D. J. 1976. Plant-loving bats, bat-loving plants. Nat. Hist. **85:**52-59.

Huffaker, C. B. 1958. Experimental studies on predation: dispersion factors and predator-prey oscillations. Hilgardia **27:**343-383.

Hutchinson, G. E. 1959. Homage to Santa Rosalia, or why are there so many kinds of animals? Am. Nat. **93:**145-159.

Kerster, H. W. 1968. Population age structure in the prairie forb, Liatris aspera. Bioscience **18:**430-432.

Lack, D. 1971. Ecological isolation in birds. Blackwell Scientific Publications, Ltd. Oxford, England.

Laws, R. M., I. S. C. Parker, and R. C. B. Johnstone, 1975. Elephants and their habitats. The Clarendon Press, London.

MacArthur, R. H. 1958. Population ecology of some warblers of northeastern coniferous forests. Ecology **39:**599-619.

May, R. M. 1976. Models for two interacting populations. In R. M. May (ed.). Theoretical ecology: principles and applications. W. B. Saunders Co. Philadelphia.

May, R. M. and R. H. MacArthur. 1972. Niche overlap as a function of environmental variability. Proc. Nat. Acad. Sci. **69:**1109-1113.

Miller, G. T., Jr. 1975. Living in the environment: concepts, problems, and alternatives. Wadsworth Publishing Co., Inc. Belmont, Calif.

Miller, R. S. 1976. Models, metaphysics, and long-lived species. Bull. Ecol. Soc. Am. **57:**2-6.

Papageorgis, C. 1975. Mimicry in neotropical butterflies. Am. Sci. **63:**522-532.

Park, T. 1962. Beetles, competition, and populations. Science **138:**1369-1375.

Perrins, C. M. 1964. Survival of young swifts in relation to brood size. Nature **201:**1147-1149.

Pianka, E. R. 1970. On r- and K-selection. Am. Nat. **104:**592-597.

Pimental, D. 1961. Animal population regulation by the genetic feedback mechanism. Am. Nat. **95:**67-79.

Powell, J. R. and T. Dobzhansky. 1976. How far do flies fly? Am. Sci. **64:**179-185.

Root, R. B. and A. M. Olson. 1969. Population increases of the cabbage aphid, Brevicoryne brassicae, on different host plants. Can. Entomol. **101:**786-793.

Schaller, G. B. 1972. The Serengeti lion: a study of predator-prey relations. University of Chicago Press. Chicago.

Schoener, T. W. 1968. The Anolis lizards of Bimini: resource partitioning in a complex fauna. Ecology **49:**704-726.

Smythe, N. 1970. Relationships between fruiting seasons and seed dispersal methods in a neotropical forest. Am. Nat. **104**:25-35.

Snow, D. W. 1976. The web of adaptation. Quadrangle/The New York Times Book Co. New York.

Southwood, T. R. E. 1962. Migration of terrestrial arthropods in relation to habitat. Biol. Rev. **37**: 171-214.

Terborgh, J. 1971. Distribution on environmental gradients; theory and a preliminary interpretation of distributional patterns in the avifauna of the Cordillera Vilcabamba, Peru. Ecology **52**:23-40.

Terborgh, J. and J. Faaborg. 1973. Turnover and ecological release in the avifauna of Mona Island, Puerto Rico. Auk **90**:759-779.

Terborgh, J. and J. Weske. 1975. The role of competition in the distribution of Andean birds. Ecology **56**:562-576.

Watts, C. R. and A. W. Stokes. 1971. The social order of turkeys. Sci. Am. **224**:112-118.

ADDITIONAL READINGS

Wilson, E. O. 1971. Insect societies. Belknap Press. Cambridge, Mass.

Wilson, E. O. 1975. Sociobiology: the new synthesis. Belknap Press. Cambridge, Mass.

11 Urban ecology

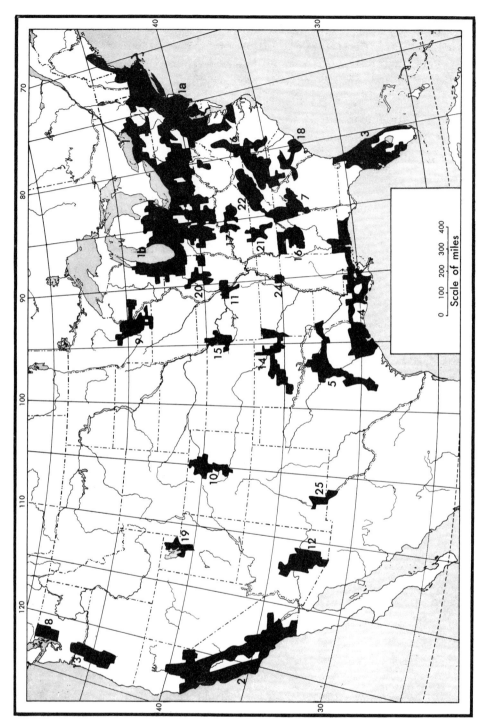

Fig. 11-1. For legend see opposite page.

THE URBANIZATION TREND

More Americans live in an urban environment than ever before. As the nation's population increases, there is a growing percentage living in the city and its suburbs. In 1790, the year of the first United States census, about 5% of the population were classified at that time as town and city residents. Less than a century later this figure had risen to 25% of the population, and 50 years afterward nearly 50% of the nation's people were urbanites. Today nearly 75% of us live in cities and, according to present projections, by the end of the century 85% of the population will be drawn together in a cohesive maze of urban centers and expanding peripheral areas. The United States today is among the most urbanized of nations.

What qualifies an area as "urban"? As a nation's population grows, the criteria also change as affected by technological advances and economic expansion. In the United States a concentration of dwellings housing at least 2500 people is classified as an urban area and for the past 25 years has been the basis of delineation between urban and rural peoples. Obviously such a classification includes numerous small communities that occur outside the corporate limits of urbanized sprawl but that become merging links with metropolitan expansion from still other urban centers. Thus urban regions have evolved. The standard metropolitan statistical area (SMSA) has entered the demographer's lexicon. It defines a county-city area, forming an integrated economic and social system with at least 50,000 people. In 1970 there were 242 SMSAs in the 48 states that included 70% of the U.S. population, and the number of SMSAs continues to grow. The combined SMSA acreage is approximately 13% of the continental United States, according to Otte. Thus the urbanized look takes on geographical proportions, transcending, modifying, even destroying whole landscapes, with here and there only isolated patches of open spaces to break the monotony of our artificial habitat. The megapolis, "Bowash," reaches from Boston to Washington, D.C.; others are forming in strategic locales along the Great Lakes as well as in the west and south. The principal urban regions of the future are shown in Fig. 11-1, indicating the pervasive growth of the city and its environs for the latter half of this century.

Fig. 11-1. Metropolitan regions of major development in the continental United States and expected growth patterns for the year 2000. The regions are designated by number as follows: metropolitan belt, 1; Atlantic seaboard, 1a; lower Great Lakes, 1b; California, 2; Florida Peninsula, 3; Gulf Coast, 4; east central Texas–Red River, 5; southern Piedmont, 6; north Georgia–southeastern Tennessee, 7; Puget Sound, 8; twin cities, 9; Colorado piedmont, 10; St. Louis, 11; metropolitan Arizona, 12; Willamette Valley, 13; central Oklahoma–Arkansas valley, 14; Missouri-Kaw, 15; northern Alabama, 16; Bluegrass region, 17; southern coastal plain, 18; Salt Lake Valley, 19; central Illinois, 20; Nashville region, 21; eastern Tennessee, 22; Oahu (not shown), 23; Memphis, 24; El Paso–Ciudad Juarez, 25. (Data from U.S. Commission on Population Growth and the American Future. 1972. Population and the American future. U.S. Government Printing Office. Washington, D.C.; base map copyright © by Denoyer-Geppert Co. Chicago; used by permission.)

The process of urbanization is universal, affecting industrial nations and less developed ones alike and, simply stated, is the trend to city living for rising proportions of a nation's populace. It is the result of a natural increase in births over deaths and of immigration, which together accelerate urban population growth compared to that in rural areas. Obviously city growth might continue even after urbanization of a nation is complete. Indeed data show that rural populations in some regions of the United States have declined in actual numbers as a result of emigration, but for the nation as a whole these populations have remained rather constant at around 45 to 50 million for approximately the last four decades. Accordingly, we see a fairly stable population classified as "rural," which in the national perspective results in an ever-declining percentage of the total population (Campbell, 1975).

In recent years, however, evidence has been emerging in the United States that urbanization per se is slowing down or even being reversed in specific instances. People are beginning to seek the countryside as a mode of living, albeit still depending on the city proper for their livelihood. There has also been a significant rise in the ownership of second homes and in the sale of country lots for weekend recreation. According to Berg, the number of second homes has increased five times since 1940. Furthermore for every second home actually built, there are six lot purchases whose new owners eventually plan to develop or perhaps (hopefully) to maintain as places of peace and quiet for the personal enjoyment of nature. Not only has suburbia grown and expanded significantly in the last 30 years, but new *exurbias* are proliferating everywhere into the farthest reaches of open country, into regions higher and higher on the mountain, and ever farther into the desert. In these places *beyond the suburbs,* new homes, country estates, condominiums, and, indeed, self-contained developments complete with lakes and golf courses are adding to the demise of open spaces and the integrity of natural areas at an ever-accelerating rate. This far-flung expansion in human habitation is an expensive energy determinant in commuter transportation, particularly in our continued use of private automobiles.

CAUSAL FACTORS

The trend toward urban living is historically related, at least in modern times (since the advent of the industrial revolution), principally to a nation's economic growth and expanding technology. People traditionally are drawn to the urban-industrial center for economic, social, and cultural reasons. Urbanization and national economic growth are shown to be strongly and positively correlated. The city offers employment opportunities, educational advantages, living conveniences, and so on that affect the daily existence of almost everyone. Per capita income on a national scale is higher in the city than in the rural environment and generally reflects that nation's level of urbanization. Data from the Population Reference Bureau, Inc. in Washington, D.C. show these relationships on a global basis (Fig. 11-2). Those countries with the least economic development and relatively large rural populations exhibit the lowest average incomes on a per capita basis. It is of interest to note from current U.N. sources that although the underdeveloped countries (UDCs) have a lesser degree of urbanization at present, urbanization is occurring more rapidly in recent times. Rates of increase for these countries greatly surpass those of

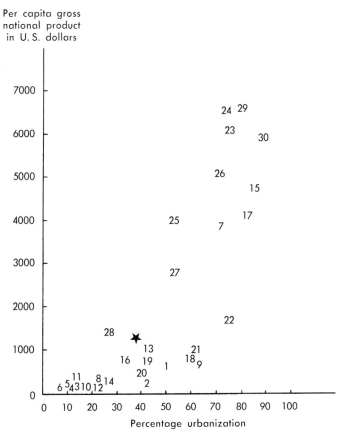

Fig. 11-2. Relationship between 1976 per capita gross national product expressed as U.S. dollars and percentage urbanization for selected industrialization and less developed nations, numbered as follows: Algeria, *1;* Egypt, *2;* Sudan, *3;* Ethiopia, *4;* Kenya, *5;* Tanzania, *6;* Japan, *7;* China, *8;* Taiwan, *9;* Burma, *10;* Thailand, *11;* India, *12;* Iran, *13;* Pakistan, *14;* Australia, *15;* Fiji, *16;* New Zealand, *17;* Brazil, *18;* Costa Rica, *19;* Ecuador, *20;* Mexico, *21;* Venezuela, *22;* Canada, *23;* United States, *24;* Austria, *25;* France, *26;* Italy, *27;* Portugal, *28;* Sweden, *29;* West Germany, *30.* The star represents the world average for all nations. (Graph based on data from World Population Data Sheet published by Population Reference Bureau, Inc., Washington, D.C.)

industrial nations. The following data for the 10-year period from 1963 to 1973 for India and the United States serve as representative examples in these contrasting cultures:

	YEAR	URBAN POPULATION	URBAN (%)	URBAN INCREASE (%)
India	1963	78,835,939	18.0	—
	1973	120,581,000	20.6	53.0
United States	1963	125,282,783	69.9	—
	1973	149,324,930	73.5	19.2

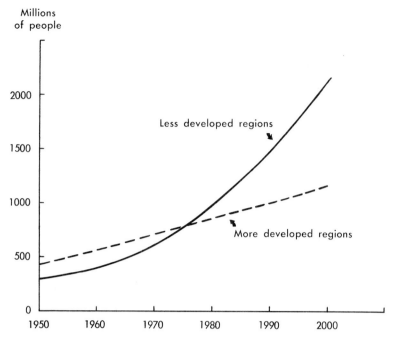

Fig. 11-3. Comparison of urban growth patterns to the year 2000 for the developed and less developed nations of the world. (Redrawn from Donella H. Meadows, Dennis L. Meadows, Jørgen Randers, and William W. Behrens III. *The Limits to Growth,* 2nd Edition, A Potomac Associates Book [New York: Universe Books, Inc.], 1975.)

A comparison of urbanization trends to the year 2000 as a composite average of developed and less developed nations is shown in Fig. 11-3.

Despite rapid increases in the urban populations of the underdeveloped countries, the Population Reference Bureau, Inc. projects that these countries will still be dominated by large and ever-growing rural populations at the end of this century. The issue of population control in these nations will become ever more critical in the years ahead, as attempts are made to eliminate poverty, to improve social conditions, and to effect an orderly transition to an urban-industrial society.

The economic and social incentives that historically are associated with urbanization in the industrialized society are extremely limited in an emerging nation. Yet the trends indicate that large urban centers or concentrations of people may develop in these nations, obviously for different reasons. In the less advanced nations such as India, a burgeoning rural population unable to support itself is an underlying factor. Many of its members leave the farms and small villages each year for the large city, seeking what limited job opportunities it has to offer. This exodus is heightened by crop failure during years of unfavorable weather and by outbreaks of disease, both common occurrences in a primitive agricultural technology. Such mass movements create extensive slums and pockets of poverty and can only make the problems of expanding cities in the poorer nations more crucial. In times of crop failure such countries must rely on foreign aid for assistance and, in some cases, even for survival for large segments of the affected population. The traditional result of demographic change from agricultural to industrial occupation, the latter with its higher standards of living, improved social conditions, and lower population growth rate, is failing to materialize in many underdeveloped countries today. The most valid reason is that agricultural production and economic expansion cannot keep pace with the explosive growth in human populations in either urban or rural environments. With few exceptions the countries with the greatest poverty are in Asia, Africa, and Latin America, which also have the highest rates of population increase (see Appendix C). Most of the nations experiencing rapid population growth within a prevailingly rural social structure must import grain for domestic consumption. It should be pointed out, however, that many industrialized nations such as the Netherlands must also import food. The output of goods and services, however, provides a balance of payments for the foods and cereals that must be imported because of limited space for growing crops in what, too, is an overpopulated situation.

EFFECTS OF URBANIZATION

In the following sections let us examine the impact of urbanization as an artificial human influence on open spaces, where it produces a shrinking countryside; on the physical environment, where we see such phenomena as the growing need for water, the urban heat island, air pollution, and "acid rain"; and on natural resources, particularly urban wildlife. In the pervasive rush for ever-expanding economies in the context of finite resources what, in addition, are the problems of today's city for its human inhabitants, and, more importantly, what are the solutions?

Shrinking countryside

The rapid urbanization taking place today is in evidence everywhere—in expanding subdivisions, industrial parks, and transportation systems spreading into open spaces once far beyond the city limits (Fig. 11-4). Incorporation of large tracts of countryside is taking place; these areas will come under control of municipalities or city-county governments for purposes of future expansion. One may wonder what these large-scale conversions of the natural setting mean in terms of actual land area in the United States. Exclusive of Alaska and the Hawaiian Islands, which are undergoing exploitive processes peculiar to their own special resources, the continental United States comprises about 3 million square miles, or over 2 billion acres.

Some history and projections for the future serve to underscore the changes taking place, which are expected to continue. In 1960 approximately 200,000 square miles were inside urban boundaries. This value is expected to double by 1980 to about 400,000 square miles, according to the U.S. Commission on Population Growth and the American Future. By the year 2000 it is projected that 500,000 square miles will be converted to metropolitan classification. This average exceeds one sixth of the total land area of the continental United States. These projections are made from U.S. Bureau of the Census data and are based on the two-child family. Obviously greater or less fertility would alter

Fig. 11-4. Effects of urban development, City Point, New Haven, Conn. The recent photograph, **B**, was taken in 1972. (From Trefethen, J. B. The American landscape: 1776-1976. Courtesy Judy Stockdale.)

the land conversion process accordingly. It is apparent that some regions are undergoing more rapid change than others, with the Atlantic seaboard (especially the northeast area) showing the greatest effects. The amount of farmland here has dwindled the most rapidly; New Jersey is a case study and a leading example. Fully 25% of its land area is tied to urban use. Once a leading truck garden region, New Jersey presently has less than 1 million acres left for the growing of forage and food plants. Today 99.5% of its population is nonagrarian. According to Kolesar and Scholl, preservation of farmland should be part of a state-wide planning process. Such lands would be designated as agricultural preserves. Paraphrasing the authors, ''an acre saved today can be developed tomorrow, but an acre developed today cannot be saved tomorrow.''

Urbanization is a concentrating mechanism for human living. Those lands most amenable to expansion, including farmland or those having future agricultural potential, tend to become developed. This includes flat terrain, low-lying ground, and alluvial soils with high fertility characteristic of river floodplains. Today floodplain development is becoming a national issue as increasingly more suburban growth occurs. Not only is agricultural land removed as a potential food source, but housing developments and industrial enterprises are subject to the hazard of periodic flooding. Crops and wildlife habitats are renewable, compared to cities built in the wrong places. Although flooding on the average may be relatively infrequent, the public must bear the financial burden through federal aid when disaster does occur.

Today 13% of all agricultural lands in the United States occur within the 242 SMSAs mentioned earlier. This includes some of the most fertile soil for growing crops, which produces 60% of the vegetables, 45% of the fruits and nuts, and 17% of all corn produced in the continental United States. Clearly a national land-use policy for preservation of our food-producing potential is needed. The problem is worldwide. According to Ingraham, an estimated 14 million hectares, or 35 million acres, of the earth's farmlands were lost to urbanization in 1975 alone! This is occurring at a time when food shortages are becoming more crucial and human populations continue to expand.

Physical environment

It has been shown that for every acre in city expansion, there are 3 additional acres outside the urban boundary that are being cleared, drained, or irrigated to support that growth. The city organism is a growing colossus whose osmoticlike influences are felt directly or indirectly far beyond its corporate limits. On the average the water used by one of eight urban dwellers is brought from a source 120 kilometers from the city in which he/she lives. The city per se cannot be viewed as an independent entity marked by distinct boundaries but rather in its holistic or ecosystem perspective as *tied to* and *dependent on* the surrounding region and even the nation as a whole. The problems of New York City, for example, are the nation's problems. As aptly stated by White and White, ''this functional extension is a most significant characteristic of the city.''

Growing need for water. Like the African water hole during the dry season, some American cities are running out of water. Continuous expansion and rapid growth since World War II have raised questions about dependable sources of clean water for the future. The increase in the use of water in some regions of the United States has been phenomenal during recent years, particularly in the west, a fact compounded by rainfall patterns being less dependable than in other regions of the United States. The problem is also growing more crucial because of the ever-present pollution of those very supplies required for human consumption and recreation on which increasing demands are being made. There is less water available per capita as more people must share a finite resource. Most citizens probably are not aware of how vulnerable the cities and towns of America

Table 11-1. Freshwater use and projected requirements for rural and urban-industrial areas to the year 2000 in the United States

USE TYPE	AMOUNT USED (billions of gallons/day)			% INCREASE IN 2000 OVER 1965
	1965	PROJECTED		
		1980	2000	
Rural				
Domestic	2.4	2.5	2.9	20.8
Livestock	1.8	2.5	3.4	88.8
Irrigation	110.8	135.8	149.8	35.2
Total	115.0	140.8	156.1	36.9
Urban-indus-trial				
Municipal	23.7	33.6	50.7	113.9
Industrial	46.4	75.0	127.4	174.5
Total	70.1	108.6	178.1	154.1
Steam-electric power plants	62.7	134.0	259.2	313.5

From Economic Research Service. 1974. Our land and water resources. Current and prospective supplies and uses. Miscellaneous Publication No. 290. U.S. Department of Agriculture. Washington, D.C.

are to critical shortages when less than normal precipitation occurs. These problems will become even more serious as urbanization continues. Here is a "catch-22" situation in which economic growth and the associated needs of water by industry are triggering a fallout of ecological deterrents to improving those very standards of living a growing economy supposedly produces. Despite the economic paradigm of growth and more growth, high-quality water and its availability may well decide the future of our cities, their growth, and where this growth will be possible. Table 11-1 shows projected water use in the United States to the year 2000. The largest increases over current use are by municipalities, industry, and steam-electric power plants. Among the largest users are power plants, the benefits being apportioned to both rural and urban consumers. Compared to 1965, needs are expected to more than quadruple by the year 2000. Based on future urban growth, approximately 80% of this water requirement is allocated to the urban dweller. Similarly a major portion of the rural water for crop irrigation would be assigned indirectly to the urban share of the total water budget.

Water tables are dropping everywhere as we sink wells ever deeper into the earth for irrigation and city use. If water is removed at rates exceeding natural replacement, hydrostatic pressure decreases, causing subsidence. One example of a sinking landform is under the city of Houston, where a drop of 4 feet has been measured during the past three decades. Drilling deeper wells also increases the possibilities of saltwater intrusion, particularly in coastal zones where urbanization and industrial development are well advanced and continue to expand. Fig. 11-5 diagrams the general effect of receding supplies of fresh groundwater. As fresh water decreases, saltwater seepage advances inland to contaminate water zones previously salt free and usable for human consumption.

Urban heat island. Cities are noted for being warmer than the open country. In

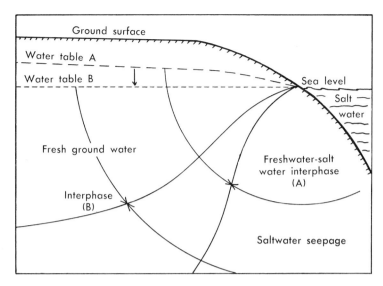

Fig. 11-5. Intrusion of salt water into freshwater substrate. As water tables continue to drop, pressures exerted by fresh water decreases against salt water moving inland, causing an equilibrium shift from *A* to *B* and the contamination of freshwater supplies.

summer solar radiation is soaked up by city buildings, streets, and parking lots usually devoid of trees and other forms of plant cover. The maximum heat difference between city and country usually occurs at night. Cooling after sundown in the city occurs more slowly than in the open country. In the presence of structures made of concrete, steel, and asphalt with relatively high specific-heat values, the surrounding air is warmed by emissions of long-wave radiation from these materials. Generally the city is also somewhat warmer in winter than the open country, but the differences are less marked. In addition to high-density artifacts that replace natural surfaces and change the thermal properties of the environment, other factors come into play that affect a city climate. Positive cooling effects are measurable from water, trees, grass, and soil as a result of evaporation and transpiration. According to Chandler, the cooling effect is considerable. In a modeling study using different percentages of surface area covered by evaporating and transpiring media, a maximum effect of 6° C (100% cover) was demonstrated. This relationship points up the significance of natural cover preserved or retained within urban developments to modern local conditions. Even with only a 30% "water-losing" surface, approximately two thirds of the maximum cooling was achieved, suggesting a nonlinear relationship. As our cities expand and more modification of the natural habitat takes place, greater attention should be given to their design in order to effect the best possible environment for the inhabitants.

Meteorologists have been able to show some variations in rainfall between city and country. Studies of summer rainfall conducted at St. Louis reveal increases in duration and intensity for downwind environs. The amounts averaged 20% to 30% more for the

Fig. 11-6. Atmospheric pollution. Approximately 50% of particulate emissions in the United States come from industrial sources. (Courtesy Environment, St. Louis, Mo., and Robert Charles Smith.)

1971 and 1972 seasons, as compared with its windward environs. The higher temperatures associated with the urban-industrial influence cause increased convective activity and cloud formation. Short-term measurements for individual storms also revealed more rainfall, from 3 to 10 times as much for the city as compared with the open countryside. This increase is mainly a result of the greater size of the rainfall cells forming on a given day. A principal effect of urban development on surface-water yield, however, is increased runoff because of the greatly enhanced area of impervious surface. In cities, therefore, less precipitation actually enters the groundwater system as a storage source for future use.

Air pollution. There are many pollutants of the urban atmosphere, varying in amount with the degree of traffic congestion and industrial activity (Fig. 11-6). These include particulates, hydrocarbons, oxides of nitrogen and sulfur, carbon monoxide, heavy metals such as lead, and photooxidants, a principal one being ozone. Congressional legislation controlling atmospheric pollution such as The Clean Air Act of 1970 provided some hope of abatement of these problems, which have become, in some cases, hazards to human health. As of 1976, however, the phenomenon of healthier air has been slow to materialize. Regarding some pollutants, the condition continues to worsen, particularly in the

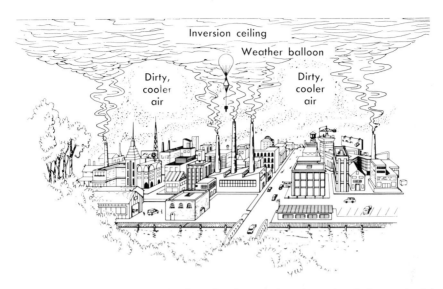

Clear, warm air

Inversion ceiling

Weather balloon

Dirty, cooler air

Dirty, cooler air

Fig. 11-7. Diagrammatic representation of a temperature inversion being probed by a weather balloon. (National Oceanic and Atmospheric Administration, U.S. Department of Commerce. National Weather Service News. NOAA 71-133, pp. 1-4, Sept. 14, 1971.)

case of organic chemicals and aerosol compounds. Another signal of deterioration is that country air is getting dirtier, which means that the "diluting effect" on urban pollution will become less with time if conditions continue as they are.

As more automobiles jam the freeways and metropolitan thoroughfares, several specific problems should be noted. The internal combustion engine is responsible for 60% of the hydrocarbons and over 50% of the oxides of nitrogen. The fine particulates of unburned fuel classified in the 1 micron* *or smaller sizes* are difficult to control and are an important factor in reducing visibility. For example, particles 0.5 micron in size are *25 times* more effective in haze formation than are particles 5 microns in diameter. These smaller particles are most actively involved in smog formation, so often observed during periods of air stagnation, or inversions (see Chapter 2), over large cities (Fig. 11-7). Fig. 11-8 shows characteristic changes in clarity of the air when conditions leading to pollution near the ground and to sharp reduction in visibility are present. The photochemical-chemical reactions are complex, but they can be outlined in simplified form as follows:

Sunlight + oxides of nitrogen → ozone (and other photooxidants) + hydrocarbons → "smog"

This mixture of atmospheric contaminants, including ozone, besides creating the characteristic haze, also causes damage to plants, cracking of rubber, and harmful irritation to

---

*1 micron = $^1/_{1000}$ of a millimeter.

Fig. 11-8. Air pollution over Montreal following a period of atmospheric stagnation. (Reproduced by permission of the National Research Council of Canada from the Canadian Journal of Botany, Volume 48, pp. 1485-1496, 1970.)

eyes and lungs. Emergency conditions can and do develop, causing debilitating effects
and even death in sensitive individuals.

Power plants and other stationary facilities that use coal for energy are the principal
contributors of oxides of sulfur as well as particulates of unburned fuel. As operating
temperatures of these plants (especially in the future) are increased, the unburned residues
of finer particles (1 micron or less) will also increase, thus adding to atmospheric contamination in general. As mentioned earlier, the finer particles are more difficult to arrest,
and even with electrostatic devices it is projected that these particles released to the
atmosphere will continue to increase. Fig. 11-9 shows projected increases to the year
2000. A principal factor, of course, is the ever-increasing demand for energy by urban
society, counteracting legislative efforts to curtail pollution. Other solutions are clearly
needed, including less per capita use of energy-producing fossil fuels and fewer automobiles in the city. Whereas automobiles are the main offenders in the release of oxides
of nitrogen, coal-using power plants are responsible for more than half of the sulfur oxides
being released into the atmosphere.

Acid rain. As the oxides of sulfur and nitrogen rise into the atmosphere, what is their
fate? Studies show that the amount of nitrous oxide delivered to the atmosphere each year
is approximately 20 times its normal occurrence there. Yet oxides of nitrogen show little

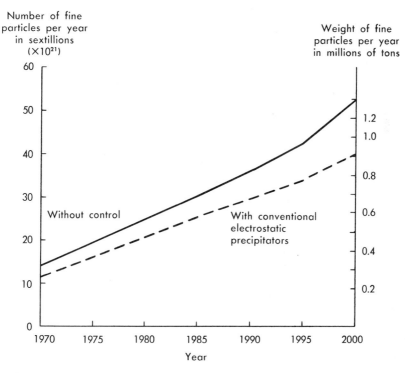

Fig. 11-9. Projected trends in fine particle (0.2 to 1.0 micron) emissions from coal-fired
utility plants. (Redrawn from Brodine, V. 1972. Environment **14**:2-11, 52.)

change in the atmospheric content. The reason is that these compounds are converted to nitric acid upon oxidation and hydrolysis and are returned to earth as ''acid rain.'' Similar reactions occur with sulfurous compounds released through pollution and subsequently ''rained out'' as sulfuric acid. Obviously this acid rain phenomenon is most characteristic of urban-industrial regions such as western Europe and the northeastern United States. A clue to increasing acidity is demonstrated by making pH measurements of the rainwater. The pH value ranges from 0 to 14 on a $\log_{10}$ scale* and, by definition, is the reciprocal of the hydrogen ion concentration expressed in moles of free hydrogen per liter of water. The actual hydrogen ion concentration can be shown as a range from $\frac{1}{10}0$ to $\frac{1}{10}14$. Thus a low pH value indicates high acidity. A pH of 7 indicates a molar concentration of $\frac{1}{10}7$ and is midway on the logarithmic scale. It represents a neutral condition of the solution being measured; pure water is an example, being neither acid nor alkaline. Normal rainfall, uncontaminated by oxides of sulfur or nitrogen, has a pH of about 5.7. This slightly acid condition is a result of the reaction of rainwater with carbon dioxide, producing carbonic acid. (This is a weak acid in contrast to sulfuric and nitric acid, which are strong and, therefore, highly ionized.) Measurements of glacial ice from the Cascade Range interpreted as a kind of historical composite prior to the ''Pollution Age'' showed a pH of 5.6. In recent years, however, rainwater has shown acidities 10 to 100 times that of unpolluted rainfall, with pH values as low as 2.8 being recorded. Remember that each unit of pH is a tenfold change in actual acidity. A pH of 4.0 is 100 times more acid than a pH of 6.0.

Although the long-range effects of acid rain are not fully understood, current observations suggest adverse biological effects such as reduced hatching success of salmon spawn in streams of highly industrialized areas of Scandinavia. Acid rain also has a greater dissolving power than normal rainfall. We can expect more rapid chemical weathering of parent rock materials and greater leaching losses of nutrients such as calcium from the upper soil profile. Decreases in plant and animal diversity of aquatic systens and lower biological productivity are among the probably long-range results of acid rain if allowed to continue on a sustained basis.

Urban wildlife

Changes in wildlife populations are inevitable with advancing urbanization. The loss of open space and the destruction of plant habitat as food and cover sites are the main factors leading to the decline of native species. Pollution and traffic noise also add to the disappearance of wildlife in urban areas. As diversity diminishes with continuing pressure on the natural habitat, certain species more tolerant of these impacts may eventually become more numerous in modified situations.

This so-called dominance-diversity interaction has an inverse relationship (see Chapter 4) and is demonstrated in the urbanization effects on bird life in Tucson, Arizona. In this rapidly expanding region of the desert Southwest, fewer bird species occurred as a result of urban development. Dominance expressed as biomass of some species' popula-

---

*Log to the base 10.

tions, however, actually increased. These latter species showed high tolerance of urbanizing influences. The effects of urban changes were compared with similar but undeveloped desert land. Comparisons in number of species and estimates of living biomass were made according to feeding guilds, that is, whether principal foraging was on the ground, on trunks of trees, in foliage, or in the air and whether seeds or insects were the mainstay of the total diet. The seedeaters had the greatest increase and reflected in large part the influence of home bird feeders, where commercial seeds were made available. Thus although the species diversity of the original avian community was simplified, there was an increase in some individual species' dominance. Exotic species became most abundant, including the house sparrow, starling, and Inca dove. Twelve species of desert birds failed to breed in the city. It was also shown that water-dependent species increased at the expense of water-independent species. Similar observations of the exotic species outnumbering the original populations have been made in other urban areas.

Data on breeding bird populations in Columbia, Maryland, a new town development, provide an excellent example of shifts in diversity and species dominance, as shown in Table 11-2. We note that the greatest "increasers" are the house sparrow and starling, which generally benefit the most where the impact of change is the greatest. As people disrupt the native habitat, newly created niches often favor the unwanted species. Parallels can also be demonstrated with other so-called "weed" species, including mammals and other animals as well as plants.

More studies are needed to understand the impact of urbanization on wildlife behavior and the survival process. As native species and their requirements in the city are better

Table 11-2. Changes in bird populations between 1966 and 1971 during the period of urban development, Columbia, Maryland*†

SPECIES	1966	1967	1968	1969	1970	1971
Decreasing						
Bobwhite	37.0	36.5	35.3	28.0	23.0	9.0
Mourning dove	17.0	12.0	11.3	7.3	4.7	4.3
Wood thrush	16.0	15.0	5.8	5.3	4.3	2.3
Eastern meadowlark	16.0	12.5	12.0	9.7	3.0	6.3
Red-winged blackbird	9.0	12.0	8.2	6.7	2.3	1.3
Indigo bunting	14.0	13.0	8.5	5.3	3.0	2.3
Grasshopper sparrow	15.0	17.0	10.0	8.7	2.7	2.0
Increasing						
House wren	0.0	0.0	1.0	1.7	2.3	2.7
Mockingbird	1.0	4.0	5.0	11.0	11.0	10.0
Starling	1.0	8.0	15.0	13.0	37.0	31.7
House sparrow	0.0	0.0	2.8	15.3	25.7	20.0
Chipping sparrow	0.0	1.0	1.2	8.7	7.3	12.3
Song sparrow	4.0	6.0	9.3	11.7	14.3	12.3

*From Geis, A. D. 1974. In Wildlife in an urbanizing environment. Planning and Research Division, Series 28, Holdsworth Natural Resources Center, University of Massachusetts. Amherst, Mass.
†The values are averages of daily counts during selected periods of the year.

understood, we can initiate sensible planning for city growth and renovation to provide suitable habitats for desirable species. Management of wildlife in the city is a necessity. It is important not only to provide habitat but also to control the opportunists such as starlings and other pest species. Forests, prairie tracts, and wetlands should become an integral part of the planning process as a conceptualized interpretation of what a city should be for man as well as other species. Rather than subjecting the native habitat to hazardous urban sprawl and unplanned development, a fractional portion of the area undergoing urbanization should be legislated as wildlife preserves. In this way we can make urban living more enjoyable for all and at the same time maintain biotic diversity at the highest level. The educational and aesthetic benefits of such programs would far outweigh the cost of the original investment. The presence of the bird feeder in the backyard, the planting of landscape material in the home environs, and the development of plant cover generally suggest that urban man still is reluctant to lose contact with green plants and wildlife and the food and cover they afford.

Recent studies show that in the establishment of trees in cities certain species are more adapted to urban conditions than others. Some are quite tolerant of polluted air, including the native sugar maple. In a study conducted by LeBlanc in Montreal it was shown that leaves taken from trees in areas of the city with severe air pollution were different in certain anatomical features from those growing in relatively clean air. Thus various native species show adaptability to environmental changes. On the other hand, epiphytic lichens and mosses have been shown to be extremely sensitive indicators of air pollution effects. In the industrial areas of Europe and the eastern United States numerous species of these plant forms have disappeared or have declining densities. In the design of cities with its incorporation of vegetal cover, species should be selected for both aesthetic and pragmatic reasons to offset the human impact on the habitat. The Pinchot Institute System for Environmental Forestry Studies, created in 1970, is an organization of the U.S. Forest Service to aid in the improvement of human environments, specifically in the heavily populated northeastern United States. Forest vegetation, where it can and should be maintained, provides the following benefits:

1. Stabilizes living conditions for humans

Table 11-3. Comparison of nonwhite population percentages between the inner city and the suburbs for metropolitan areas of 1 million or more (1970) for several regions in the United States*

REGION	INNER CITY			SUBURBAN RING		
	1950	1960	1970	1950	1960	1970
Northeast	10.4	17.6	26.6	3.6	3.8	4.8
North central	12.6	20.3	27.6	3.2	2.6	3.3
South	24.6	30.8	38.6	13.0	9.7	8.3
West	6.1	8.4	12.5	2.8	2.8	4.2

*Adapted from Chinitz, B. 1976. In M. Chatterji (ed.). Space location and regional development. Proceedings of the University Symposium on Regional Science. 1974. Binghamton, N.Y.

2. Furnishes sites for recreation
3. Protects water supplies
4. Provides wildlife sanctuaries
5. Screens industrial developments
6. Abates urban noise
7. Reduces summertime temperatures
8. Filters out dust and other particulates
9. Enhances the landscape for aesthetic enjoyment

Detailed discussion of urban design and harmony with nature in any landscape is beyond the scope of this brief treatment, but the interested reader should consult the works of Ian McHarg.

Human condition

An "inner city" and its suburban ring characterize the growth and development of most metropolitan areas. Each is an environmental composite, having distinct social and economic characteristics. The former depicts the old origins of the city, now often stagnant, congested, with decaying structures beside high rises, and the latter a dynamic, horizontally expanding perimeter of new dwellings, shopping centers, and industrial parks. Those with mobility and financial advantages join in the exodus to the suburbs, the poor and indigent being left behind. The latter are locked into an economic structure from which it is difficult or impossible to depart. Besides the sociological implication of depreciating neighborhoods, failing tax revenues, and limited employment opportunities, the inner city almost invariably exhibits great ecological deterioration as well.

Downtown is a city of contrasts: slum and tenements coexist with office buildings housing commercial and financial interests. In a recent survey white-collar jobs actually increased, but blue-collar opportunities declined, contributing to the unemployment of the poor and less educated. The plight of the disadvantaged is self-perpetuating under these conditions. This lack of opportunity for employment, education, and a healthier environment affects the minorities to the greatest degree. In Table 11-3 data are presented to show increasing percentages of the total population in the inner city being composed of non-

Table 11-4. Comparison of white and nonwhite family incomes in the inner city as percentages of total population for metropolitan areas of 1 million or more (1970)*

REGION	LESS THAN $5,000/YR		MORE THAN $15,000/YR	
	WHITE	NONWHITE	WHITE	NONWHITE
Northeast	22	35	18	9
North central	20	32	20	11
South	25	39	19	7
West	17	26	25	15

*Adapted from Chinitz, B. 1976. In M. Chatterji (ed.). Space location and regional development. Proceedings of the University Symposium on Regional Science. 1974 Binghamton, N.Y.

whites during the 20-year period from 1950 to 1970. These trends are similar in every region of the country. Lacking educational opportunities, these same groups also make up the larger segment in the lowest income bracket. Supporting data are provided in Table 11-4. Mass transit systems would not only alleviate the pollution problem of the private automobile but would also provide access to jobs outside the city core.

The environmental deterioration of the inner city is a greater hazard to human health than any hazards in the suburbs. Automobile congestion on the freeways, power plants, and old factories make air pollution a major problem. Air-borne lead, especially from automobile exhaust, is a serious pollutant. A revealing study shows different lead concentrations in the blood based on place of residence and occupation. Data assembled by the Center for Biology of Natural Systems are as follows:

SUBJECT	LEAD POISONING (μg lead/100 ml blood)
Urban infants	22.1
Suburban infants	18.3
Adults in rural United States	10-16
Adults in urban United States	16-21
Cincinnati traffic police and garage workers	30-31

The reduction in private use of cars would reduce pollution and would also save energy. Hawley presents data to show that a single-track mass transit system could carry 40,000 passengers at peak rush hour capacity compared to 13 lanes of cars, each averaging 1.5 passengers. Furthermore the capital investment for highways per person trip during peak period is $1670 compared to $440 for the subway and $140 for elevated rail lines. Clearly much of the urban cost of energy and depreciation of the atmospheric environment could be reduced by curtailed use of private transportation in the metropolitan area.

SUMMARY

Urbanization traditionally follows a nation's economic expansion and technological development. The advanced countries of the world are characterized by highly developed and complex urban-industrial societies in which most of the population lives in cities. More recently, less developed countries are exhibiting rapid urbanization rates, exceeding those of the advanced nations. The reasons are intricately interwoven, but a salient feature providing impetus to urbanization is explosive growth in the rural population. Land per capita is diminishing; that fact coupled with outdated agricultural operations and even threats of periodic famine have resulted in immigration to the city to seek better living conditions. In the affluent nations and particularly in the United States where heretofore land for expansion has always been available, the emergence of extensive metropolitan regions typifies growth in the latter part of the twentieth century. These regions assimilate and modify the open countryside. The impact of congestion, exhibited by proliferating highways, expanding suburbia, and the inner city itself, has led to a deteriorating environment, extirpation of wildlife species, and even threats to human health from pollution. Usable water will become a limiting factor in some regions in the coming decades. The city core and its suburban ring constitute an environmental composite with distinctive modes of human existence. Besides social and economic depreciation, the older sections of the city exhibit the greatest ecological deterioration and also the greatest hazards to health, especially for the residents of slums. Air pollution is a major problem. Areas with large concentrations of people require the preservation and maintenance of open spaces and parks for a more livable habitat. Efficient mass transit systems are a necessity to provide economic opportunity for the disadvantaged, curtail the use of private automobiles, conserve energy, and alleviate noise and atmospheric pollution. Lastly, our dwellings and other artifacts that characterize the urban existence should provide ecological compromises in the holistic sense to blend and incorporate natural values as much as possible.

DISCUSSION QUESTIONS

1. Explain the historical factors promoting expanding urbanization in rich and poor nations.
2. Discuss the extravagant use of land in developing suburbia. What should we do to conserve land for wildlife and agricultural production?
3. What can be done to make the inner city more livable for its permanent inhabitants and to arrest the flight to the suburbs?

REFERENCES

Adams, R. 1966. The evolution of urban society. Aldine-Atherton Publishing Co., Inc. Chicago.

Berg, N. A. 1974. Matching resources and people. Water Spectrum 6:19-25.

Boorstin, D. 1976. Rural immigration. In H. Gimlin (ed.). Editorial research reports on the American future. Congressional Quarterly, Inc. New York.

Bowman, G. A. (ed.). 1976. Land use: issues and research needs for planning policy and allocation. Agricultural Research Committee, National Association of State Universities and Land Grant Colleges. Washington State University. Pullman, Wash.

Brodine, V. 1972. Running in place. Environment 14:2-11, 52.

Caldwell, J. C. 1976. Toward a restatement of demographic transition theory. Pop. Dev. Rev. 2:321-366.

Caldwell, W. A. (ed.). 1973. How to save urban America. The New American Library, Inc. New York.

Campbell, R. R. 1975. Beyond the suburbs: the changing rural scene. In H. H. Hawley and V. P. Rock (eds.). Metropolitan America in contemporary perspective. Holsted Press. New York.

Center for Biology of Natural Systems. 1973. Clinical and subclinical effects of lead poisoning in children. Note 6. pp. 12-19. Washington University. St. Louis, Mo.

Center for Biology of Natural Systems. 1973. The SST and ozone. Notes 1-9. Washington University. St. Louis, Mo.

Chandler, E. (ed.). 1974. City's land subsides as water table falls. Water Spectrum 6:48.

Chandler, T. J. 1976. Urban climatology and its relevance to urban design. Technical Note No. 149. World Meteorological Organization. Geneva.

Chavooshian, B. B. 1975. Growth management: a new planning approach. Urban Land 34:22-27.

Chinitz, B. 1976. Metropolitan growth patterns in the United States in the post World War 2 period. In M. Chatterji (ed.). Space location and regional development. Proceedings of the University Symposium on Regional Science. 1974. Binghamton, N.Y.

Christianson, C. M. 1975. Mountain scenery under city lights. Parks Rec. 9:28-29.

Clarke, J. F. and J. T. Peterson. 1973. An empirical model using eigenvectors to calculate the temporal and spatial variations of the St. Louis heat island. J. Appl. Meteor. 12:195-210.

Clawson, M. 1974. Suburban land conversion in the United States. An economic and governmental process. A study conducted for Resources for the Future, Inc. The Johns Hopkins University Press. Baltimore.

Cramp, S. and H. D. Tomkins. 1966. The birds of inner London. Br. Birds 59:209-233.

Davis, K. 1965. The urbanization of the human population. In Man and the ecosphere, readings from Scientific American. W. H. Freeman & Co., Publishers. San Francisco.

DeGraaf, R. M. and J. W. Thomas. 1974. A strategy for wildlife research in urban areas. In Wildlife in an urbanizing environment. Planning and Research Division, Series 28, Holdsworth Natural Resources Center, University of Massachusetts. Amherst, Mass.

Eaton, R. L. 1971. The animal parks: the new and valuable biological resource. Bioscience 21:810-811.

Eckholm, E. 1976. Losing ground. Environment **3**:7.

Economic Research Service. 1974. Our land and water resources. Current and prospective supplies and uses. Miscellaneous Publication No. 290. U.S. Department of Agriculture. Washington, D.C.

Emlen, J. T. 1974. An urban bird community in Tucson, Arizona: derivation, structure, vegetation. Condor **76**:184-197.

Ewing, R. H. 1972. Potential relief from extreme urban air pollution. J. Appl. Meteor. **11**:1342-1345.

Geis, A. D. 1974. Effects of urbanization and types of urban development on bird populations. In Wildlife in an urbanizing environment. Planning and Research Division, Series 28, Holdsworth Natural Resources Center, University of Massachusetts. Amherst, Mass.

Hage, K. D. 1972. Nocturnal temperatures in Edmonton, Alberta. J. Appl. Meteor. **11**:123-129.

Hawley, A. H. 1971. Urban society: an ecological approach. The Ronald Press Co. New York.

Hoben, J. E. 1975. The costs of sprawl. Challenge **6**:24-26.

Huff, F. A. and P. T. Schickedanz. 1974. METROMEX: rainfall analysis. Bull. Am. Meteor. Soc. **55**:90-92.

Ingraham, E. W. 1976. Lead time for assessing land use: a case study. Science **194**:17-22.

Keyfitz, N. 1966. Population density and the style of social life. Bioscience **16**:868-873.

Keyfitz, N. 1976. World resources and the world middle class. Sci. Am. **235**:28-35.

Kolesar, J. and J. Scholl. 1975. Saving farmland. The Center for Analysis of Public Issues, Inc. Princeton, N.J.

LeBlanc, F. and J. De Sloover. 1970. Relation between industrialization and the distribution and growth of epiphytic lichens and mosses in Montreal. Can. J. Bot. **48**:1485-1496.

Lowry, W. P. 1967. The climate of cities. Sci. Am. **217**:15-23.

Marston, L. 1973. Land use: water. Spectrum **5**:7-11.

McHarg, I. L. 1969. Design with nature. The American Museum of Natural History. Natural History Press. Garden City, N.Y.

McPherson, M. B. 1972. Hydrologic effects of urbanization in the United States. Office of Water Resources Research. National Technical Education Service. Springfield, Va.

Meadows, D. H. et al. 1975. The limits to growth, ed. 2. A Potomac Associates Book; Universe Books. New York.

Meadows, D. L. et al. 1974. Dynamics of growth in a finite world. Wright-Allen Press, Inc. Cambridge, Mass.

Meir, R. L. 1976. A stable urban ecosystem. Science **192**:962-968.

Mitchell, J. M., Jr. 1971. The effect of atmospheric aerosols on climate with special reference to temperature near the earth's surface. J. Appl. Meteor. **10**:703-714.

National Oceanic and Atmospheric Administration. 1974. Denver "brown cloud" study. Bull. Am. Meteor. Soc. **55**:1501.

Ogden, S. R. (ed.). 1969. America the vanishing rural life and the price of progress. Stephen Greene Press. Brattleboro, Vt.

Otte, R. C. 1974. Farming in the city's shadow. Agricultural Economics Research Service. Report No. 250. U.S. Department of Agriculture. Washington, D.C.

Pinchot Institute. 1973. The Pinchot Institute system for environmental forestry studies. U.S. Forest Service General Technical Department NE-2. Northeastern Forest Experimental Station. Upper Darby, Pa.

Reilly, W. K. (ed.). 1973. The use of land: a citizen's policy guide to urban growth. Thomas Y. Crowell Co., Inc. New York.

Ryan, D. A. and J. S. Larson. 1976. Chipmunks in residential environments. Urban Ecol. **2**:173-178.

Saarinem, T. F. 1976. Environmental planning perception and behavior. Houghton Mifflin Co. Boston.

Schmid, A. A. 1968. Converting land from rural to urban uses. A study conducted for Resources for the Future, Inc. The Johns Hopkins University Press. Baltimore.

Shomon, J. J. 1971. Open land for urban America. The Johns Hopkins University Press. Baltimore.

Stearns, F. and T. Montag (eds.). 1974. The urban ecosystem: a holistic approach, Part I. Dowden, Hutchinson & Ross, Inc. Stroudsburg, Pa.

Trefethen, J. B. The American landscape: 1776-1976. Wildlife Management Institute. Washington, D.C.

Turner, F. C. 1976. The rush to the cities in Latin America. Science **192**:955-962.

U.S. Commission on Population Growth and the American Future. 1972. Population and the American future. U.S. Government Printing Office. Washington, D.C.

Uthe, E. E. and P. B. Russell. 1974. Experimental study of the urban aerosol structure and its relation to urban climate modification. Bull. Am. Meteor. Soc. **55**:115-121.

White, M. and L. White. 1962. The intellectual versus the city. Mentor Books. New York.

White, W. H. et al. Formation and transport of secondary air pollutants: ozone and aerosols in the St. Louis urban plume. Science **194:**187-189.

World Population Data Sheet. 1976. Population Reference Bureau, Inc. Washington, D.C.

ADDITIONAL READINGS

Auerbach, J. S. 1976. Unequal justice. Lawyers and social change in modern America. Oxford University Press, Inc. New York.

Barnes, P. (ed.). 1975. The people's land. Rodale Press. Emmaus, Pa.

Commoner, B. 1976. The poverty of power: energy and the economic crisis. Alfred A. Knopf, Inc. New York.

Gist, N. P. and S. F. Flava. 1974. Urban society. Thomas Y. Crowell Co., Inc. New York.

Haar, C. M. 1972. The end of innocence. A suburban reader. Scott, Foresman & Co. Glenview, Ill.

Havlick, S. W. 1974. The urban organism. Macmillan Inc. New York.

Higbee, E. 1967. The squeeze, cities without space. William Morrow & Co., Inc. New York.

Keil, C. 1966. Urban blues. University of Chicago Press. Chicago.

Lynch, K. 1960. The image of the city. The M.I.T. Press. Cambridge, Mass.

Still, H. 1968. Man: the next 30 years. Hawthorn Books, Inc. New York.

Stratford, A. H. 1973. Airports and the environment. St. Martin's Press, Inc. New York.

Whyte, W. H. 1968. The last landscape. Doubleday & Co., Inc. New York.

Wolf, P. 1974. The future of the city. Watson-Guptill Publications. New York.

PART FIVE

Postscript

12 Imperatives

People adopt a *laissez-faire* attitude toward population because they refuse to face the hard choices of population control . . . a change of our own choosing can be pleasanter than the unavoidable change that will be forced upon us if we refuse to make a choice.*

GARRET HARDIN

Much has been written about our environmental problems and the reasons underlying them. Nature's resilience and power of renewal do indeed have bounds. As stated earlier, our earth is a finite system of checks and balances. Its resources are not without limits. It is true that we can acquire, redistribute, concentrate, and use these resources with varying degrees of efficiency. The fact remains, however, that what one of us gains another must give up. In the global perspective this axiom applies to governments as well as to individuals and corporations. When there was a frontier, the finite nature of the earth had not yet been realized. Then people moved on to the next valley or open place, to new lands and fresh opportunities. What once was is no longer so. In terms of suitable living conditions for over 4 billion people, we have now reached, if not exceeded, the carrying capacity of our planet.

What would have happened during the Irish potato famine of the nineteenth century if 2 million people over several decades had not been able to emigrate to a relatively unpopulated America? As it was, another 2 million did die of malnutrition and hunger. Doubtless many more would have starved, all because of the potato blight fungus, known by its scientific name as *Phytophthora infestans*. This mass egress from a western culture is paralleled in other countries as well for a number of reasons. Today we are much more crowded than people were a century ago during the potato famine; there is literally no place to which we can escape. There is more human frustration and, inevitably, more social strife. At the same time a growing sameness, ugliness, and monotony associated with the megapolis existence is becoming ever more apparent. Space and resources are being hard pressed on all fronts.

Some may say that the United States, with an average population density of 55 persons per square mile, has no space crisis compared to other parts of the world. Thriving Holland might be cited as an example, with 975 persons per square mile. These comparisons are ill founded. We cannot, in the ecosystem view of the earth, isolate and compartmentalize ourselves and our problems. Without food imports, Hollanders would soon starve, as would the people of many other populous, highly industrialized countries such as Japan and Great Britain. The resources, food, and energy from less populated regions of the earth are often the mainstay of support for the so-called advanced nation with a per capita deficiency of arable land. The industrial society is an osmotic force that sequesters the resources and potential wealth of less technologically advanced nations, often for its own benefit and at their expense.

In this mechanical age the impact of human influences on fragile ecosystems is occurring at an increasing pace and with ever greater power and finality. We can conceive of

---

*Hardin, G. 1972. Exploring new ethics for survival: the voyage of the spaceship Beagle. Penguin Books, Inc. Baltimore.

affluent sportsmen in helicopters tracking down polar bears on the Arctic ice pack, motor-cyclists making erosion tracks in an alpine tundra, a giant coal combine stripping away whole counties, or still another shopping center being built anywhere and everywhere. What should be done to reverse this trend or to at least decelerate it so as to reach an accommodation with our environment?

Almost as a matter of tradition we have come to accept bigness as a measure of success. The gross national product (GNP) is an index of growth. A larger GNP in each succeeding year is a source of national pride and a symbol of progress. But is such growth, measured in terms of financial transactions for goods and services, indeed representative of true social progress and human betterment? Is it a real criterion, for example, when the costs of combating unnecessary pollution created by industry are included in the total assessment? Obviously pollution prevention is less costly than pollution remedies. The reduction in environmental quality arising from pollution is not measurable on such a yardstick. How do we determine the loss of ponderosa pine forest caused by Los Angeles smog, the industrial eutrophication of Lake Erie, the large-scale acid pollution caused by the surface mining of coal in Kentucky, or the destruction of our estuaries and intertidal zones by oil, industrial, and real-estate interests?

Pollution and destruction of open spaces are as great a threat to human existence as threats of starvation, if not in fact greater. Our food resources can be expanded at the expense of biotic diversity and environmental stability so that many more can live a sub-marginal life on a crowded planet. But should this be our main goal? Yet such artifacts as the GNP and increasing production are frequently misused as a measure of the well-being of a nation and its people. Indeed it is safe to predict that over time the growing volume of goods and services is inversely related to ecological stability. Other, more sensitive criteria are needed to gauge social progress and the condition of our surroundings. The reader is urged to examine in detail the thesis of E. F. Schumacher concerning the adverse effects of materialism. Challenging questions are posed about the desirability of continuous economic growth.

Although the technological interdiction of ecosystem processes may serve the short-range needs of society, they can only weaken the environmental fabric in the long run. Even though we enjoy the amenities and comforts of an industrialized society and would find it difficult to give them up, we have them at the cost of resource depletion and poisoned environments. Somewhere there must be a compromise, to trade materialism and increasing demands for food for environmental integrity. We know that the ecosystem has been damaged, in some instances severely. Obviously one of the first corrective measures in a concerted program of environmental improvement is to clean up the mess we have already made. A second measure is to prevent further deterioration. Stringent controls are needed to minimize pollution of the air, water, and land. This means legislative action and a proper set of inducements to prevent recurrences. Some progress has been made in the lawmaking process, but much remains to be done to stem the tide of degradation. The annual report (1976) by the editors of *National Wildlife* indicates little if any improvement in the following six major indices of the environment:

1. Air quality

2. Water quality   5. Forest cutting versus regeneration
3. Soil protection   6. Open space
4. Wildlife preservation

If our environment is to be kept whole, the philosophy of affluence that we have come to accept as the basis for our way of life should be questioned. Can we own snowmobiles and power boats, summer homes, and an endless list of mechanical gadgets and still naively expect to have the environment as we would like it maintained? Obviously we cannot. A government official once commented on DDT for a select readership as follows: ". . . farming isn't going to be very successful if we're forced to ban another necessary pesticide every time a few birds drop dead." Although DDT is not the issue it once was, at least in this country, its manufacture still continues for foreign export. Nonetheless whether it is DDT or some other chemical, the attitude toward the use of those chemicals on the part of many in this country remains unchanged.

Ecology provides us with an irrevocable set of rules—something must be given for something else. This exchange applies to the sanctimoniously guarded province of family size and the rights of parents. When there was a frontier and a crude society, the more hands the better in the development of the natural wealth. In this context perhaps this attitude was relevant to the public good. But is such an attitude relevant today, in the midst of ghettos, poisoned lakes, oil-soaked birds, dead fish, congested parks, and highways leading everywhere but nowhere?

Clearly, and most pervasively, the real question is how far we want to push the earth's carrying capacity at the expense of environmental quality and the human condition. Should the carrying capacity be 4 billion, 10 billion, or even some smaller value? According to Hulett, the optimum population of the earth with a level of affluence similar to ours in the United States would be about *1 billion* persons. The more there are of us, the less quality and dignity there will be in the life of the average person. It is true that some, by virtue of privileged position within the social structure, will live well, but as populations expand, the finite aspect of our planetary resources dictates that the gulf between the haves and the have-nots will widen. The hungry people of the world are multiplying twice as fast as those who have plenty of food. Compulsory control of births will become necessary if personal initiative does not stem the tide, not only in the poor countries but in the affluent ones as well.

Thomas Paine, in *Rights of Man,* wrote many decades ago that "My country is the world. . . ." In these words is the essence of our ecology, the interdependence of natural systems, and the inescapable truism that we cannot identify for security and convenience solely with a given piece of this planet while exploiting and controlling other areas from afar. Religious parochialism, intense nationalism, and tribal inclinations to control resources, dominate territory, and accumulate wealth are subtle, pervasive, and often attractive influences. Yet these very influences depress human values, create war and civil strife, and threaten the very environment on which we depend for survival. The environment is everyone's concern. We cannot afford to be exclusionists who would, with philistine independence, use it without sacrificing materialism in a Thoreau-like fashion and developing a personal commitment to protect our environment. In *A Sand County almanac,* Aldo Leopold wrote the following long before environmental problems became national issues:

> There is, as yet, no sense of shame in the proprietorship of a sick landscape. We tilt windmills in behalf of conservation in convention halls and editorial offices, but on the back forty we disclaim owning even a lance.

The real ecological challenge is to change traditional attitudes toward our only environment and its unique biotic diversity that has evolved over millions of years and, in the Olympian sense, toward even ourselves.

DISCUSSION QUESTIONS

1. Discuss the carrying capacity of the earth for the human population. What are the implications concerning an actual carrying capacity for the earth?
2. As human populations continue to expand, finite resources and sound management are threatened. Explain.
3. Our environmental problems are planetary in scope. What must be done to achieve social stability in the global perspective? Explain.

REFERENCES*

Boulding, K. E. 1970. Fun and games with the gross national product. In H. W. Helfrick (ed.). The environmental crisis. Yale University Press. New Haven.

Boulding, K. E. 1973. The shadow of the stationary state. In M. Olson and H. H. Landsberg (eds.). The no-growth society. W. W. Norton & Co., Inc. New York.

Callahan, D. 1972. Ethics and population limitation. Science 175:487-494.

Chasteen, E. R. 1971. The case for compulsory birth control. Prentice-Hall, Inc. Englewood Cliffs, N.J.

Commoner, B. 1971. The closing circle. The Viking Press, Inc. New York.

Dale, E. L., Jr. 1970. A Nixon advisor doubts overpopulation. The New York Times. 29 March: Sect. 1, p. 75.

Drucker, P. F. 1972. Saving the crusade/the high cost of protecting our future. Harpers 244:66-71.

Ehrlich, P. R. 1970. Famine 1975: fact or fallacy. In H. W. Helfrick, Jr. (ed.). The environmental crisis. Yale University Press. New Haven, Conn.

Elder, F. 1970. Crisis in Eden; a religious study of man and environment. Abingdon Press. Nashville, Tenn.

Farvar, M. T. and J. P. Milton. 1972. The careless technology. Natural History Press. Garden City, N.Y.

*Additional readings in socioecological problems and issues are listed in Appendix A.

Frome, M. 1962. Whose woods these are: the story of the natural forests. Doubleday & Co., Inc. New York.

Gofman, J. W. 1971. Nuclear power and ecocide: an adversary of new technology. Bull. Atom. Sci. 27:28-32.

Hardin, G. 1972. Exploring new ethics for survival: the voyage of the spaceship Beagle. Penguin Books, Inc. Baltimore.

Harris, C. 1971. Man in nature: model for a new radicalism. In T. R. Harney and R. Disch (eds.). The dying generation. Dell Publishing Co., Inc. New York.

Hauser, P. M. 1963. The population dilemma. Prentice-Hall, Inc. Englewood Cliffs, N.J.

Heilbroner, R. L. 1974. An inquiry into the human prospect. W. W. Norton & Co., Inc. New York.

Hulett, H. R. 1970. Optimum world population. Bioscience 20:160-161.

Huxley, A. 1970. The politics of ecology: the question of survival. In C. E. Johnson (ed.). Eco-crisis. John Wiley & Sons, Inc. New York.

Keyfitz, N. 1971. On the momentum of population growth. Demography 8:71-81.

Large, E. C. 1962. The advance of the fungi. Dover Publications, Inc. New York.

Leopold, A. 1949. A Sand County almanac. Oxford University Press. Oxford, England.

Leyhausen, P. 1969. The social dilemma of man. Science J. 5:60-65.

Pope Paul VI. 1968. Humanae vitae. The New York Times. 30 July: 20-21.

Randers, J. and D. Meadows. 1973. The carrying capacity of our global environment: a look at the ethical alternatives. In I. G. Barbour (ed.). Western man and environmental ethics. Addison-Wesley Publishing Co., Inc. Reading, Mass.

Schumacher, E. F. 1973. Small is beautiful. Economics as if people mattered. Harper & Row, Publishers. New York.

Stewart, G. 1968. Not so rich as you think. Houghton Mifflin Co. Boston.

Washington report. 1970. Pesticide crackdown. Successful Farming 68:7.

APPENDICES

Selected readings on environmental issues

There are numerous works dealing with the human impact on the environment, including anthologies and book-length treatises. In Part One excerpts or brief comments are provided from a limited list. Part Two is a listing of additional readings.

PART ONE

Anderson, P. K. (ed). 1971. Omega. William C. Brown Co., Publishers. Dubuque, Iowa.

The title is apocalyptic but perhaps with good reason when we consider the lack of foresight exhibited in some quarters of our society. Prime Minister Sato of Japan, for example, is quoted as saying that his country's birthrate must be increased to offset a worsening labor shortage that threatens the nation's economic growth.

Auerbach, J. S. 1976. Unequal justice. Oxford University Press, Inc. New York.

The author effectively reviews the historical basis for the gap between law and justice in America. As society becomes more urbanized and increasingly complex in social structure, the legal profession must become more than a private club whose services are limited to those with the ability to pay. Justice, to be served, should be addressed to the welfare of society as a whole regardless of the economic circumstance of its members.

Beckmann, P. 1973. Eco-hysterics and the technophobes. The Golem Press. Boulder, Colo.

This author minimizes, even ridicules, the fears of those environmentalists who continue to voice publicly their concern about overpopulation, impending food shortages, and ecological degradation such as oil spills on the high seas. The author's rationale is that more technology, not less, is needed to clean up the environment and safeguard it against future catastrophes; yet it is the advanced technology of industrialized nations that has had, up to now at least, the greatest impact on the global environment. He debunks dangers to the ozone layer, minimizes the deleterious effects of pesticides on nontarget species in food chains, and states that there is no danger of worldwide famine if only we adopt certain measures to ensure new supplies of energy.

Benarde, M. A. 1970. Our precarious habitat. W. W. Norton & Co. Inc. New York.

The author draws on a large reservoir of information dealing with environmental deterioration and human health. He cites as examples the upset of ecological balance in the Nile by the construction of the Aswan High Dam and how the disease bilharziasis will increase as the range of infected snails is expanded into new habitats.

Blair, S. D. and J. B. Rodenbeck (eds.). 1971. The house we live in. Macmillan Inc. New York.

A collection of essays prepared and organized for college classes in English composition, containing selections by a wide range of authors, including Henry Thoreau, Aldo Leopold, Albert Camus, Lewis Mumford, Barry Commoner, Rene Dubos, Garret Hardin, Kenneth Boulding, Margaret Mead, and Ralph Nader.

Brown, L. R. 1974. In the human interest. W. W. Norton & Co., Inc. New York.

The author brings into sharp focus the problem of overpopulation and the need to stabilize growth if the world is to socially progress. In the developing nations any increase in food is negated by a corresponding expansion in population so that the well-being of society in these countries is not being improved.

301

Editors of *The Progressive*. 1970. The crisis of survival. Scott, Foresman & Co. Glenview, Ill.

A collection of papers selected for their relevance, constructively aimed at the development of creative programs and the realignment of national priorities in the public interest.

George, C. J. and D. McKinley. 1974. Urban ecology/in search of an asphalt rose. McGraw-Hill Book Co. New York.

The general thrust of this book is that cities as perceived today are bad places for man as well as other species. Thus the historical view of cities as centers of civilization is brought into question. With expanding urbanization superimposed on finite resources, the human condition worsens. The authors propose a return of nature to the city environment, a provision for wildlife and natural habitats, and a recycling of our wastes as self-sustaining ecosystems.

Graham, F., Jr. 1970. Since silent spring. Houghton Mifflin Co. Boston.

When Rachel Carson completed *Silent Spring* in 1962, some manufacturers attempted to prevent its publication. After 7 years of controversy, Miss Carson's position has been vindicated; yet chemicals of all types are still being released into our environment, and apologists for their continued use are as vocal as ever.

Hardin, G. 1972. Exploring new ethics for survival/the voyage of the spaceship Beagle. The Viking Press, Inc. New York.

Technology cannot solve the population problem. There are no "easy answers" to the human condition, only those that will effect revolutionary changes in our traditional view of finite nature and, accordingly, our utilization of nature. The author develops his argument for population control through family responsibility tied to the power of reproduction as a prerequisite for maintenance of a "stable commons." State interference is a possible threat to personal decision making concerning family size if societal attitudes fail to effect control of future population growth on spaceship Earth.

Harney, T. R. and R. Disch (eds.). 1971. The dying generations. Dell Publishing Co., Inc. New York.

This anthology of essays deals with man and nature, how he views and exploits it. Following is a quote from the section on the politics of survival: "The idea of an infinitely expanding Gross National Product on an isolated sphere, a finite system, an island in space, is complete nonsense, or to put it the way I personally perceive it, may be, together with population growth, the most dangerous tendency in the world today."

Harrington, M. 1976. The twilight of capitalism. Simon & Shuster, Inc. New York.

For its continued existence, capitalism depends on a maldistribution of wealth. It is also self-destructive. The top one half of 1% of Americans own over one fourth of the nation's wealth; the bottom one fourth own none of it. The rise in welfare legislation is seen as a transitional phenomenon between capitalism and its successor, collectivism.

Heilbroner, R. L. 1974. An inquiry into the human prospect. W. W. Norton & Co., Inc. New York.

Is there hope for man? The question is founded in the anxiety occurring in many advanced nations today. Corruption in government, continuing racial and religious strife, and unabated world poverty prompt these fears and lack of self-confidence in directing our social destiny. There is also the pervasive awareness of the deterioration of our surroundings. The "quality of life," despite governmental assistance and high hopes in technology and economic growth, continues its downward path. The external dangers to the human prospect are objectively defined, including continuing population growth at an exponential rate and threats of nuclear disasters. Regardless of the socioeconomic system, planning is seen to initiate future problems in the long view. Short-term solutions are always uppermost in allaying human predicaments. Despite the gloomy outlook, the author views the human prospect as a formidable array of challenges that will be met by either man's foresight or nature's contravention. In either case a postindustrial age is foreseen that will permit the continuance of human life but within a social structure much different from the one we now know.

Helfrick, H. W., Jr. (ed.). 1970. The environmental crisis. Yale University Press. New Haven.

A lecture series based on a symposium concerning issues in the environmental crisis held at the Yale School of Forestry. As our technological ability grows, it is accompanied by a steady and seemingly inexorable deterioration of our environment.

Henkin, H., M. Merta, and J. Staples. 1971. The environment, the establishment, and the law. Houghton Mifflin Co. Boston.

An account of the "DDT trial" held in Madison, Wisconsin. The court ruling arising from volumes of testimony stated that DDT and its analogs are environmental pollutants.

Horsfield, B. and P. B. Stone. 1972. The great ocean business. The New American Library, Inc. New York.

A most readable book clearly stating the issues of the ocean as a physical and biotic resource. The ocean must be viewed in the global perspective with international rights in sharing these resources by both the advanced and the less advanced nations. Intense nationalism and overfishing will transcend ecological thinking, running higher risks of species extirpation, environmental degradation, and even armed confrontation on the high seas.

Johnson, H. D. (ed.). 1970. No deposit—no return. Addison-Wesley Publishing Co., Inc. Reading, Mass.

A long list of contributions, systematically cataloged, that deal with environmental problems, natural resources, and constructive action on the part of individuals and institutions. A recurrent theme is building quality into human life and revamping priorities in the interests of a better environment.

Linton, R. M. 1970. Terracide. Paperback Library. New York.

Traditionally we have dealt with environmental hazards to our welfare only when a situation has reached crisis proportions. The author sees the urgent need to make decisions in order to prevent future deteriorations, rather than seeking solutions to problems as they arise.

Marine, G. 1969. America the raped. Simon & Schuster, Inc. New York.

The technological blitzkrieg that is destroying the environment is being confused with social progress. The author says, "We need a sense of responsibility not only toward land and rivers, but a whole way of thinking that embraces ghettos as well as mountains. . . ."

Means, R. 1969. The ethical imperative. Doubleday & Co., Inc. New York.

The dominant tendency of the social sciences has been to delineate culture and nature. This exclusionist attitude makes it possible to regard the natural world as a subject of indiscriminate exploitation. This approach opposes the inclusionist position, which holds that man is a part of nature.

Mostert, N. 1974. Supership. Alfred A. Knopf, Inc. New York.

The author discusses the global transport of oil in tankers routinely more than a 1000 feet long with over one quarter of a million tons of cargo. The integrity of marine and coastal systems everywhere is threatened by possible oil spills from these gargantuan vessels. This threat to natural ecosystems is heightened as ships become ever larger and as safety standards become less critical in the industrialized countries' rush for greater and ever-greater amounts of energy.

Nicholson, M. 1970. The environmental revolution. McGraw-Hill Book Co. New York.

An in-depth study of the changing environment and of changing attitudes concerning it. The author's concluding remark: "Before straying away to other planets let us first learn the facts of life and discover who we are by what we do on this one."

Parker, D. H. 1970. Schooling for what? McGraw-Hill Book Co. New York.

The author reveals an educational crisis. "With each turn of the money machine, oligarchical monopolistic capitalism may be undermining its own structure as more and more of the young refuse to be stuffed into the corporate sausage grinder."

Potter, Van R. 1971. Bioethics, bridge to the future. Prentice-Hall, Inc. Englewood Cliffs, N.J.

The author dedicates his work to the late Aldo Leopold, whose writings more than a quarter of a century ago urged the necessity of man's moral responsibility to the land and its wildlife, soil, and water.

Revelle, R. and H. H. Lansberg (eds.). 1970. America's changing environment. Houghton Mifflin Co. Boston.

A collection of essays on environmental protection, national goals, and the rights of society. Unfortunately, private interests often benefit at the public expense.

Ridgeway, J. 1970. The politics of ecology. E. P. Dutton & Co., Inc. New York.

The ecology movement includes a number of established pollutors who have capitalized on the environmental crisis. Some of our largest industrial combines have profit-taking enterprises in the control of pollution.

Roloff, J. G. and R. C. Wylder (eds.). 1971. There is no "away." Glencoe Press. Beverly Hills, Calif.

An anthology of action-oriented essays. It provides timely reading on a number of issues ranging from agricultural pesticides and the AEC's nuclear experiments to overcrowding and starvation. Senator George McGovern, in "Hungry Every Night," says, "We are producing people faster than we can feed them."

Shepard, P. and D. McKinley (eds.). 1969. The subversive science: essays toward an ecology of man. Houghton Mifflin Co. Boston.

A collection of papers whose central theme is that man's well-being extends beyond material goods. A healthy, attractive environment is also required. One of the best and earliest preparations on environmental issues.

Shepard, P. and D. McKinley (eds.). 1971. Environmental essays on the planet as a home. Houghton Mifflin Co. Boston.

The editors bring together another set of articles emphasizing the human aspects of the environment. The opinion is voiced that we have a right to ecological information: "We are not limited to a choice between empty political activism and disinterested ivory-tower intellectualism."

Snyder, E. F. 1971. Please stop killing me! The New American Library, Inc. New York.

A book that, by the author's admission, hopefully will frighten people into action. He says, "People cannot go on polluting the biosphere, exploiting the environment, defacing nature, and still act as though they are the chosen creatures of a benevolent God."

Toffler, A. 1970. Future shock. Random House, Inc. New York.

The author covers a wide gamut of social change that will affect our future mode of living. Fed by a proliferating technology and its scientific base, life moves at an ever-faster, albeit transient, pace with more options in life-style. Yet such changes are wasteful of energy and resources, emphasizing throwaways, short-duration artifacts, and high-cost services. Drastic changes for the future are attributed to rapid urbanization, exploding population growth, and the human capacity for innovation. Although it is impossible to return to a "state of nature," we must control future technology in the long-range perspective to minimize future shock on the human system in the superindustrial age.

PART TWO

Bates, M. 1961. Man in nature. Prentice-Hall, Inc. Englewood Cliffs, N.J.

Bernarde, M. A. 1968. Race against famine. Macrae Smith Co. Philadelphia.

Borgstrom, G. 1969. Too many, a study of earth's biological limitations. Macmillan Inc. New York.

Borgstrom, G. 1973. World food resources. Intext Educational Publishers. New York.

Boughey, A. S. 1976. Strategy for survival/an exploration of the limits of further population and industrial growth. W. A. Benjamin, Inc. Menlo Park, Calif.

Boulding, K. E. 1968. Beyond economics. University of Michigan Press. Ann Arbor, Mich.

Brown, H. 1954. The challenge of man's future. An inquiry concerning the condition of man during the years that lie ahead. The Viking Press, Inc. New York.

Caras, R. 1966. Last chance on earth. A requiem for wildlife. Chilton Book Co. Philadelphia.

Carefoot, G. and E. R. Sprott. 1967. Famine on the wind; man's battle against disease. Rand McNally & Co. Chicago.

Carr, D. E. 1976. Energy and the earth machine. W. W. Norton & Co., Inc. New York.

Carson, R. 1962. Silent spring. Houghton Mifflin Co. Boston.

Chermayeff, S. and C. Alexander. 1963. Community and privacy; toward a new architecture of humanism. Doubleday & Co. New York.

Clark, C. 1958. World population. Nature **181:** 1235-1236.

Commoner, B. 1966. Science and survival. The Viking Press, Inc. New York.

Commoner, B. 1971. The closing circle. Alfred A. Knopf, Inc. New York.

Dasmann, R. F. 1963. The last horizon. Macmillan Inc. New York.

de Bell, G. 1970. The environmental handbook. Ballantine Books, Inc. New York.

Domhoff, G. W. 1967. Who rules America? Prentice-Hall, Inc. Englewood Cliffs, N.J.

Drucker, P. 1968. The age of discontinuity. Harper & Row, Publishers. New York.

Dubos, R. J. 1968. So human an animal. Charles Scribner's Sons. New York.

Ehrlich, P. 1968. The population bomb. Ballantine Books, Inc. New York.

Elder, F. 1970. Crisis in Eden. Abingdon Press. Nashville, Tenn.

Faltermayer, E. K. 1968. Redoing America. Macmillan Inc. New York.

Farb, P. 1959. Living earth. Harper & Row, Publishers. New York.

Farvar, M. T. and J. P. Milton. 1972. The careless technology/ecology and international development. Natural History Press. Garden City, N.Y.

Friedman, M. 1962. Capitalism & freedom. University of Chicago Press. Chicago.

Frome, M. 1962. Whose woods these are: the story of the natural forests. Doubleday & Co. New York.

Fuller, R. B. 1973. Earth, Inc. Anchor Press. Garden City, N.Y.

Galbraith, J. 1969. The affluent society. Houghton Mifflin Co. Boston.

Glick, W. (ed.). 1969. The recognition of Henry David Thoreau; selected criticism since 1848. University of Michigan Press, Ann Arbor, Mich.

Graham, F., Jr. 1966. Disaster by default; politics and water pollution. J. B. Lippincott Co. Philadelphia.

Hardin, G. J. (ed.). 1964. Population, evolution and birth control; a collage of controversial readings. W. H. Freeman & Co., Publishers. San Francisco.

Hauser, P. M. 1963. The population dilemma. Prentice-Hall, Inc. Englewood Cliffs, N.J.

Leopold, A. 1949. A Sand County almanac, and sketches here and there. Oxford University Press. New York.

Marsden, R. W. (ed.). 1975. Politics, minerals, and survival. University of Wisconsin Press. Madison, Wis.

McCallum, H. D. and F. T. McCallum. 1965. The wire that fenced the west. University of Oklahoma Press. Norman, Okla.

McHarg, I. 1969. Design with nature. Natural History Press. Garden City, N.Y.

Meadows, D. H. et al. 1972. The limits to growth. Universe Books. New York.

Morgan, A. E. 1971. Dams and other disasters/a century of Army Corps of Engineers in civil works. Porter Sargent Publisher. Boston.

Morris, D. 1967. The naked ape. Dell Publishing Co., Inc. New York.

Mumford, L. 1961. The city in history: its origins, its transformations, and its prospects. Harcourt Brace Jovanovich, Inc. New York.

Olson, M. and H. H. Landsberg. (ed.). 1973. The no-growth society. W. W. Norton & Co., Inc. New York.

Ophuls, W. 1977. Ecology and the ecology of scarcity. W. H. Freeman & Co., Publishers. San Francisco.

Osborn, F. 1948. Our plundered planet. Little, Brown & Co. Boston.

Paddock, W. and P. Paddock. 1967. Famine—1975! America's decision: who will survive? Little, Brown & Co. New York.

Reich, C. A. 1971. The greening of America. Bantam Books, Inc. New York.

Revelle, R., A. Khosha, and M. Vinovskis. (eds.). 1971. The survival equation: man, resources and his environment. Houghton Mifflin Co. Boston.

Ridgeway, J. 1973. The last play. The struggle to monopolize the world's energy resources. The New American Library, Inc. New York.

Rienow, R. and L. T. Eienow. 1967. Moment in the sun, a report on the deteriorating quality of the American environment. The Dial Press. New York.

Rudd, R. L. 1964. Pesticides and the living landscape. University of Wisconsin Press. Madison, Wis.

Scheler, M. 1961. Man's place in nature. The Noonday Press, Inc. New York.

Sears, P. B. 1935. Deserts on the march. University of Oklahoma Press. Norman, Okla.

Sommer, R. 1969. Personal space; the behavioral basis of design. Prentice-Hall, Inc. Englewood Cliffs, N.J.

Stewart, G. 1968. Not so rich as you think. Houghton Mifflin Co. Boston.

Storer, J. H. 1968. Man in the web of life. The New American Library, Inc. New York.

Theobald, R. 1961. Challenge of abundance. Clarkson N. Potter, Inc. New York.

Thomas, W. 1956. Symposium on man's role in changing the face of the earth. University of Chicago Press. Chicago.

Udall, S. L. 1963. The quiet crisis. Holt, Rinehart & Winston, Inc. New York.

Vogt, W. 1948. Road to survival. William Sloane Associates, Inc. New York.

Wickenden, L. 1949. Make friends with your land. The Devin-Adair Co. Old Greenwich, Conn.

Zurhorst, C. 1970. The conservation fraud, Cowles Book Co., Inc. New York.

B Endangered species*

The following lists of fish, amphibians, reptiles, birds, and mammals represent the current status of those species that are in most urgent need of human protection. For additional information concerning estimated numbers, breeding rates, reasons for decline, and so forth, the *Red Data Book* should be consulted. As a matter of convenience, only common names are used here; obscure provinces, states, and islands have been identified by the more familiar designations of country, subcontinent, or ocean. Species are arranged according to approximate geographic locations.

FISH (Pisces)

Gila topminnow	Arizona
Greenback cutthroat trout	Colorado
Pahrump killifish	Nevada
Woundfin	Utah
Ala balik	Zamanti River (Turkey)
Cicek	Lake Egridir (Turkey)

AMPHIBIANS (Amphibia)

Desert slender salamander	Hidden Palm Canyon (Santa Rosa Mountains, California)
Santa Cruz long-toed salamander	Santa Cruz County (California)
Texas blind salamander	Hayes County (Texas)
Houston toad	South-central Texas
Vegas Valley leopard frog	Nevada
Orange toad	Costa Rica
Italian spade-foot toad	Northern Italy and Switzerland
Israel painted frog	Eastern shore of Lake Huleh (Israel)

REPTILES (Reptilia)

San Francisco garter snake	San Francisco area
Black lizard	Monterey Peninsula (California)

---

*From International Union for Conservation of Nature and Natural Resources. Vol. 1, 2, 3, and 4, 1972, 1966, 1975, and 1969, respectively. Morges, Switzerland; personal communication, Warren B. King, International Council for Bird Preservation, Smithsonian Institution, Washington, D.C.

REPTILES (Reptilia)—cont'd

Blunt-nosed lizard	California
Morelet's crocodile	Vera Cruz province (Mexico)
Atlantic Ridley turtle	Gulf of Mexico
Cuban crocodile	Zapata Peninsula
Puerto Rican boa	Puerto Rico and Virgin Islands
Rodriguez day gecko	Rodriguez
St. Croix ground lizard	St. Croix
American crocodile	Central America
Magdalena caiman	Central America
Keel-scaled boa	Round Island
Round Island boa	Round Island
South American red-lined turtle	Colombia
Rio Apaporis caiman	Southeastern Colombia
Orinoco crocodile	Colombia and Venezuela
South American river turtle	Northern South America
Black caiman	Peru, Ecuador, and Brazil
Spectacled caiman	North-central South America
Galapagos giant tortoise *	Galapagos Islands
Broad-nosed caiman	Southern South America
Paraguay caiman	Southern South America
Hawksbill turtle	Atlantic, Pacific, and Indian Oceans
Leathery turtle	Tropics
Green turtle	Tropical seas
Geometric tortoise	South Africa
African slender-snouted crocodile	Western and Central Africa
Dwarf crocodile	Western Africa from south of the Sahara to north of Congo basin
Central Asian cobra	Transcaspia to Afghanistan
Marsh crocodile	Pakistan to Burma
Siamese crocodile	Thailand
Indian gavial	Nepal, Bhutan, and Bangladesh
River terrapin	Bengal to Sumatra (Southeast Asia)
False gavial	Sumatra and Borneo
Fiji banded iguana	Fiji and Tonga Islands
China alligator	Yangtze River
Short-necked turtle	Southern West Australia

BIRDS (Aves)

California condor	California
Everglade kite	Florida
Attwater's prairie chicken	Texas, coastal prairies
Whooping crane	Canada to Texas
Kirtland's warbler	Michigan and Bahamas
Eskimo curlew	Eastern North America
Ivory-billed woodpecker (probably extinct)	Texas to Florida
Bachman's warbler	South Carolina
Hawaiian duck or koloa	Hawaiian Islands

*The Galapagos tortoise consists of several subspecies endemic to particular islands of the archipelago, the status of which is more critical for some than others.

BIRDS (Aves)—cont'd

Hawaiian crow	Hawaiian Islands
Puaiohi	Hawaiian Islands
Kauai OO	Hawaiian Islands
Maui nukupuu	Hawaiian Islands
Nukupuu	Hawaiian Islands
Kauai akialoa	Hawaiian Islands
Molokai creeper	Hawaiian Islands
O U	Hawaiian Islands
Junin grebe	Peru
Atitlan or giant pied-billed grebe	Guatemala
Grenada hook-billed kite	Grenada (West Indies)
Red-billed curassow	Brazil
Puerto Rico plain pigeon	Puerto Rico
Grenada dove	Grenada (West Indies)
Puerto Rico parrot	Puerto Rico
Imperial woodpecker	Mexico
Cuba ivory-billed woodpecker	Cuba
Euler's flycatcher	Grenada (West Indies)
St. Lucia wren	West Indies
Martinique brown trembler	Martinique (West Indies)
Martinique white-breasted thrasher	Martinique (West Indies)
St. Lucia white-breasted thrasher	West Indies
Isle of Pines solitaire	Cuba
Sempler's warbler	West Indies
Slender-billed grackle	Mexico
Grand Cayman troupial	Grand Cayman Islands (Caribbean)
Arabian ostrich (once thought extinct, may still persist)	Saudi Arabia
Cahow	Castle Harbor Island (Atlantic Ocean)
Spanish imperial eagle	Spain
Azores wood pigeon	Azores
Long-toed pigeon	Madeira, Tenerife Islands (Atlantic Ocean)
São Miguel bullfinch	Azores
Anjouan Island sparrow hawk	Anjouan Islands (Indian Ocean)
Seychelles kestrel	Seychelles (Indian Ocean)
Mauritius kestrel	Mauritius (Indian Ocean)
Western tragopan	West Pakistan, India
Mauritius ring-necked parakeet	Mauritius (Indian Ocean)
Seychelles owl	Seychelles (Indian Ocean)
Soumagne's owl	Madagascar (Malagasy)
Anjouan Scops owl	Anjouan (Indian Ocean)
Réunion cuckoo shrike	Réunion Islands (Indian Ocean)
Seychelles magpie-robin	Seychelles (Indian Ocean)
Rodriguez warbler	Rodriguez Islands (Indian Ocean)
Seychelles warbler	Seychelles (Indian Ocean)
Japanese crested ibis	Japan
Campbell Island flightless teal	Campbell Islands (Pacific Ocean)
Monkey-eating eagle	Philippines
Auckland Island rail	Auckland Island (Pacific Ocean)
Australian night parrot	Western Australia
New Zealand laughing owl	New Zealand
Tristram's woodpecker	Korea
New Zealand bush wren	New Zealand

BIRDS (Aves)—cont'd

Noisy scrub-bird	Western Australia
South Island kokako	New Zealand
Piopio	New Zealand
Cebu black shama	Philippines
Molokai thrush	Molokai Island (Pacific Ocean)
Eryean grass wren	South Australia

MAMMALS (Mammalia)

Salt-marsh harvest mouse	San Francisco area
Morro Bay kangaroo rat	California
Cedros Island deer	Cedros Island (off Baja California)
Columbia white-tailed deer	Washington and Oregon
Northern Rocky Mountain wolf	British Columbia to South Dakota and south to Idaho
Eastern cougar	Eastern Canada
Delmarva Peninsula fox squirrel	Pennsylvania, Maryland, and Virginia
Florida cougar	Florida
Texas ocelot	Southern Texas and northeastern Mexico
Cuban solendon	Cuba
Haitian solendon	Dominican Republic
Volcano rabbit	Mexico
Central American tapir	Central America
La Plata otter	Southern Brazil, Paraguay, northern Argentina, and Uruguay
Marine otter	Coastal South America
Southern river otter	Chile and Argentina
Golden lion tamarin	Brazil
Mountain tapir	Eastern Andes Mountains
Barbary hyena	Algeria, Tunisia, and Morocco
Rio de Oro Dama gazelle	Spanish Sahara (West Africa)
Cameroon clawless otter	Republic of Cameroon
Jentink's duiker	Liberia and Ivory Coast
African wild ass	Sudan and Ethiopia
Tora hartebeest	Ethiopia, Sudan, and Egypt
Barbary deer	Algerian/Tunisian border
Barbary leopard	Morocco, Algeria, and Tunisia
Corsican red deer	Corsica and Sardinia
Tana River red colobus	Kenya
Mountain gorilla	Uganda and Zaire
Verreaux's sifaka	Madagascar
Indri	Madagascar
Sanford's lemur	Madagascar
Sclater's lemur	Madagascar
Black lemur	Madagascar
Red-fronted lemur	Madagascar
Red-tailed sportive lemur	Western Madagascar
White-footed sportive lemur	Southern Madagascar
Zanzibar suni	Zanzibar
Black-faced impala	Angola and Southwest Africa
Pelzeln's gazelle	Somali
South Arabian leopard	Arabian Peninsula
Sand gazelle	Jordan south into Arabian Peninsula
Syrian wild ass	North of Syrian-Turkish border

MAMMALS (Mammalia)—cont'd

Arabian gazelle (oryx)	Western Arabian Peninsula
Arabian tahr	Oman (Arabian Peninsula)
Indian wild ass	Gujarat/Pakistan border
Saudi Arabian dorcas gazelle	Middle East
Kabul markhor	Afghanistan and Pakistan
Straight-horned markhor	Pakistan
Chiltan markhor	Chiltan Range, Pakistan
Asiatic lion	India
Lion-tailed macaque	Western Ghats
Great Indian rhinoceros	Northern India and Nepal
Manipur brow-antlered deer	India
Flea's muntjac	Burma, Thailand
Pileated gibbon	Thailand, Cambodia
Thailand brow-antlered deer	Thailand, Laos, Cambodia, and Vietnam
Baluchistan bear	Baluchistan
Chinese tiger	Eastern and central China
Wild yak	China
Yarkand deer	Chinese Turkestan
Przewalski's horse	Mongolian Republic to Gobi Desert
Snow leopard	Mountains of central Asia
Ryukyu rabbit	Japan
Sumatran serow	Sumatra
Bengal tiger	Southeastern Asia
Malayan tapir	Thailand to Sumatra
Pig-tailed langur	Mentawai Islands (off the coast of Sumatra)
Kloss's gibbon	Mentawai Islands
Orangutan	Sumatra and Borneo
Bridle nail–tailed wallaby	Queensland through New South Wales (Australia)
Crescent nail–tailed wallaby	Central Australia
Eastern Jerboa marsupial	Australia
Humpback whale	Two groups in Northern and Southern Hemispheres
Bowhead whale	Arctic waters
Black right whale	North Atlantic and Pacific Oceans and south of Antarctic convergence

C Population and income data

Population data and per capita income based on gross national product for representative developed and less developed countries of the world*

REGION AND SELECTED NATIONS†	POPULATION ESTIMATES FOR MID-1976 (millions)	ANNUAL GROWTH RATE (%)	POPULATION DOUBLING TIME‡(yr)	PER CAPITA INCOME (U.S. dollars)
North Africa	99.4	3.0	23.3	340
Algeria	17.3	3.3	21.2	650
Egypt	38.1	2.3	30.4	280
Sudan	18.2	3.2	21.9	150
East Africa	119.1	2.9	24.1	200
Ethiopia	28.6	2.6	26.9	90
Kenya	13.8	3.7	18.9	200
Tanzania	15.6	3.0	23.3	140
Zambia	5.1	3.2	21.9	480
East Asia	1153.3	2.3	30.4	710
Japan	112.3	1.1	63.6	3880
Peoples Republic of China	836.8	2.4	29.2	300
Taiwan	16.3	2.0	35.0	720
Southeast Asia	338.4	2.7	25.9	220
Burma	31.2	2.4	29.2	90
Indonesia	134.7	2.3	30.4	150
Singapore	2.3	1.4	50.0	2120
Thailand	43.3	3.1	22.6	300
Middle South Asia	893.8	2.7	25.9	160
Bangladesh	76.1	2.7	25.9	100
India	620.7	2.6	26.9	130
Iran	34.1	3.1	22.6	1060
Pakistan	72.5	3.6	19.4	130

*Data from 1976 World Population Data Sheet prepared by The Environmental Fund, 1302 Eighteenth Street, N.W., Washington, D.C.

†The countries selected represent only a portion of that region's population; hence, with the exception of North America, their total population does not equal the total for the region.

‡Data from The Environmental Fund; calculations by the author. *Continued.*

Population data and per capita income based on gross national product for representative developed and less developed countries of the world—cont'd

REGION AND SELECTED NATIONS	POPULATION ESTIMATES FOR MID-1976 (millions)	ANNUAL GROWTH RATE (%)	POPULATION DOUBLING TIME (yr)	PER CAPITA INCOME (U.S. dollars)
Southwest Pacific	21.5	1.8	38.9	3800
Australia	13.8	1.4	50.0	4760
Fiji	0.6	2.1	33.3	720
New Zealand	3.2	2.2	31.8	4100
Latin America	333.5	2.6	26.9	940
Brazil	110.2	2.9	24.1	900
Costa Rica	2.0	2.3	30.4	790
Ecuador	6.9	3.2	21.9	460
Haiti	4.6	2.1	33.3	140
Mexico	62.3	2.2	31.8	1000
Venezuela	12.3	3.4	20.6	1710
North America	245.3	1.3	53.8	6580
Canada	23.1	1.3	53.8	6080
United States	222.2	1.3	53.8	6640
Europe	475.8	0.6	116.7	3680
Austria	7.5	0.0		4050
France	53.1	0.8	87.5	5190
Italy	56.3	0.7	100.0	2770
Portugal	8.5	0.4		1540
Sweden	8.2	0.4	175.0	6720
West Germany	62.1	0.1	700.0	5890
World	4240.7	2.2	31.8	1360

D Glossary

abiotic Referring to the nonliving factors of the ecosystem, for example, temperature, light, nutrients, water, and pH.

adaptation Genetically controlled trait of an organism having survival value in a given environment.

allochthonous Indicating external origin, or produced elsewhere (the opposite of *autochthonous*).

anaerobic Condition lacking free oxygen.

autochthonous Indicating origin in place, or produced within the system (the opposite of *allochthonous*).

autotrophic Literally self-nourishing (synthesizing own food), with specific reference to chlorophyllous, or green plants, and also including certain bacteria that are either pigmented (photosynthetic) or those that can synthesize usable energy from inorganic compounds (chemosynthetic).

Batesian mimicry Condition whereby one species resembles in color or pattern a harmful or distasteful species to gain protection from potential predators.

benthos Collective term for bottom-dwelling organisms in a body of water.

biological control One species controlling the numbers of another species that is considered a pest.

biological oxygen demand (BOD) Referring to the oxygen requirement in the microbial breakdown of organic materials deposited in reservoirs or natural waters.

biotic Referring to all organisms of the ecosystem.

capillary capacity Value indicating maximum amount of water held in the soil by capillary tension and often expressed as a percentage of the oven-dry weight of the soil (same as *field capacity*).

carnivore Consumer level utilizing other live animals in the food chain.

carrying capacity Number of individuals a given environment can support, designated as K in logistic models.

climax Referring to the terminal stage or community in a sere (succession) and interpreted as a stable, generally self-perpetuating combination of species.

cohort Group of individuals studied throughout the life cycle to record age at reproduction and death.

commensalism Two-species interaction whereby one species benefits whereas the other is unaffected.

community Assemblage of species occurring together over a given period of time.

competitive exclusion principle Concept that no two distinct species can live in the same place at the same time.

coprophagy Eating of feces.

density Number of individuals per unit area.

density-dependent factors Factors that influence an increasing percent of the population as density increases.

diffuse competition Term applying to situations in which distantly related species may have important competitive effects on a given species.

direct nutrient cycling Process of nutrient uptake in tropical wet forests from a mycorrhizal (host/root-fungus relation) development in the soil surface, thus bypassing nutrient release to soil colloid and subsequent absorption by the root system.

dispersal Movement of individuals away from birthplace or areas of high population density.

dispersion Spatial distribution pattern of a species, for example, *random, clumped* or *aggregated,* or *regular,* at one period in time.

dominance Expression of high importance of a species or species group measured in numbers, biomass, energy production, and so on.

ecological isolation Separation of coexisting species effected by differences in their ecological requirements.

ecosystem Concept of biotic and abiotic integration as a structural and functional whole.

edaphic Factors referring to the soil environment.

emigration Movement of individuals out of a population or area.

evapotranspiration Collective water losses from both plants and soil.

exploitation competition Utilization of a resource by one species to the exclusion of other species.

exponential growth Growth of a population in which there is the same proportional increase in number per time unit. Such growth results in a curve with an ever-increasing slope when numbers are plotted against time.

herbivore Consumer level utilizing plants in the food chain.

heterotrophic Literally ''other-nourishing'' and referring to the consumers of the ecosystem.

home range Area in which an organism normally lives, exclusive of migration or erratic wanderings.

homeostasis Condition of equilibrium achieved in a system when growth and species change cease.

homeotherm (homoiotherm) ''Warm-blooded'' organisms, or those maintaining a constant body temperature through metabolic processes.

immigration Movement of individuals into a population or area.

instantaneous rate of increase (r) Rate at which a population is growing at a given moment expressed as the proportional increase of the population per unit of time.

interference competition Direct interaction of competitors, including actual defense of resources.

interspecific competition Competition for resources among members of different species, as opposed to intraspecific competition in the same species.

K selection Selection type favoring equilibrium species with strategies for energy allocations in biomass structure and competition (see *r selection*).

lentic Referring to ponds, lakes, and so on, or still water.

life table Enumeration of the numbers and proportion of individuals dying at a given age and the number of progeny produced at each age.

littoral Environment of the seashore, that is, the intertidal zone.

logistic equation Mathematical description of population growth and equilibrium in a resource-limited environment (see *carrying capacity*):

$$K = \frac{dN}{dt} = rN\,\frac{K - N}{N}$$

lotic Referring to streams, and so on; or running water.

Müllerian mimicry Condition of convergence in distasteful or harmful organisms toward standard patterns of warning coloration.

mutualism Interaction of two species that benefits both.

mycorrhizae Symbiotic relationship of absorbing roots of host plant and a fungus.

natality, age-specific Number of daughters produced by an average female at age x, designated as m_x.

neritic Zone of inshore water over the continental shelf, as opposed to *oceanic,* referring to the deeper basins.

net ecosystem production Total energy or biomass production for all trophic levels of the ecosystem after respiration. When NEP = 0, there is no further growth, and the ecosystem is in equilibrium with the available resources.

net replacement rate Number of daughters produced by an average female over her lifetime, designated as R_0.

niche Functional role of the species within a given community structure.

oceanic Zone of deep water beyond the continental shelf.

parasitism Interaction between two species in which one benefits at the expense of the other without killing it.

pelagic Open ocean beyond the intertidal zone.

permanent wilting percent Critical moisture level at which most plants become permanently wilted and will not recover turgor when water is restored.

plankton Collective term for both free-floating plants and animals, usually of microscopic dimensions, distinguished by the prefix ''phyto'' for plants (algae) and ''zoo'' for animals.

poikilotherm ''Cold-blooded'' organisms, or those whose body temperatures vary within limits as a function of ambient conditions.

population Interbreeding individuals of a single species occurring in a given area.

predation Interaction between two species in which one benefits by harvesting the other.

productivity Production of energy or biomass on a time-rate basis, for example, grams per square meter per year.

r selection Selection for pioneer or seral species with energy allocations for high seed production and dispersal and relatively less for structure and competition (see *K-selection*).

respiration Metabolic expenditure of available energy for growth and maintenance.

stability Equilibrium condition resisting normal perturbations in the environment and capacity for self-perpetuation.

stable age distribution Proportional distribution of ages characteristic of a given l_x and m_x schedule that does not change as the population varies in size provided l_x and m_x are stable.

succession Sequential pattern of species replacement through time, terminating in a stable community.

survivorship, age-specific Proportion of newborn individuals that will live to age x, designated as l_x.

territory Any defended area that allows the holder exclusive use.

urbanization National or regional process affected by social and economic conditions whereby an increasing proportion of the total population live and work in the city.

Index